Epigenetics
in
Human Reproduction and Development

Epigenetics
in
Human Reproduction and Development

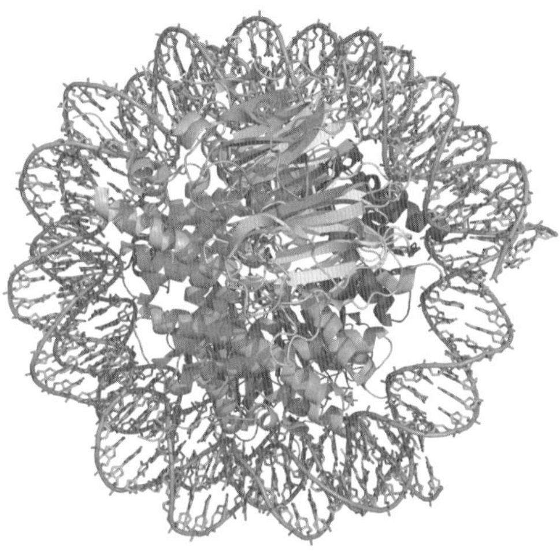

Editors

Anna K Naumova

Teruko Taketo

McGill University, Canada

World Scientific

NEW JERSEY • LONDON • SINGAPORE • BEIJING • SHANGHAI • HONG KONG • TAIPEI • CHENNAI • TOKYO

Published by

World Scientific Publishing Co. Pte. Ltd.
5 Toh Tuck Link, Singapore 596224
USA office: 27 Warren Street, Suite 401-402, Hackensack, NJ 07601
UK office: 57 Shelton Street, Covent Garden, London WC2H 9HE

Library of Congress Cataloging-in-Publication Data
Names: Naumova, Anna K., editor. | Taketo, Teruko, editor.
Title: Epigenetics in human reproduction and development / [edited by]
Anna K Naumova, Teruko Taketo.
Description: New Jersey : World Scientific, 2016. | Includes bibliographical references and index.
Identifiers: LCCN 2016035691 | ISBN 9789813144262 (hardback : alk. paper)
Subjects: | MESH: Epigenesis, Genetic | Reproduction--genetics | Human Development
Classification: LCC RB155 | NLM QU 475 | DDC 616/.042--dc23
LC record available at https://lccn.loc.gov/2016035691

British Library Cataloguing-in-Publication Data
A catalogue record for this book is available from the British Library.

Typeset by Stallion Press
Email: enquiries@stallionpress.com

Printed in Singapore

Preface

Epigenetic phenomena have been studied since early works in chromosomal biology and chromatin that date back to the beginning of the 20th century. The term "epigenetics" was coined by Conrad Waddington in 1942. In the recent years epigenetics studies have undergone a renaissance due to the advent of new technologies and the realization that neither genetic variation nor environment alone explain the aetiology of human disease. The availability of new technologies and the massive body of data generated by the Human and Mouse Genome Projects, as well as the Encyclopedia of DNA Elements (ENCODE) and the Roadmap Epigenomics projects permitted answering many research questions that were not possible to address in the past. It has been demonstrated that different somatic cell types, different stages of development and different diseases show differences in chromatin states and DNA methylation levels throughout the genome. Sex, age and environmental factors are believed to modulate epigenetic states and thereby gene expression.

The initial focus of epigenetics research was on the relationship between gene and phenotype, however in recent years more emphasis is given to the role of environmental factors and the perceived role of epigenetic factors as a mode of communication and adaptation between the genome and environment. In this book we tried to embrace the different facets of epigenetics and its roles in the interaction of the genome with both intrinsic and extrinsic factors. As open systems that have to react to environmental changes to adapt and survive, living organisms respond to environmental cues by changes in the epigenome and consequently modifications in gene expression patterns. Such external events may or may not have lasting effects on the epigenome. Why certain changes are fully reversible while other leave a robust epigenetic footprint is an important question for which the answer has yet to be found. Furthermore, emerging evidence suggests that certain environmental influences may elicit epigenetic changes that affect the epigenome of the next generation.

The epigenetic processes that occur in the parental germ cells, such as marking DNA for recombination, genomic imprinting, massive repackaging of the genome into different types of histones or protamines in spermatogenesis, etc. are critical for the formation of normal gametes and the health of the future embryo. These epigenetic reprogramming events that occur in the germ line and in the embryo after fertilization are of paramount importance for the survival of the species and health of the individual. In this book, the fascinating epigenetic events that take place during gametogenesis are covered in the first four Chapters. Chapter 1 by Sofia Malki and Alex Bortvin reviews the role of epigenetic factors in maintaining genome integrity in the germ cells. Chapter 2 by Rabindranath De La Fuente, Claudia Baumann, and Maria M. Viveiros focuses on the pivotal role of epigenetic modifications in oogenesis. Chapter 3 by Fatima Smagulova provides an overview of the process of meiotic recombination in humans and the role of histone modifications in directing crossover formation. Chapter 4 by Debbie Montjean, Antoine D Rolland, and Celia Ravel examines the role of epigenetic factors in spermatogenesis and changes that lead to male infertility.

Chapters 5 and 6 review essential epigenetic processes that take place during embryonic/fetal development, X chromosome inactivation and genomic imprinting. Chapter 5 by Bernhard Payer provides an update on the current understanding of the phenomenon of X chromosome inactivation and discusses certain mechanistic aspects of X-inactivation that remain elusive. Chapter 6 by Courtney Hana and Wendy Robinson focuses on the role of genomic imprinting in placental development.

Studies of epigenetic variation in humans show that the same types of somatic cells from different individuals may have different epigenetic profiles and that such inter-individual variation is part of the norm rather then abnormality. Furthermore, recent epigenome-wide studies (EWAS) conducted in large cohorts demonstrated that genetic variation influences DNA methylation levels in *cis* and *trans*. The vast majority of these data are based on studies of human peripheral blood cells. However, DNA methylation levels at certain genomic regions change with age and may depend on individual's sex. Such large-scale studies have yet to be done in human germ cells and will be extremely important for determining the extent of normal epigenetic variation in the human population and the significance of certain epigenetic profiles for reproductive outcomes. To date, very few studies of this nature have been reported. Most of them focus on a specific pathology or a candidate genomic region and are often conducted on a rather small number of samples. Therefore inter-individual epigenetic variation in the human germ line and its link to genetic variation remains a vast unchartered territory waiting to be explored. Chapter 7 by Celia Greenwood and colleagues examines how the interplay between genetic variants and DNA methylation at an imprinted region influences birth weights and other developmental parameters.

The importance of the remodelling of the epigenome during gametogenesis and embryonic development raises the question of time points and specific conditions when the epigenome becomes vulnerable to external influences. Certain environmental factors acting at these points of vulnerability may elicit changes that have a lasting impact on health and even behaviour of the individual. These aspects of the early environmental impacts on the epigenome are discussed in the three last chapters of our book. Chapter 8 by Renata H. Bahous, Karen E. Christensen, and Rima Rozen provides a review of the mechanisms by which genetic variation and maternal diet during pregnancy influence folate metabolism and impact the epigenome of the offspring. Chapter 9 by Anke Hoffmann and Dietmar Spengler examines the epigenetic programming of the hypothalamic-pituitary-adrenal axis by maternal stress and its impact on mental health. Chapter 10 by Debra Fulton, Seang Lin Tan, and Teruko Taketo examines the effects of assisted reproduction techniques on the embryonic epigenome.

We hope that this book will be helpful for researchers and students in their studies of human reproduction and development.

We would like to express our gratitude to all contributors for making this project possible.

Anna K. Naumova
McGill University, Canada

Contents

Epigenetics and Transposon Control in the Mammalian Germline

1

Safia Malki and Alex Bortvin

Department of Embryology,
Carnegie Institution for Science,
Baltimore, MD USA

Germ cells bridge generations and are critical for the successful propagation of species. Early in development, a small group of embryonic cells embarks on a long but fascinating journey that will take them through most dramatic changes in their genomic architecture, epigenetic makeup and cellular morphology (Lesch and Page 2012). The germ cell lineage will multiply, protect and prepare a few original copies of the newly created embryonic genome for the ultimate purpose of fertilization. Failure to precisely fulfill these functions at any level of cellular organization leads to deleterious mutations, chromosomal numerical and structural abnormalities, subfertility or infertility of an organism.

Integrity of germ cells' genomes is critical for reproduction. In addition to numerous exogenous factors constantly bombarding somatic and germ cells of any organism throughout its life cycle, each of those cells' genomes has a dark secret in the form of massive amounts of apparently non-functional DNA sequences. While not coding for any of the regulatory or structural proteins that make up a successful gamete, these DNA sequences possess a tremendous potential power to harm the germ cell and its genome. This chapter will introduce these DNA sequences in some detail and describe the complex love–hate relationship between them and germ cells.

1.1 What are Transposons?

Barbara McClintock's discovery of "controlling elements" in maize in the late 1940s marked the critical departure from the earlier vision of the constant character of the genome (Comfort 1999, 2001, McClintock 1948, 1951, 1956, 1984). Now known as transposable elements (TEs) or transposons, these discrete segments of DNA capable of integrating into a new location in the genome remain phenotypically invisible unless specifically looked for or presented with a chance to cause an observable change in an organism such as variegated coloring of maize kernels studied by McClintock. Since then, owing to the incredible advances in molecular biology, we have gained an extensive knowledge of the immense variety of plant and animal transposons, their structures and mechanisms of spreading throughout host genomes (Craig *et al.* 2015, Kazazian 2011). This knowledge empowers ongoing and future exploration of TEs' roles in genome structure and evolution. Furthermore, transposons have become critical tools in every day biological research and have a great promise in human gene therapy (Hou *et al.* 2014, Ivics and Izsvak 2006, Rubin and Spradling 1983).

As we learn more about evolution and mechanisms of transposition of transposons, their classification continues to evolve (Finnegan 1989, Piégu *et al.* 2015, Wicker *et al.* 2007). According to their dependency on an RNA intermediate for insertion to a new location in the host genome, any TE belongs to one of two broad classes. A *DNA transposon* generally uses a "cut and paste" mechanism whereby it physically excises its entire body and inserts it at a new site in the genome (Craig *et al.* 2015, Pace and Feschotte 2007). This transposition process is possible because DNA transposons encode a specific enzyme "transposase", which binds in a sequence-specific manner to terminal inverted repeats (TIRs) flanking transposon sequence and catalyzes both the excision and integration reactions of transposition (Hickman *et al.* 2005, Yuan and Wessler 2011). Some DNA transposons, such as miniature inverted-repeat transposable elements (MITEs), lack transposases and jump around the genome in a non-autonomous manner using transposases of other elements (Fattash *et al.* 2013).

In contrast, RNA intermediate-dependent elements transpose by a "copy and paste" mechanism (Goodier and Kazazian 2008). Here, DNA sequence of an element remains intact at the original genomic site but its RNA transcript templates the synthesis of DNA by reverse transcriptase. Based on their absolute dependency on reverse transcriptase, these TEs are classified as *retrotransposons*. There are two principle types of retrotransposons — those flanked by long terminal repeats (*LTR retrotransposons*) and those lacking LTRs altogether (*non-LTR retrotransposons*). The LTR retrotransposons (also known as Endogenous RetroViruses, ERVs) encode GAG and POL and, in some cases, ENV proteins and therefore, might represent ancestors of true retroviruses (Dewannieux and Heidmann 2013, Maksakova *et al.* 2006, Ribet *et al.* 2008, Weiss 2005). In mice, low copy number ERVs bearing similarity to murine leukemia virus (MLV) exhibit significantly higher activity than copious Intracisternal A particle (IAP) elements. In addition to hundreds of full-length copies of ERVs, the genome contains large numbers of solitary LTRs that arise by way of homologous recombination between the 5′ and 3′ LTRs of the full-length ERV. It is also quite possible that ERVs encoding the ENV protein have retained infectivity and propagate, at least in part, by infecting germ cells (Maksakova *et al.* 2006).

The non-LTR retrotransposons are subdivided into autonomous elements called long interspersed nuclear element (LINEs) and non-autonomous short interspersed nuclear element (SINEs) (Han 2010, Kramerov and Vassetzky 2010, Ohshima and Okada 2004). Full-length LINEs such as those in the LINE-1 (L1) super-family encode an RNA-binding chaperon ORF1p and a reverse transcriptase/endonuclease ORF2p (Babushok and Kazazian 2007, Feng *et al.* 1996, Martin 2006, 2010, Martin and Bushman 2001, Mathias *et al.* 1991, Moran *et al.* 1996). Although the mechanism of LINE-1 retrotransposition requires reverse transcription, it is distinct from that of LTR retrotransposons. Instead of forming a cytoplasmic virus-like particle, LINE-1-encoded ORF1p and ORF2p proteins form cytoplasmic ribonucleoprotein particles (RNPs) with their L1 mRNA that must reach the nuclear genome not only to integrate but also to create LINE-1 cDNA in the first place (Doucet *et al.* 2010, Martin 1991, 2006). Once in the nucleus, endonuclease activity of ORF2p nicks genomic DNA (predominantly at 5′TTAAAA/3′AA|TTTT sequences) to provide a T-rich template sequence to anneal LINE-1 mRNA poly(A)-tail and to initiate synthesis of the cDNA by reverse transcriptase activity of ORF2p (Cost and Boeke 1998; Feng *et al.* 1996, Jurka 1997). Details of later steps in this retrotransposition mechanism (known as Target-Primed Reverse Transcription or TPRT) remain await further elucidation (Gilbert *et al.* 2005, Han 2010, Kojima 2010).

Non-autonomous SINEs have originated from tRNA, 7SL RNA or 5S rRNAs. SINE-B1 (~140 bp) and Alu (~300 bp) are predominant elements, respectively, in mouse and human species (Kramerov and Vassetzky 2010). Transcription of gene-embedded SINEs occurs by RNA polymerase II while polymerase III transcribes individual SINEs utilizing internal promoter. The presence of a 3′-terminal L1-common sequence and an A-rich tail facilitates recognition of SINE RNA by L1 reverse transcriptase and retrotransposition (Dewannieux *et al.* 2003, Dewannieux and Heidmann 2005, Wallace *et al.* 2008). In addition, the human genome encodes ~2700 copies of another non-LTR retrotransposon — 700–4,000 bp-long composite SINE-VNTR-Alu (SVA) elements (Hancks and Kazazian 2010). Like SINEs, these "repeat of repeats" elements rely on L1 reverse transcriptase but possess sequence features most consistent with their RNA polymerase II transcription using internal or upstream promoters.

1.2 How Many Transposons are There in a Mammalian Genome?

As early as 1968, Britten and Kohne demonstrated a high repeat content of plant and animal genomes (Britten and Kohne 1968). Over the past two decades, genome sequencing projects revealed that much of that repeat content is of transposon origin. For example, at least 42% of the human genome is made up of sequences of intact and mutated transposable elements (19% of LINE-1, 12% of SINEs, 8% LTR and 3% DNA transposons) (Chalopin *et al.* 2015, Lander *et al.* 2001). Likewise, the mouse genome contains 31% of transposon sequences (15% LINE, 6% SINE, 9% LTR and 1% DNA TEs) while transposons account for 54% of the Zebrafish genome (3.6% LINE, 1.8% SINE, 5% LTR and 41% DNA TEs) (Chalopin *et al.* 2015, Waterston *et al.* 2002). While these are already very impressive numbers, they likely underestimate the true contribution of TEs to genomes due to the challenge of genome assembly from highly similar repeated elements (de Koning *et al.* 2011).

Most TEs populating modern genomes have lost their ability to transpose due to truncation and mutations introduced during transposition or accumulated over time. For example, there is no evidence of active DNA transposons since ~40 million years in mouse and human genome (Goodier and Kazazian 2008, Pace and Feschotte 2007, Ray *et al.* 2008, Smit and Riggs 1996). A census of DNA transposons of the Hg19 human genome assembly listed 383,000 DNA sequences derived from DNA transposons (Pace and Feschotte 2007). These elements are subdivided within 125 families counting 100 to 1,000 copies per family. With mouse genome that counts 110,000 DNA transposon elements, the human genome contains the highest copy number among eukaryotic species (Chalopin *et al.* 2015, Lander *et al.* 2001). Loss of DNA transposons activity in mammals throughout evolution (with the exception for PiggyBat elements recently described in Bat) is puzzling with respect to their quantitative importance in the genome (Mitra *et al.* 2013, Ray *et al.* 2008).

Only about 100 of 500,000 LINE-1 elements in the human genomes remain competent for retrotransposition but just 6 of those could account for most of the ongoing LINE-1 activity (Brouha *et al.* 2003, Sassaman *et al.* 1997). The mouse genome encodes upwards of 10,000 full-length LINE-1 elements belonging to evolutionarily "young" families (DeBerardinis *et al.* 1998, Goodier *et al.* 2001). LTR transposons are still active in the mouse, for example IAP elements, but not in the human genomes (Horie *et al.* 2007, Lander *et al.* 2001, McCarthy and McDonald 2003, Qin *et al.* 2010, Smit and Riggs 1996).

Transposons reside throughout mammalian genomes with no clear preference for specific locations yet LINE-1 appears to amass in G-banding regions and the X chromosome while SINEs favor R-banding regions (Sandmeyer *et al.* 1989, Wichman *et al.* 1992) and both sequence and chromatin features could play significant role in targeting integration of transposons (Gao *et al.* 2008). Interestingly, both human and mouse genomes possess transposon-free regions some of which include *Hox* gene clusters (Simons *et al.* 2006a, 2006b).

1.3 Transposons — Genomic Junk or Agents of Change?

What purpose do transposons have in the genome — is it solely that of selfish propagation or are they accidental building blocks or, more precisely, the clay of evolution? The functionality of transposable elements has been and remains one of the most exciting and contentious questions that go back to the original ideas of McClintock. In fact, contrary to the wide-spread belief it was not the mere existence of transposons but rather McClintock's vision of them as "controlling elements" that received an overwhelmingly cold reception by contemporaries (Comfort 1999, 2001). In the early days of the molecular biology era, evolutional considerations and the *C*-value paradox of marked disparity between size of the genome and morphological complexity of the organism raised a strong argument that transposable elements are most likely "selfish", "parasitic" or "junk" bits of modern genomes (Doolittle and Sapienza 1980, Orgel and Crick 1980). While such view did not exclude sporadic advantageous roles of junk DNA, little experimental evidence at the time could argue against it otherwise. Yet, careful consideration of these features leads to a conclusion that not a single description could account for the complexity of interactions between transposons and their host genomes (Kidwell and Lisch 2000). Over the years, ample although not yet systematic evidence of transposon impact on genome structure and gene expression has erode to some extent the purely negative perception of transposons (Britten 1996, Britten and Davidson 1971, Bucheton 1990, Corces and Geyer 1991, Fedoroff 1989, Kazazian *et al.* 1988, Pardue *et al.* 1996, Smit 1996, Spradling 1994, Weiner *et al.* 1986, Wichman *et al.* 1992). With the emergence of sequenced genomes and next generation sequencing technologies over the last decade, transposons continued to gain further recognition as "potent agents of change" (Kidwell and Lisch 2000). The recently completed

Encyclopedia of DNA Elements (ENCODE) project has claimed a functional role of 80% of the human genome sequence thus reigniting an enthusiastic debate on this subject (Doolittle 2013, Dunham 2012, Eddy 2012). Whether or not transposons will remain qualified as junk DNA or be upgraded to a more attractive status, this discourse underscores an important fact that we are still in the very active phase of understanding the structure and function of the genome (Doolittle *et al.* 2012, Stencel and Crespi 2013). The ongoing exploration of the world of short and long non-coding RNAs promises to further expand the boundaries of our knowledge and it is without any doubt that transposable elements will be an important part of the new vision of the genome (Hangauer *et al.* 2013, Kelley and Rinn 2011, Rory and Roderic 2014). In all likelihood, future studies of vast numbers of eukaryotic genomes will illuminate additional roles of transposons in the evolution of species thus proving correct once again Dobzhansky's aphorism that "Nothing in biology makes sense except in the light of evolution" (Dobzhansky 1973).

1.4 Consequences of Transposition

Transposable elements possess a remarkable potential to alter the structure and expression capacity of the genome (Burns and Boeke 2012, Cordaux and Batzer 2009, Deininger *et al.* 2003, Han and Boeke 2005, Hancks and Kazazian 2010, 2012, Huang *et al.* 2012, Kazazian, 2000, 2004, Kazazian and Moran 1998). For example, a newly inserted DNA sequence can be used as a DNA recombination substrate leading to duplications, deletions, non-allelic exchanges and gene conversion depending on interacting sequences (Burns and Boeke 2012, Deininger *et al.* 2003, Huang *et al.* 2012). New transposon insertions have the capacity to influence the expression potential of the surrounding DNA territory both quantitatively and qualitatively (Han and Boeke 2005, Han *et al.* 2004). Insertions could change spacing between functional elements or introduce additional regulatory sequences with a capability of changing every step in initiation and termination of transcription, mRNA splicing and polyadenylation (Bourque *et al.* 2008, Han and Boeke 2005, Han *et al.* 2004, Pierre-Étienne *et al.* 2013, Savage *et al.* 2013, Sundaram *et al.* 2014, Varagona *et al.* 1992). Frequently retrotransposon mRNAs extend into downstream coding and non-coding sequences resulting in their inclusion and transduction along with the original element (Moran *et al.* 1999, Tubio *et al.* 2014). Because of the concerted efforts of several mechanisms of transposon regulation and silencing, new transposon insertions nucleate assembly of repressive chromatin states that involve DNA methylation and histone modification, and lead to reduced accessibility of DNA to transacting factors (Castel and Martienssen 2013,

Holmquist *et al.* 1998, Pezic *et al.* 2014, Rangan *et al.* 2011, Spradling 1994). A newly integrated DNA sequence corresponding to a full-length or partial transposable element could serve as a useful marker for the purpose of understanding of genomic diversity and evolution of species (Akagi *et al.* 2008, Xing *et al.* 2007, 2009). In addition to insertion-mediated effects, reverse transcriptase of retrotransposon L1 is also responsible for mobilization of non-autonomous elements as well as creating a whole universe of processed pseudogenes and functional retrogenes throughout the genome (Dewannieux *et al.* 2003, Dewannieux and Heidmann 2005, Doucet *et al.* 2015, Esnault *et al.* 2000, Ewing *et al.* 2012, Garcia-Perez *et al.* 2007).

Transposons have greatly contributed to various aspects of mammalian reproduction. Retrotransposons exhibit high levels of expression at various times of germ cell development and differentiation resulting in new integrations in germ cells and early embryos (Kano *et al.* 2009, van den Hurk *et al.* 2007). Transposons participated extensively in the species-specific rewiring of core transcriptional networks in mammalian embryos (Kunarso *et al.* 2010). Pluripotency transcription factors OCT4 and NANOG play key roles in mouse and human embryonic stem cells yet exhibit only minor (5%) homologous occupancy of their sites between the two species. At least a quarter of novel sites have been contributed by specie-specific integration events that have occurred since their divergence. In addition, in the oocyte and the embryo, transcription of many genes initiates not at normal promoters but within MT transposons that provide alternative promoters and exons (Peaston *et al.* 2004). Furthermore, reverse transcriptase activity of L1 has been critical for the creation of autosomal retrogene copies of several house-keeping and germ-cell specific sex-linked genes thus creating a functional backup during their transcriptional inactivation in gametogenesis (Wang *et al.* 2005). Consistent with a role of transposons in evolution of epigenetic control of the genome, remnants of a transposon embedded into a differentially methylated regulatory region of an imprinted *Rasgrf1* bring its *de novo* DNA methylation under the control of piRNAs (Watanabe *et al.* 2011). In the female germline, L1 appears to play a role in determining the size and quality of the ovarian reserve by regulating the process of massive elimination of fetal oocytes following the initiation of meiotic prophase I (Malki *et al.* 2014).

1.5 When and Where Does Retrotransposon Expression Occur?

Given the transformative power of retrotransposition on the host genome's structure and expression, when and where do retrotransposons mobilize, if at all, during the life cycle of an organism? A most precise answer to this question should come from studies identifying *de novo* insertions in the host genome. The main challenge of detecting retrotransposition lies in the uniqueness of every insertion and historical lack of sufficiently sensitive tools to pinpoint such events at a single genome level (Burns and Boeke 2012). In the absence of direct unequivocal evidence of retrotransposition, expression of retrotransposon RNA and proteins can be used as a proxy of retrotransposon activity. It is important to remember that while transcription of a retrotransposon is an obligatory step for transposition, it is not a guarantee of the appearance of new insertions. The opposite is also true — lack of apparent expression of an element does not rule out a possibility of ongoing retrotransposition. Based on these considerations, best examples of *de novo* retrotransposition events have been described in those tissues where a cell carrying such an insertion had an opportunity to multiply to allow for its detection. Therefore, it comes as no surprise that most studies of *de novo* insertions deal with those occurring in the germline or early embryonic cells, particularly if they are associated with phenotypic changes or disease. This germ-cell bias does not imply that transposons prefer the germline to somatic cells but likely reflects on the difficulty of analysis of somatic retrotransposition *in vivo*. Because of the cell type complexity and renewal of most somatic tissues, comparable analysis of *de novo* retrotransposition in the soma has been and still remains technically challenging. However, some cancers (particularly of the epithelial lineage) offer an exciting opportunity to identify *de novo* retrotransposon insertions and understand their contribution to oncogenesis (Babatz and Burns 2013, Carreira *et al.* 2013, Rodic *et al.* 2014). The fact that many cells in a tumor possess the same new insertion point to the involvement of cancer stem cells in the process raises an important question whether stem cell properties are conducive to retrotransposition. Interestingly, retrotransposon expression is apparent in mouse and human embryonic carcinoma (EC) cells, embryonic stem (ES) cells and induced pluripotent stem (iPS) cells (Marchetto *et al.* 2013, Martin 1991). Furthermore, some studies in rodents and human brain tissues and derived cultured cells uncovered evidence of L1 expression and retrotransposition in neuronal stem cells (Baillie *et al.* 2011, Muotri *et al.* 2005). Together with extensive evidence of retrotransposon expression in germ and embryonic cells, these findings suggest that some aspect of stem cell biology makes them permissive for retrotransposon expression.

1.5.1 *Erasure of DNA methylation in the embryonic germline*

Epigenetic reprogramming is an obligatory event in the development of germ cells that completes the restoration of their developmental potential and replaces the original bi-parental epigenetic makeup with the one corresponding to the sex of the embryo (Messerschmidt *et al.* 2014, Seisenberger *et al.* 2013). Most of our current knowledge of epigenetic reprogramming in mammals comes from studies in mice. While some aspects of mouse germ cell development and cell biology are clearly different from that of humans, it is an excellent experimental model organism amenable to a broad spectrum of genetic and molecular approaches to guide further exploration of reproduction in humans. Studies over the past two decades have revealed a complex picture of the gradual loss of DNA methylation. DNA methylation and histone modification reprogramming encompasses imprinted and germ cell-specific genes as well as genes with meiotic functions (Hackett *et al.* 2012, Yamaguchi *et al.* 2012). In addition, since DNA methylation plays a major role in transposon repression, epigenetic reprogramming is accompanied by activation of retrotransposons, most prominently L1 elements (Popp *et al.* 2010, Seisenberger *et al.* 2012, Walsh *et al.* 1998, Yoder *et al.* 1997). As a whole, LTR transposons reactivate to a lesser extent owing to an additional layer of repression conferred by histone modifications (Lane *et al.* 2003, Leung and Lorincz 2012, Popp *et al.* 2010, Seisenberger *et al.* 2012).

Erasure of pre-existing bi-parental DNA methylation landscape in the mouse begins in primordial germ cells (PGCs) during their migration towards the developing genital ridges (Hajkova *et al.* 2002, Seisenberger *et al.* 2013, Seki *et al.* 2007). Both passive (replication-dependent) and active demethylation mechanisms contribute to the process of replacement of methylated cytosines with unmodified bases in the germ cell genome (Hackett *et al.* 2013, Kagiwada *et al.* 2013, Ohno *et al.* 2013, Ooi and Bestor 2008, Yamaguchi *et al.* 2013). Active DNA demethylation implies the existence of an enzymatic mechanism for the removal of methylated cytosine or the methyl group from DNA. A leading candidate process under extensive investigation at present is oxidative modification of the methyl group of cytosines by Ten-Eleven Translocation (TET) proteins (Iyer *et al.* 2009, Pastor *et al.* 2013). In the current view of this mechanism, three mammalian dioxygenases TET1, 2 and 3 have the capacity to gradually transform methylated cytosine to three successive oxidized forms. These could be further removed by DNA repair mechanisms such as base excision repair or passively through DNA replication in proliferating cells. Whether and how TET enzymes regulate transposons is an important question that awaits detailed investigation.

In addition to reprogramming of DNA methylation of their genomes, migratory and gonadal PGCs undergo extensive turnover of histone modifications and histones subtypes. For example, a survey of migratory mouse PGCs (embryonic days E8.5–E10.5) revealed loss of H3K9me2, a repressive mark that also appears to be involved in guidance of DNA methylation (Hajkova *et al.* 2008, Seki *et al.*

2007). This agrees with the early wave of passive DNA demethylation in migratory PGCs. In a parallel mechanism, symmetrical di-methylation of arginine residues (R3) in H2A and H4 tails by arginine methyltransferase PRMT5 protects germ cell genomes from damage by L1 and LTR elements (Kim *et al.* 2014). PGC entry into the embryonic gonad is accompanied by prominent chromatin changes that involve temporary disappearance of chromocenters, linker histone H1, H3K9me3 and H3K27me3 modifications and HP1 proteins that bind these repressive marks in constitutive and facultative heterochromatin (Hajkova *et al.* 2008). Interestingly, gonadal PGCs also lack H2A/H4R3me2 marks as their writer PRMT5 becomes exclusively cytoplasmic (Kim *et al.* 2014, Li *et al.* 2015). Therefore, E12.5–E13.5 PGCs appear to lack most efficient epigenetic mechanisms preventing expression of retrotransposons and enter the most vulnerable period of their development.

1.5.2 *De novo DNA methylation of germ cells*

Erasure of DNA methylation is only the first critical step of epigenetic reprogramming. Establishment of new DNA methylation patterns is just as critical and perhaps, even a more complex process as its temporal and spatial precision will have a long-lasting impact on the quality of gametes and a newly formed embryonic genome following fertilization (Smallwood and Kelsey 2011). A central role in *de novo* DNA methylation of imprinted gene and repeats in the germ cell genome belongs to *de novo* methyltransferase DNMT3a and a related but enzymatically inactive DNMT3l protein (Bourc'his and Bestor 2004, Bourc'his *et al.* 2001, Kaneda *et al.* 2004, Kato *et al.* 2007, Webster *et al.* 2005). Another *de novo* methyltransferase DNMT3b modifies satellite repeats and some retrotransposons. Unlike DNMT1 maintenance methyltransferase, *de novo* methyltransferases do not require hemi-methylated DNA substrates and could, in principle, methylate any CpG dinucleotide. Yet, despite this apparent simplicity and anticipated promiscuity of *de novo* methylation, this process is finely regulated. Notably, DNMT3l plays a key role in this regulation by fine-tuning where and when *de novo* methylation should occur and accomplishes this role in a sex-specific manner (Bourc'his and Bestor, 2004, 2006; Bourc'his *et al.* 2001). One mechanism by which DNMT3l guides *de novo* DNA methylation involves binding to unmethylated lysine 4 residues in the tail of histone H3 (Ooi *et al.* 2007). Furthermore, unmethylated H3K4 residue also activates DNMT3a (Guo *et al.* 2015). Consistent with these *in vitro* observations, genetic analysis showed that *de novo* methylation of maternally imprinted genes requires the activity of H3K4 histone demethylase KDM1B (Ciccone *et al.* 2009). Another mechanism involves the control of the catalytic activity of DNMT3a through the formation of tetramers with DNMT3l resulting in a pattern of methylated CpG dinucleotides with an 8–10 bp periodicity in maternally imprinted genes (Jia *et al.* 2007). In addition to the canonical CpG methylation, mitotically arrested fetal and postnatal spermatogonia possess appreciable levels of non-CpG methylation, preferentially in hemimethylated CpApG or CpApC contexts (Ichiyanagi *et al.* 2012). However, despite the dependence on DNMT3l, the significance of this methylation is unclear as it disappears upon mitotic re-entry of postnatal spermatogonia.

Under normal physiological conditions, genomes of male fetal germ cells regain DNA methylation by birth and abrogation of DNMT3l prevents *de novo* methylation of imprinted genes and retrotransposons in fetal germ cells (Bourc'his and Bestor 2004, Webster *et al.* 2005). As a result, postnatal spermatogonia are hypomethylated and delay entry into meiotic prophase I while spermatocytes express high levels of retrotransposons and exhibit a complex phenotype eloquently described as "meiotic catastrophe", which causes arrest of gametogenesis and infertility. In contrast, neither DNMT3a nor DNMT3l are expressed in non-growing oocytes, which resist forced *de novo* DNA methylation, and it is not until the fully-grown oocyte stage that DNA methylation reaches its maximum (Hara *et al.* 2014, Smallwood *et al.* 2011). Like non-cycling prospermatogonia, oocytes also exhibit high levels of non-CpG methylation of unknown functional significance and which generally follows the pattern of CpG methylation (Shirane *et al.* 2013).

1.5.3 *A big role for small RNAs in transposon control in male germ cells*

In the male germline, *de novo* DNA methylation of truncated and inactive retrotransposons precedes that of full-length and potentially active elements from evolutionarily young families (Molaro *et al.* 2014). Interestingly, efficient and timely targeting of the *de novo* DNA methylation to full-length elements requires an RNA interference (RNAi)-related mechanism. In RNAi and related small RNA-mediated mechanisms, the fundamental principle of nucleic acid complementarity guides the activity of associated effector complexes equipped with short (19–34 nt long) RNAs towards their targets (Fire *et al.* 1998, Ghildiyal and Zamore 2009). Two most common outcomes of activity of such small RNA-Induced Silencing Complexes (RISCs) are transcriptional and posttranscriptional gene silencing (TGS and PTGS, respectively). As the name implies, TGS involves prevention of transcription of an element while PTGS results in degradation or translational repression of mRNA. Both of these mechanisms contribute to retrotransposon silencing in the male germline. The primary guides in these processes are specialized small RNAs known as Piwi-interacting RNAs (or piRNAs) for their association with Piwi proteins. An evolutionarily conserved ARGONAUTE/Piwi protein family serves as the effector core of all RNAi-related pathways (Juliano *et al.* 2010, Ku and Lin 2014, Weick and Miska 2014). piRNAs have an evolutionarily conserved task of protecting germ cell genomes from the threat posed by transposable elements. piRNAs are distinct from other types of small RNAs such as microRNAs (miRNAs) or endogenous short interfering RNAs (endo-siRNAs) in several aspects of their biogenesis and function. The unique features include their different genomic origins, their larger size of 24–32 nt (depending on the specific interacting Piwi protein), independence of their biogenesis on the processing machinery involved in the production of miRNAs and endo-siRNAs and, as the name implies, their association with Piwi rather than with Ago proteins.

piRNAs and numerous proteins involved in piRNA biogenesis and function comprise "the piRNA pathway" (Bortvin 2013, Castaneda *et al.* 2011, Ku and Lin 2014). The mouse genome encodes three Piwi proteins (MIWI/PIWIL1, MILI/PIWIL2 and MIWI2/PIWIL4) while the human genome has four (Chuma and Nakano 2013). In addition, animal genomes encode a large group of evolutionarily conserved accessory proteins with essential roles in the piRNA pathway. Tudor domain containing proteins (TDRDs) feature prominently this system and serve as scaffolds for assembly of large complexes containing piRNAs, piRNA precursors and piRNA pathway proteins (Pek *et al.* 2012). Tudor domains bind specific symmetrically dimethylated arginine residues in histone tails and on Piwi proteins providing an intriguing mechanistic link between the two types of transposon control that is further supported by studies of PRMT5 arginine methyltransferase mentioned above (Kim *et al.* 2014, Kirino *et al.* 2009, 2010, Vagin *et al.* 2009). Other piRNA pathway factors such as MOV10L1, MVH/DDX4, Zucchini/MitoPLD, HEN1 and Maelstrom possess RNA binding, unwinding, modifying or ribonuclease activities consistent with their essential functions for piRNA biogenesis (Genzor and Bortvin 2015, Ipsaro *et al.* 2012, Kirino and Mourelatos 2007, Kuramochi-Miyagawa *et al.* 2010, Matsumoto *et al.* 2015, Vourekas *et al.* 2015).

Mouse Piwi proteins exhibit strong germ-cell specificity and their temporal pattern of expression defines distinct classes of piRNAs (Bortvin 2013, Ku and Lin 2014, Weick and Miska 2014). Interestingly, DNA methylation reprogramming of PGC genomes not only derepresses retrotransposons but also activates expression of piRNA factors (Hackett *et al.* 2012). Initially distributed throughout the cytoplasm, piRNA pathway proteins begin to accumulate in characteristic germinal granules, cumulatively known as nuage (for *cloud* in French), in fetal prospermatogonia

(Chuma *et al.* 2009, Eddy 1974). MILI partners with TDRD1 in close proximity to mitochondria to form electron dense "intermitochondrial cement" or pi-bodies (Aravin *et al.* 2009, Chuma *et al.* 2006, Hosokawa *et al.* 2007, van der Heijden *et al.* 2010). MILI levels and pi-body assembly depend on the function of GASZ protein, which is also a component of this form of nuage and required for piRNA biogenesis (Ma *et al.* 2009). MIWI2 and its Tudor domain containing partner TDRD9 reside in germinal granules known as piP-bodies since they contain piRNA factors MIWI2, TDRD9, MAEL and TDRD4/RFN17 and proteins characteristic of processing bodies such as GW182 and DCP1a (Aravin *et al.* 2009, Castaneda *et al.* 2014). Furthermore, initially exclusively cytoplasmic, MIWI2 and TDRD9 become nuclear and this transition correlates with *de novo* DNA methylation of L1 elements. Interestingly, MIWI2 and TDRD9 fail to localize to piP-bodies entirely and delay nuclear entry in the absence of MAEL. MAEL possesses a unique RNA binding HMG-box domain and a potential ssRNA endonuclease domain, and associates with retrotransposon mRNAs (Chen *et al.* 2015, Genzor and Bortvin 2015, Matsumoto *et al.* 2015). Taken together with nucleo-cytoplasmic shuttling of MAEL and its association with cytoskeletal and cargo proteins, it is quite possible that MAEL facilitates trafficking of piRNA precursors to their processing sites such as piP-bodies (Castaneda *et al.* 2014, Findley *et al.* 2003, Sato *et al.* 2011, Soper *et al.* 2008). MIWI2-TDRD9 accumulation at piP-bodies could reflect on the ongoing processing of piRNAs whose loading onto MIWI2 would allow its nuclear localization and transposon silencing.

The above dynamics of piRNA pathway factor expression parallels the appearance and accumulation of fetal piRNAs. A significant share of fetal piRNAs maps to retrotransposons and other repeats (Aravin *et al.* 2008, Kuramochi-Miyagawa *et al.* 2008). Therefore, piRNA biogenesis is a PTGS process in itself as it consumes transcripts originating from genomic repeats. However, how the piRNA machinery identifies RNAs for the production of so-called primary piRNAs in the first place is an open question. Interestingly, the piRNA-based defense does not stop at degrading transposon RNAs but takes advantage of the frequent occurrence of anti-sense transposon transcripts. Sequence analyses of piRNAs have hinted at the existence of an intricate amplification mechanism of piRNA biogenesis that allows sense piRNAs to sample the transcriptome for anti-sense transcripts and generate anti-sense piRNAs that subsequently target nuclear MIWI2 effector complexes to nascent RNA transcripts (Aravin *et al.* 2008, De Fazio *et al.* 2011, Wang *et al.* 2014).

While early studies envisioned that piRNA machinery targeted transposons irrespectively of their capacity for retrotransposition, recent work suggested that piRNAs focus predominantly on full-length and, therefore, potentially active, elements (Molaro *et al.* 2014, Pezic *et al.* 2014). As a result, *de novo* DNA methylation of most repeated sequence in the genome does not require piRNAs. In contrast, piRNAs appear to target *de novo* DNA methylation and histone modification machinery to full-length retrotransposons. However, at this point, direct mechanistic links between piRNAs and DNA methylation or histone methylation remain unknown.

De novo DNA methylation of the male germline is set in place by birth and MIWI2 expression and piRNA levels wane off (Aravin *et al.* 2008). MILI remains the sole Piwi protein in undifferentiated and differentiating spermatogonia in prepubertal and adult testes. Spermatogenesis in mice and men is a cyclical process maintained by the pool of spermatogonial stem cells. Following a period of proliferation and differentiation, spermatogonia complete the last pre-meiotic round of DNA replication and enter meiosis. Interestingly, despite the critical role of fetal piRNAs in repression of active retrotransposons, fetal and adult spermatogonial populations do not exhibit immediate accumulation of DNA damage or cytotoxicity normally associated with retrotransposon activation. In all likelihood, an explanation to this lack of obvious phenotype lies in redundancy of piRNAs, DNA methylation and repressive histone modifications in spermatogonia (Di Giacomo *et al.* 2013a, 2013b).

Remarkably, an acute demise of piRNA-deficient male germ cells does not occur until the pachytene substage of meiotic prophase I (Ollinger *et al.* 2010, Zody *et al.* 2006). A prolonged meiotic prophase I is a highly specialized stage of meiosis devoted to a series of complex and precisely orchestrated events leading to homologous chromosome pairing and recombination. It is at this point in meiotic prophase I, piRNA-deficient spermatocytes arrest and apoptose. Most likely, the principle causes of death of piRNA-deficient spermatocytes are unrepaired DNA breaks and wide-spread asynapsis of homologous chromosomes. Even though spermatocytes normally possess hundreds of programmed meiotic DNA breaks created by a specialized SPO11 protein, homologous recombination machinery of the wild-type germ cell repairs these by mid-pachynema. Interestingly, spermatocytes lacking both a piRNA factor MAEL and SPO11 exhibit abundant non-meiotic DNA breaks generated most likely in the course of retrotransposition of L1 (Soper *et al.* 2008).

Expression of a third Piwi protein in the mouse, MIWI, is under the control of A-MYB transcription factor and starts in the pachynema substage of meiotic prophase I (Li *et al.* 2013). MIWI-interacting piRNAs (known as pachytene piRNAs) are highly abundant in the adult testes but their functional role remains controversial. Unlike fetal and pre-pachytene piRNAs most of which are processed from transposon messages, pachytene piRNAs originate from over a hundred of long non-coding RNA transcripts with low levels of transposon sequences (Li *et al.* 2013). Consequently, even though the deletion of MIWI arrests spermiogenesis, the lack of clear RNA targets for pachytene piRNAs precludes the straightforward prediction of their function. Possibly, numerous large pachytene piRNA genes present on practically every mouse and human chromosomes function as traps for new transposon insertions. Alternatively, pachytene piRNAs might impact translation of meiotic and spermatid RNAs (Castaneda *et al.* 2014, Grivna *et al.* 2006, Vourekas *et al.* 2012, Zhao *et al.* 2013).

Taken together, the above findings clearly demonstrate that small RNA-mediated retrotransposon control plays a critical role in spermatogenesis both at transcriptional and post-transcriptional levels. It also identifies a large group of piRNA pathway factors as potentially fertility factors and promising candidate targets for male contraceptives.

1.5.4 *The curious case of retrotransposon activity in oogenesis*

In contrast to the overwhelming evidence of the significance of piRNAs in control of transposons in the male germline, Piwi-deficient female mice are fertile (Carmell *et al.* 2007, Deng and Lin 2002, Kuramochi-Miyagawa *et al.* 2004, Lim *et al.* 2013). It appears that, at least in the mouse, the piRNA pathway functions redundantly with an additional small RNA-based mechanism of transposon regulation. Small RNA studies of growing, full-grown and metaphase II oocytes from adult wild-type ovaries uncovered abundant endo-siRNAs targeting transposon and oocyte mRNAs (Tam *et al.* 2008, Watanabe *et al.* 2008). Deletions of RNase III enzyme DICER, a critical factor in miRNA and endo-siRNA biogenesis, or its oocyte-specific isoform in growing oocytes lead to infertility with characteristic meiotic spindle defects (Flemr *et al.* 2013, Ghildiyal and Zamore 2009, Murchison *et al.* 2007). Likewise, a catalytic mutant of AGO2, which functions downstream of DICER, exhibits a similar phenotype (Stein *et al.* 2015). Importantly, the above mutants exhibit upregulation of transposon and oocyte mRNAs and, therefore, implicate endo-siRNAs in regulation of both classes of transcripts. However, how perturbation of endo-siRNAs levels leads to meiotic failure in the mature oocyte remains unclear. One attractive model envisions formation of double-stranded RNA precursors between maternal mRNAs encoding cytoskeletal proteins and anti-sense RNAs produced from pseudogenes that could be processed by DICER–AGO2 pathway. Interestingly, recent analysis of human, primate and bovine oocyte piRNAs revealed abundant expression of piRNAs that could be linked to the presence of an additional Piwi protein (PiwiL3), which is lacking in the mouse (Roovers *et al.* 2015).

The above findings provide an important insight into transposon control in the adult ovary but what about the fetal oogenesis where DICER is dispensable (Hayashi *et al.* 2008)? Fetal and primordial oocytes express many piRNA pathway proteins and possess electron-dense nuage structures reminiscent of those observed in the male germline (Lim *et al.* 2013). Yet, even though early postnatal ovaries of *Mili*, *Mvh* and *Gasz* mutants express elevated levels of L1 and IAP retrotransposons, none of these mutants exhibit significant reduction in fertility. However, an important insight into the problem of retrotransposon control comes from a recent study of retrotransposon L1 expression in the context of fetal oocyte attrition (FOA), an evolutionarily conserved process of the loss of 80% of the starting oocyte pool (Baker 1963, 1966, Burgoyne and Baker 1985, Malki *et al.* 2014). FOA is of particular interest because all fetal oocytes initiate meiosis while demethylated and will not re-methylate their genomes until after birth (Kobayashi *et al.* 2013, Seisenberger *et al.* 2012, Shirane *et al.* 2013). This late *de novo* methylation and lack of repressive H2A/H4Rme2 modifications create a window of opportunity for retrotransposon expression, particularly of L1. Indeed, detailed analysis revealed that while all fetal oocytes express L1, FOA selectively eliminates those oocytes overexpressing L1 (Malki *et al.* 2014) (Figure 1.1(a)). Furthermore, L1 activity appears to generate two distinct triggers of oocyte death. Endonuclease activity of L1ORF2p protein is likely responsible for generating wide-spread DNA damage and asynapsis that trigger pachytene checkpoint and oocyte apoptosis. Unexpectedly, treatment of pregnant females with reverse transcriptase (RT) inhibitor AZT protected early meiotic prophase I oocytes from FOA suggesting that an RT-dependent intermediate, such as L1 RNA-cDNA hybrids or another cellular process affected by AZT, profoundly impacts viability of fetal oocytes (Figure 1.1(b)). Interestingly, excessive L1 activity also interferes with crossover formation in both male and female germ cells (Castaneda *et al.* 2014, Malki *et al.* 2014). However, unlike in males where such defective spermatocytes apoptose, chiasmata-deficient oocytes reach ovulation resulting in embryonic aneuploidy and lethality. Intriguingly, an RT-dependent intermediate of L1 retrotransposition appears to underlie this chiasmata defect since AZT treatment has a strong protective effect and even increases chiasmata numbers in the wild-type mice. An important implication of this work is that a newborn's ovaries might already contain primordial follicles deficient for chiasmata. In this regard, an age-related decline in oocyte quality and increased incidence of aneuploidy could stem from preferential ovulation of unaffected oocytes but as their numbers decline, ovulation of aneuploid eggs becomes prominent.

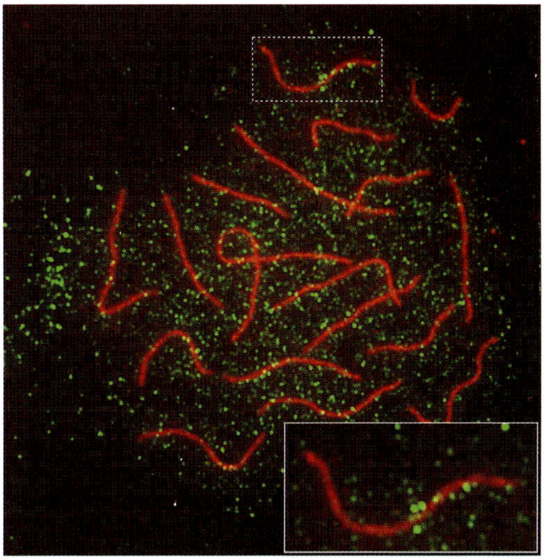

(a)

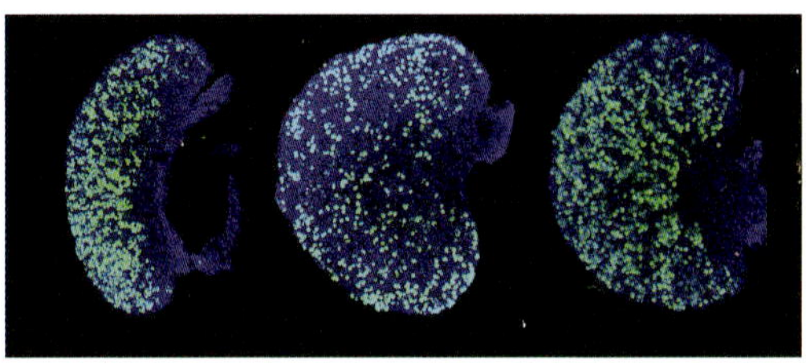

(b)

Figure 1.1. (a) Vast amount of L1 ribonucleoprotein particles in the wild-type mouse pachytene oocyte. Shown is an example of meiotic nuclear spread of pachytene stage fetal oocyte (embryonic day 18.5) labeled with antibodies to L1-encoded ORF1p protein (green) and synaptonemal complex protein SYCP2 (red). Inset — a close-up image of a single synaptonemal complex with associated L1ORF1p signal indicative of ongoing L1 retrotransposition into meiotic chromosome. (b) Reverse transcriptase inhibitor AZT has a profound protective effect on viability of mouse fetal oocytes. Embryonic day 15.5 and 18.5 wild-type ovaries are labeled with germ-cell specific marker TRA98 (green). All fetal oocytes in the E15.5 ovary are alive and will survive to E18.5 if treated with AZT but only 50% will live in the untreated E18.5 control (extracted from Malki *et al.* 2014, Figure 7(c)).

Taken together, these observations suggest that L1 elements could be important determinants of both quantitative and qualitative aspects of oogenesis in mammals. Furthermore, these findings reveal different strategies of retrotransposon control between male and female germ cell lineages. In males, transposon silencing is set in place before spermatogonial stem cell specification and is backed up by stringent checkpoints to maximize production of large numbers of high quality gametes. In contrast, the female germline appears to lack the urgency of transposon silencing thus uncovering those oocytes with most active retrotransposon content. Differential L1 activity could arise from new insertions, differential epigenetic reprogramming of germ cell genomes or redistribution of L1 mRNA or RNPs among oocytes comprising a germ cell cyst (Lei and Spradling 2013, Pepling and Spradling 1998). However, irrespective of the specific mechanism behind differential L1 expression in fetal oocytes, it appears that the female germline begins the process of selecting best oocytes for the next generation as early as in the fetal stage of oogenesis (Malki *et al.* 2014).

1.6 Conclusion

This chapter outlined the complex interplay between retrotransposons and their mammalian germline hosts. Great advancements in our ability to survey genomic sequences, characterize gene expression and place specific proteins in various states of their modification throughout the genomes of wild-type and mutant mice have opened an unprecedented opportunity to identify and characterize all participants in the arms race between mobile genetic elements and germ cell genomes. We have only recently learned about major pathways involved in transposon control and began to appreciate the great contribution of junk DNA to evolution of genomes and their hosts. Germ cells and their genomes occupy the central role in the circle of life and future work will uncover even more exciting examples of good, bad and neutral effects of transposons on the human genome. Finally, further studies of these new mechanisms and participating proteins, and RNAs will without a doubt provide important insights into mammalian gametogenesis and reproduction.

References

Akagi, K., Li, J., Stephens, R. M., Volfovsky, N. and Symer, D. E. (2008). Extensive variation between inbred mouse strains due to endogenous L1 retrotransposition. *Genome Research* 18, 869–880.

Aravin, A. A., Sachidanandam, R., Bourc'his, D., Schaefer, C., Pezic, D., Toth, K. F., Bestor, T. and Hannon, G. J. (2008). A piRNA pathway primed by individual transposons is linked to de novo DNA methylation in mice. *Molecular Cell* 31, 785–799.

Aravin, A. A., van der Heijden, G. W., Castaneda, J., Vagin, V. V., Hannon, G. J. and Bortvin, A. (2009). Cytoplasmic compartmentalization of the fetal piRNA pathway in mice. *PLoS Genetics* 5, e1000764.

Babatz, T. D. and Burns, K. H. (2013). Functional impact of the human mobilome. *Current Opinion in Genetics & Development* 23, 264–270.

Babushok, D. V. and Kazazian, H. H., Jr. (2007). Progress in understanding the biology of the human mutagen LINE-1. *Human Mutations* 28, 527–539.

Baillie, J. K., Barnett, M. W., Upton, K. R., Gerhardt, D. J., Richmond, T. A., De Sapio, F., Brennan, P. M., Rizzu, P., Smith, S., Fell, M. *et al.* (2011). Somatic retrotransposition alters the genetic landscape of the human brain. *Nature* 479, 534–537.

Baker, T. G. (1963). A quantitative and cytological study of germ cells in human ovaries. *Proceedings of the Royal Society B: Biological Sciences* 158, 417–433.

Baker, T. G. (1966). A quantitative and cytological study of oogenesis in the rhesus monkey. *Journal of Anatomy* 100, 761–776.

Bortvin, A. (2013). PIWI-interacting RNAs (piRNAs) — a mouse testis perspective. Biochemisty (Moscow) 78, 592–602.

Bourc'his, D. and Bestor, T. H. (2004). Meiotic catastrophe and retrotransposon reactivation in male germ cells lacking Dnmt3L. *Nature* 431, 96–99.

Bourc'his, D. and Bestor, T. H. (2006). Origins of extreme sexual dimorphism in genomic imprinting. *Cytogenet Genome Research* 113, 36–40.

Bourc'his, D., Xu, G. L., Lin, C. S., Bollman, B. and Bestor, T. H. (2001). Dnmt3L and the establishment of maternal genomic imprints. *Science* 294, 2536–2539.

Bourque, G., Leong, B., Vega, V. B., Chen, X., Lee, Y., Srinivasan, K. G., Chew, J.-L., Ruan, Y., Wei, C.-L. and Ng, H. (2008). Evolution of the mammalian transcription factor binding repertoire via transposable elements. *Genome Research* 18, 1752–1762.

Britten, R. J. (1996). DNA sequence insertion and evolutionary variation in gene regulation. *Proceedings of the National Academy of Sciences of the United States of America* 93, 9374–9377.

Britten, R. J. and Davidson, E. H. (1971). Repetitive and non-repetitive DNA sequences and a speculation on the origins of evolutionary novelty. *The Quarterly Review of Biology* 46, 111–138.

Britten, R. J. and Kohne, D. E. (1968). Repeated sequences in DNA. Hundreds of thousands of copies of DNA sequences have been incorporated into the genomes of higher organisms. *Science* 161, 529–540.

Brouha, B., Schustak, J., Badge, R. M., Lutz-Prigge, S., Farley, A. H., Moran, J. V. and Kazazian, H. H., Jr. (2003). Hot L1s account for the bulk of retrotransposition in the human population. *Proceedings of the National Academy of Sciences of the United States of America* 100, 5280–5285.

Bucheton, A. (1990). I transposable elements and I-R hybrid dysgenesis in Drosophila. *Trends Genetics* 6, 16–21.

Burgoyne, P. S. and Baker, T. G. (1985). Perinatal oocyte loss in XO mice and its implications for the aetiology of gonadal dysgenesis in XO women. *Journal of Reproduction and Fertility* 75, 633–645.

Burns, K. H. and Boeke, J. D. (2012). Human transposon tectonics. *Cell* 149, 740–752.

Carmell, M. A., Girard, A., van de Kant, H. J., Bourc'his, D., Bestor, T. H., de Rooij, D. G. and Hannon, G. J. (2007). MIWI2 is essential for spermatogenesis and repression of transposons in the mouse male germline. *Developmental Cell* 12, 503–514.

Carreira, P. E., Richardson, S. R. and Faulkner, G. J. (2013). L1 retrotransposons, cancer stem cells and oncogenesis. *The FEBS Journal* 281, 63–73.

Castaneda, J., Genzor, P. and Bortvin, A. (2011). piRNAs, transposon silencing, and germline genome integrity. *Mutation Research* 714, 95–104.

Castaneda, J., Genzor, P., van der Heijden, G. W., Sarkeshik, A., Yates, J. R., 3rd, Ingolia, N. T. and Bortvin, A. (2014). Reduced pachytene piRNAs and translation underlie spermiogenic arrest in Maelstrom mutant mice. *The EMBO Journal* 33, 1999–2019.

Castel, S. E. and Martienssen, R. A. (2013). RNA interference in the nucleus: roles for small RNAs in transcription, epigenetics and beyond. *Nature Reviews Genetics* 14, 100–112.

Chalopin, D., Naville, M., Plard, F., Galiana, D. and Volff, J. N. (2015). Comparative analysis of transposable elements highlights mobilome diversity and evolution in vertebrates. *Genome Biology and Evolution* 7, 567–580.

Chen, K.-M.M., Campbell, E., Pandey, R. R., Yang, Z., McCarthy, A. A. and Pillai, R. S. (2015). Metazoan Maelstrom is an RNA-binding protein that has evolved from an ancient nuclease active in protists. *RNA* (New York, NY) 21, 833–839.

Chuma, S., Hosokawa, M., Kitamura, K., Kasai, S., Fujioka, M., Hiyoshi, M., Takamune, K., Noce, T. and Nakatsuji, N. (2006). Tdrd1/Mtr-1, a tudor-related gene, is essential for male germ-cell differentiation and nuage/germinal granule formation in mice. *Proceedings of the National Academy of Sciences of the United States of America* 103, 15894–15899.

Chuma, S., Hosokawa, M., Tanaka, T. and Nakatsuji, N. (2009). Ultrastructural characterization of spermatogenesis and its evolutionary conservation in the germline: germinal granules in mammals. *Molecular and Cellular Endocrinology* 306, 17–23.

Chuma, S. and Nakano, T. (2013). piRNA and spermatogenesis in mice. *Philosophical Transactions of the Royal Society of London. Series B, Biological Sciences* 368, 20110338.

Ciccone, D. N., Su, H., Hevi, S., Gay, F., Lei, H., Bajko, J., Xu, G., Li, E. and Chen, T. (2009). KDM1B is a histone H3K4 demethylase required to establish maternal genomic imprints. *Nature* 461, 415–418.

Comfort, N. C. (1999). "The real point is control": the reception of Barbara McClintock's controlling elements. *Journal of the History of Biology* 32, 133–162.

Comfort, N. C. (2001). *The Tangled Field: Barbara McClintock's Search for the Patterns of Genetic Control*. Massachusetts, Harvard University Press.

Corces, V. G. and Geyer, P. K. (1991). Interactions of retrotransposons with the host genome: the case of the gypsy element of Drosophila. *Trends in Genetics: TIG* 7, 86–90.

Cordaux, R. and Batzer, M. A. (2009). The impact of retrotransposons on human genome evolution. *Nature Reviews Genetics* 10, 691–703.

Cost, G. J. and Boeke, J. D. (1998). Targeting of human retrotransposon integration is directed by the specificity of the L1 endonuclease for regions of unusual DNA structure. *Biochemistry* 37, 18081–18093.

Craig, N. L., Chandler, M., Gellert, M., Lambowitz, A. M. and Sandmaeyer, S. (2015). *Mobile DNA III*, 3rd edn. Canada, ASM Press.

De Fazio, S., Bartonicek, N., Di Giacomo, M., Abreu-Goodger, C., Sankar, A., Funaya, C., Antony, C., Moreira, P. N., Enright, A. J. and O'Carroll, D. (2011). The endonuclease activity of Mili fuels piRNA amplification that silences LINE1 elements. *Nature* 480, 259–263.

de Koning, A. P., Gu, W., Castoe, T. A., Batzer, M. A. and Pollock, D. D. (2011). Repetitive elements may comprise over two-thirds of the human genome. *PLoS Genetics* 7, e1002384.

DeBerardinis, R. J., Goodier, J. L., Ostertag, E. M. and Kazazian, H. H., Jr. (1998). Rapid amplification of a retrotransposon subfamily is evolving the mouse genome. *Nature Genetics* 20, 288–290.

Deininger, P. L., Moran, J. V., Batzer, M. A. and Kazazian, H. H., Jr. (2003). Mobile elements and mammalian genome evolution. *Current Opinion in Genetics and Development* 13, 651–658.

Deng, W. and Lin, H. (2002). miwi, a murine homolog of piwi, encodes a cytoplasmic protein essential for spermatogenesis. *Developmental Cell* 2, 819–830.

Dewannieux, M., Esnault, C. and Heidmann, T. (2003). LINE-mediated retrotransposition of marked Alu sequences. *Nature Genetics* 35, 41–48.

Dewannieux, M. and Heidmann, T. (2005). L1-mediated retrotransposition of murine B1 and B2 SINEs recapitulated in cultured cells. *Journal of Molecular Biology* 349, 241–247.

Dewannieux, M. and Heidmann, T. (2013). Endogenous retroviruses: acquisition, amplification and taming of genome invaders. *Current Opinion in Virology* 3, 646–656.

Di Giacomo, M., Comazzetto, S., Saini, H., De Fazio, S., Carrieri, C., Morgan, M., Vasiliauskaite, L., Benes, V., Enright, A. J. and O'Carroll, D. (2013a). Multiple epigenetic mechanisms and the piRNA pathway enforce LINE1 silencing during adult spermatogenesis. *Molecular Cell* 50, 601–608.

Di Giacomo, M., Comazzetto, S., Sampath, S. C., Sampath, S. C. and O'Carroll, D. (2013b). G9a co-suppresses LINE1 elements in spermatogonia. *Epigenetics & Chromatin* 7, 24.

Dobzhansky, T. (1973). Nothing in biology makes sense except in the light of evolution. *The American Biology Teacher* 35, 125–129.

Doolittle, W. F. (2013). Is junk DNA bunk? A critique of ENCODE. *Proceedings of the National Academy of Sciences* 110, 5294–5300.

Doolittle, W. F., Fraser, P., Gerstein, M. B., Graveley, B. R., Henikoff, S., Huttenhower, C., Oshlack, A., Ponting, C. P., Rinn, J. L., Schatz, M. C. *et al.* (2012). Sixty years of genome biology. *Genome Biology* 14, 113.

Doolittle, W. F. and Sapienza, C. (1980). Selfish genes, the phenotype paradigm and genome evolution. *Nature* 284, 601–603.

Doucet, A. J., Droc, G., Siol, O., Audoux, J. and Gilbert, N. (2015). U6 snRNA pseudogenes: markers of retrotransposition dynamics in mammals. *Molecular Biology and Evolution* 32, 1815–1832.

Doucet, A. J., Hulme, A. E., Sahinovic, E., Kulpa, D. A., Moldovan, J. B., Kopera, H. C., Athanikar, J. N., Hasnaoui, M., Bucheton, A., Moran, J. V. *et al.* (2010). Characterization of LINE-1 ribonucleoprotein particles. *PLoS Genetics* 6, pii: e1001150.

Dunham, I. (2012). An integrated encyclopedia of DNA elements in the human genome. *Nature* 489, 57–74.

Eddy, E. M. (1974). Fine structural observations on the form and distribution of nuage in germ cells of the rat. *The Anatomical Record* 178, 731–757.

Eddy, S. R. (2012). The C-value paradox, junk DNA and ENCODE. *Current Biology: CB* 22, 9.

Esnault, C., Maestre, J. and Heidmann, T. (2000). Human LINE retrotransposons generate processed pseudogenes. *Nature Genetics* 24, 363–367.

Ewing, A. D., Ballinger, T. J., Earl, D., Broad Institute Genome Sequencing and Analysis Program and Platform, Harris, C. C., Ding, L., Wilson, R. K. and Haussler, D. (2012). Retrotransposition of gene transcripts leads to structural variation in mammalian genomes. *Genome Biology* 14, R22.

Fattash, I., Rooke, R., Wong, A., Hui, C., Luu, T., Bhardwaj, P. and Yang, G. (2013). Miniature inverted-repeat transposable elements: discovery, distribution and activity. *Genome/National Research Council Canada = Génome/Conseil national de recherches Canada* 56, 475–486.

Fedoroff, N. V. (1989). About maize transposable elements and development. *Cell* 56, 181–191.

Feng, Q., Moran, J. V., Kazazian, H. H., Jr. and Boeke, J. D. (1996). Human L1 retrotransposon encodes a conserved endonuclease required for retrotransposition. *Cell* 87, 905–916.

Findley, S. D., Tamanaha, M., Clegg, N. J. and Ruohola-Baker, H. (2003). Maelstrom, a Drosophila spindle-class gene, encodes a protein that colocalizes with Vasa and RDE1/AGO1 homolog, Aubergine, in nuage. *Development* 130, 859–871.

Finnegan, D. J. (1989). Eukaryotic transposable elements and genome evolution. *Trends in Genetics* 5, 103–107.

Fire, A., Xu, S., Montgomery, M. K., Kostas, S. A., Driver, S. E. and Mello, C. C. (1998). Potent and specific genetic interference by double-stranded RNA in Caenorhabditis elegans. *Nature* 391, 806–811.

Flemr, M., Malik, R., Franke, V., Nejepinska, J., Sedlacek, R., Vlahovicek, K. and Svoboda, P. (2013). A retrotransposon-driven dicer isoform directs endogenous small interfering RNA production in mouse oocytes. *Cell* 155, 807–816.

Gao, X., Hou, Y., Ebina, H., Levin, H. L. and Voytas, D. F. (2008). Chromodomains direct integration of retrotransposons to heterochromatin. *Genome Research* 18, 359–369.

Garcia-Perez, J. L., Doucet, A. J., Bucheton, A., Moran, J. V. and Gilbert, N. (2007). Distinct mechanisms for trans-mediated mobilization of cellular RNAs by the LINE-1 reverse transcriptase. *Genome Research* 17, 602–611.

Genzor, P. and Bortvin, A. (2015). A unique HMG-box domain of mouse maelstrom binds structured RNA but not double stranded DNA. *PLoS One* 10, e0120268.

Ghildiyal, M. and Zamore, P. D. (2009). Small silencing RNAs: an expanding universe. *Nature Reviews Genetics* 10, 94–108.

Gilbert, N., Lutz, S., Morrish, T. A. and Moran, J. V. (2005). Multiple fates of L1 retrotransposition intermediates in cultured human cells. *Molecular Cell Biology* 25, 7780–7795.

Goodier, J. L. and Kazazian, H. H., Jr. (2008). Retrotransposons revisited: the restraint and rehabilitation of parasites. *Cell* 135, 23–35.

Goodier, J. L., Ostertag, E. M., Du, K. and Kazazian, H. H., Jr. (2001). A novel active L1 retrotransposon subfamily in the mouse. *Genome Research* 11, 1677–1685.

Grivna, S. T., Pyhtila, B. and Lin, H. (2006). MIWI associates with translational machinery and PIWI-interacting RNAs (piRNAs) in regulating spermatogenesis. *Proceedings of the National Academy of Sciences of the United States of America* 103, 13415–13420.

Guo, X., Wang, L., Li, J., Ding, Z., Xiao, J., Yin, X., He, S., Shi, P., Dong, L., Li, G. *et al.* (2015). Structural insight into autoinhibition and histone H3-induced activation of DNMT3A. *Nature* 517, 640–644.

Hackett, J. A., Reddington, J. P., Nestor, C. E., Dunican, D. S., Branco, M. R., Reichmann, J., Reik, W., Surani, M. A., Adams, I. R. and Meehan, R. R. (2012). Promoter DNA methylation couples genome-defence mechanisms to epigenetic reprogramming in the mouse germline. *Development* 139, 3623–3632.

Hackett, J. A., Sengupta, R., Zylicz, J. J., Murakami, K., Lee, C., Down, T. A. and Surani, M. A. (2013). Germline DNA demethylation dynamics and imprint erasure through 5-hydroxymethylcytosine. *Science* 339, 448–452.

Hajkova, P., Ancelin, K., Waldmann, T., Lacoste, N., Lange, U. C., Cesari, F., Lee, C., Almouzni, G., Schneider, R. and Surani, M. A. (2008). Chromatin dynamics during epigenetic reprogramming in the mouse germ line. *Nature* 452, 877–881.

Hajkova, P., Erhardt, S., Lane, N., Haaf, T., El-Maarri, O., Reik, W., Walter, J. and Surani, M. A. (2002). Epigenetic reprogramming in mouse primordial germ cells. *Mechanics of Development* 117, 15–23.

Han, J. S. (2010). Non-long terminal repeat (non-LTR) retrotransposons: mechanisms, recent developments, and unanswered questions. *Mobile DNA* 1, 15.

Han, J. S. and Boeke, J. D. (2005). LINE-1 retrotransposons: modulators of quantity and quality of mammalian gene expression? *BioEssays: News and Reviews in Molecular, Cellular and Developmental Biology* 27, 775–784.

Han, J. S., Szak, S. T. and Boeke, J. D. (2004). Transcriptional disruption by the L1 retrotransposon and implications for mammalian transcriptomes. *Nature* 429, 268–274.

Hancks, D. C. and Kazazian, H. H. (2010). SVA retrotransposons: Evolution and genetic instability. *Seminars in Cancer Biology* 20, 234–245.

Hancks, D. C. and Kazazian, H. H., Jr. (2012). Active human retrotransposons: variation and disease. *Current Opinion in Genetics & Development* 22, 191–203.

Hangauer, M. J., Vaughn, I. W. and McManus, M. T. (2013). Pervasive transcription of the human genome produces thousands of previously unidentified long intergenic noncoding RNAs. *PLoS Genetics* 9, e1003569.

Hara, S., Takano, T., Fujikawa, T., Yamada, M., Wakai, T., Kono, T. and Obata, Y. (2014). Forced expression of DNA methyltransferases during oocyte growth accelerates the establishment of methylation imprints but not functional genomic imprinting. *Human Molecular Genetics* 23, 3853–3864.

Hayashi, K., Chuva de Sousa Lopes, S. M., Kaneda, M., Tang, F., Hajkova, P., Lao, K., O'Carroll, D., Das, P. P., Tarakhovsky, A., Miska, E. A. *et al.* (2008). MicroRNA biogenesis is required for mouse primordial germ cell development and spermatogenesis. *PloS One* 3, e1738.

Hickman, A. B., Perez, Z. N., Zhou, L., Musingarimi, P., Ghirlando, R., Hinshaw, J. E., Craig, N. L. and Dyda, F. (2005). Molecular architecture of a eukaryotic DNA transposase. *Nature Structural Molecular Biology* 12, 715–721.

Holmquist, G. P., Kapitonov, V. V. and Jurka, J. (1998). Mobile genetic elements, chiasmata, and the unique organization of beta-heterochromatin. *Cytogenetics and Cell Genetics* 80, 113–116.

Horie, K., Saito, E.-S.S., Keng, V. W., Ikeda, R., Ishihara, H. and Takeda, J. (2007). Retrotransposons influence the mouse transcriptome: implication for the divergence of genetic traits. *Genetics* 176, 815–827.

Hosokawa, M., Shoji, M., Kitamura, K., Tanaka, T., Noce, T., Chuma, S. and Nakatsuji, N. (2007). Tudor-related proteins TDRD1/MTR-1, TDRD6 and TDRD7/TRAP: domain composition, intracellular localization and function in male germ cells in mice. *Developmental Biology* 301, 38–52.

Hou, X., Du, Y., Deng, Y., Wu, J. and Cao, G. (2014). Sleeping beauty transposon system for genetic etiological research and gene therapy of cancers. *Cancer Biology & Therapy* 16, 8–16.

Huang, C. R., Burns, K. H. and Boeke, J. D. (2012). Active transposition in genomes. *Annual Review of Genetics* 46, 651–675.

Ichiyanagi, T., Ichiyanagi, K., Miyake, M. and Sasaki, H. (2012). Accumulation and loss of asymmetric non-CpG methylation during male germ-cell development. *Nucleic Acids Research* 41, 738–745.

Ipsaro, J. J., Haase, A. D., Knott, S. R., Joshua-Tor, L. and Hannon, G. J. (2012). The structural biochemistry of Zucchini implicates it as a nuclease in piRNA biogenesis. *Nature*, 491, 279–283.

Ivics, Z. and Izsvak, Z. (2006). Transposons for gene therapy! *Current Gene Therapy* 6, 593–607.

Iyer, L. M., Tahiliani, M., Rao, A. and Aravind, L. (2009). Prediction of novel families of enzymes involved in oxidative and other complex modifications of bases in nucleic acids. *Cell Cycle* (Georgetown, Tex) 8, 1698–1710.

Jia, D., Jurkowska, R. Z., Zhang, X., Jeltsch, A. and Cheng, X. (2007). Structure of Dnmt3a bound to Dnmt3L suggests a model for *de novo* DNA methylation. *Nature* 449, 248–251.

Juliano, C., Wang, J. and Lin, H. (2010). Uniting germline and stem cells: the function of Piwi proteins and the piRNA pathway in diverse organisms. *Annual Review of Genetics* 45, 447–469.

Jurka, J. (1997). Sequence patterns indicate an enzymatic involvement in integration of mammalian retroposons. *Proceedings of the National Academy of Sciences of the United States of America* 94, 1872–1877.

Kagiwada, S., Kurimoto, K., Hirota, T., Yamaji, M. and Saitou, M. (2013). Replication-coupled passive DNA demethylation for the erasure of genome imprints in mice. *The EMBO Journal* 32, 340–353.

Kaneda, M., Okano, M., Hata, K., Sado, T., Tsujimoto, N., Li, E. and Sasaki, H. (2004). Essential role for *de novo* DNA methyltransferase Dnmt3a in paternal and maternal imprinting. *Nature* 429, 900–903.

Kano, H., Godoy, I., Courtney, C., Vetter, M. R., Gerton, G. L., Ostertag, E. M. and Kazazian, H. H., Jr. (2009). L1 retrotransposition occurs mainly in embryogenesis and creates somatic mosaicism. *Genes & Development* 23, 1303–1312.

Kato, Y., Kaneda, M., Hata, K., Kumaki, K., Hisano, M., Kohara, Y., Okano, M., Li, E., Nozaki, M. and Sasaki, H. (2007). Role of the Dnmt3 family in *de novo* methylation of imprinted and repetitive sequences during male germ cell development in the mouse. *Human Molecular Genetics* 16, 2272–2280.

Kazazian, H. H. (2011). *Mobile DNA: Finding Treasure in Junk*, 1st edn. New Jersey, FT Press Science.

Kazazian, H. H., Jr. (2000). Genetics. L1 retrotransposons shape the mammalian genome. *Science* 289, 1152–1153.

Kazazian, H. H., Jr. (2004). Mobile elements: drivers of genome evolution. *Science* 303, 1626–1632.

Kazazian, H. H., Jr. and Moran, J. V. (1998). The impact of L1 retrotransposons on the human genome. *Nature Genetics* 19, 19–24.

Kazazian, H. H., Jr., Wong, C., Youssoufian, H., Scott, A. F., Phillips, D. G. and Antonarakis, S. E. (1988). Haemophilia A resulting from *de novo* insertion of L1 sequences represents a novel mechanism for mutation in man. *Nature* 332, 164–166.

Kelley, D. and Rinn, J. (2011). Transposable elements reveal a stem cell-specific class of long noncoding RNAs. *Genome Biology* 13.

Kidwell, M. G. and Lisch, D. R. (2000). Transposable elements and host genome evolution. *Trends in Ecology & Evolution* 15, 95–99.

Kim, S., Gunesdogan, U., Zylicz, J. J., Hackett, J. A., Cougot, D., Bao, S., Lee, C., Dietmann, S., Allen, G. E., Sengupta, R. *et al.* (2014). PRMT5 protects genomic integrity during global DNA demethylation in primordial germ cells and preimplantation embryos. *Molecular Cell* 56, 564–579.

Kirino, Y., Kim, N., de Planell-Saguer, M., Khandros, E., Chiorean, S., Klein, P. S., Rigoutsos, I., Jongens, T. A. and Mourelatos, Z. (2009). Arginine methylation of Piwi proteins catalysed by dPRMT5 is required for Ago3 and Aub stability. *Nature Cell Biology* 11, 652–658.

Kirino, Y. and Mourelatos, Z. (2007). The mouse homolog of HEN1 is a potential methylase for Piwi-interacting RNAs. *RNA* 13, 1397–1401.

Kirino, Y., Vourekas, A., Sayed, N., de Lima Alves, F., Thomson, T., Lasko, P., Rappsilber, J., Jongens, T. A. and Mourelatos, Z. (2010). Arginine methylation of Aubergine mediates Tudor binding and germ plasm localization. *RNA* 16, 70–78.

Kobayashi, H., Sakurai, T., Miura, F., Imai, M., Mochiduki, K., Yanagisawa, E., Sakashita, A., Wakai, T., Suzuki, Y., Ito, T. *et al.* (2013). High-resolution DNA methylome analysis of primordial germ cells identifies gender-specific reprogramming in mice. *Genome Research* 23, 616–627.

Kojima, K. K. (2010). Different integration site structures between L1 protein-mediated retrotransposition in cis and retrotransposition in trans. *Mobile DNA* 1, 17.

Kramerov, D. A. and Vassetzky, N. S. (2010). SINEs. *Wiley Interdisciplinary Reviews RNA* 2, 772–786.

Ku, H. Y. and Lin, H. (2014). PIWI proteins and their interactors in piRNA biogenesis, germline development and gene expression. *National Science Review* 1, 205–218.

Kunarso, G., Chia, N. Y., Jeyakani, J., Hwang, C., Lu, X., Chan, Y. S., Ng, H. H. and Bourque, G. (2010). Transposable elements have rewired the core regulatory network of human embryonic stem cells. *Nature Genetics* 42, 631–634.

Kuramochi-Miyagawa, S., Kimura, T., Ijiri, T. W., Isobe, T., Asada, N., Fujita, Y., Ikawa, M., Iwai, N., Okabe, M., Deng, W. *et al.* (2004). Mili, a mammalian member of Piwi family gene, is essential for spermatogenesis. *Development* 131, 839–849.

Kuramochi-Miyagawa, S., Watanabe, T., Gotoh, K., Takamatsu, K., Chuma, S., Kojima-Kita, K., Shiromoto, Y., Asada, N., Toyoda, A., Fujiyama, A. *et al.* (2010). MVH in piRNA processing and gene silencing of retrotransposons. *Genes & Development* 24, 887–892.

Kuramochi-Miyagawa, S., Watanabe, T., Gotoh, K., Totoki, Y., Toyoda, A., Ikawa, M., Asada, N., Kojima, K., Yamaguchi, Y., Ijiri, T. W. *et al.* (2008). DNA methylation of retrotransposon genes is regulated by Piwi family members MILI and MIWI2 in murine fetal testes. *Genes and Development* 22, 908–917.

Lander, E. S., Linton, L. M., Birren, B., Nusbaum, C., Zody, M. C., Baldwin, J., Devon, K., Dewar, K., Doyle, M., FitzHugh, W. *et al.* (2001). Initial sequencing and analysis of the human genome. *Nature* 409, 860–921.

Lane, N., Dean, W., Erhardt, S., Hajkova, P., Surani, A., Walter, J. and Reik, W. (2003). Resistance of IAPs to methylation reprogramming may provide a mechanism for epigenetic inheritance in the mouse. *Genesis* 35, 88–93.

Lei, L. and Spradling, A. C. (2013). Mouse primordial germ cells produce cysts that partially fragment prior to meiosis. *Development* 140, 2075–2081.

Lesch, B. J. and Page, D. C. (2012). Genetics of germ cell development. *Nature Reviews Genetics* 13, 781–794.

Leung, D. C. and Lorincz, M. C. (2012). Silencing of endogenous retroviruses: when and why do histone marks predominate? *Trends in Biochemical Sciences* 37, 127–133.

Li, X. Z., Roy, C. K., Dong, X., Bolcun-Filas, E., Wang, J., Han, B. W., Xu, J., Moore, M. J., Schimenti, J. C., Weng, Z. *et al.* (2013). An ancient transcription factor initiates the burst of piRNA production during early meiosis in mouse testes. *Molecular Cell* 50, 67–81.

Li, Z., Yu, J., Hosohama, L., Nee, K., Gkountela, S., Chaudhari, S., Cass, A. A., Xiao, X. and Clark, A. T. (2015). The Sm protein methyltransferase PRMT5 is not required for primordial germ cell specification in mice. *The EMBO Journal* 34, 748–758.

Lim, A. K., Lorthongpanich, C., Chew, T. G., Tan, C. W., Shue, Y. T., Balu, S., Gounko, N., Kuramochi-Miyagawa, S., Matzuk, M. M., Chuma, S. *et al.* (2013). The nuage mediates retrotransposon silencing in mouse primordial ovarian follicles. *Development* 140, 3819–3825.

Ma, L., Buchold, G. M., Greenbaum, M. P., Roy, A., Burns, K. H., Zhu, H., Han, D. Y., Harris, R. A., Coarfa, C., Gunaratne, P. H. *et al.* (2009). GASZ is essential for male meiosis and suppression of retrotransposon expression in the male germline. *PLoS Genetics* 5, e1000635.

Maksakova, I. A., Romanish, M. T., Gagnier, L., Dunn, C. A., van de Lagemaat, L. N. and Mager, D. L. (2006). Retroviral elements and their hosts: insertional mutagenesis in the mouse germ line. *PLoS Genetics* 2, e2.

Malki, S., van der Heijden, G. W., O'Donnell, K. A., Martin, S. L. and Bortvin, A. (2014). A role for retrotransposon LINE-1 in fetal oocyte attrition in mice. *Developmental Cell* 29, 521–533.

Marchetto, M. C., Narvaiza, I., Denli, A. M., Benner, C., Lazzarini, T. A., Nathanson, J. L., Paquola, A. C., Desai, K. N., Herai, R. H., Weitzman, M. D. *et al.* (2013). Differential L1 regulation in pluripotent stem cells of humans and apes. *Nature* 503, 525–529.

Martin, S. L. (1991). Ribonucleoprotein particles with LINE-1 RNA in mouse embryonal carcinoma cells. *Journal of Molecular Cell Biology* 11, 4804–4807.

Martin, S. L. (2006). The ORF1 protein encoded by LINE-1: structure and function during L1 retrotransposition. *Journal of Biomedicine and Biotechnology* 2006, 45621.

Martin, S. L. (2010). Nucleic acid chaperone properties of ORF1p from the non-LTR retrotransposon, LINE-1. *RNA Biology* 7, 706–711.

Martin, S. L. and Bushman, F. D. (2001). Nucleic acid chaperone activity of the ORF1 protein from the mouse LINE-1 retrotransposon. *Molecular and Cellular Biology* 21, 467–475.

Mathias, S. L., Scott, A. F., Kazazian, H. H., Jr., Boeke, J. D. and Gabriel, A. (1991). Reverse transcriptase encoded by a human transposable element. *Science* 254, 1808–1810.

Matsumoto, N., Sato, K., Nishimasu, H., Namba, Y., Miyakubi, K., Dohmae, N., Ishitani, R., Siomi, H., Siomi, M. C. and Nureki, O. (2015). Crystal structure and activity of the endoribonuclease domain of the piRNA pathway factor maelstrom. *Cell Reports* 11, 366–375.

McCarthy, E. M. and McDonald, J. F. (2003). Long terminal repeat retrotransposons of Mus musculus. *Genome Biology* 5, R14.

McClintock, B. (1948). Mutable loci in maze. *Carnegie Inst Wash* 47, 155–169.

McClintock, B. (1951). Chromosome organizatoin and genic expression. *Cold Spring Harbor Symposia on Quantitative Biology* 16, 13–47.

McClintock, B. (1956). Controlling elements and the gene. *Cold Spring Harbor Symposia on Quantitative Biology* 21, 197–216.

McClintock, B. (1984). The significance of responses of the genome to challenge. *Science* 226, 792–801.

Messerschmidt, D. M., Knowles, B. B. and Solter, D. (2014). DNA methylation dynamics during epigenetic reprogramming in the germline and preimplantation embryos. *Genes & Development* 28, 812–828.

Mitra, R., Li, X., Kapusta, A., Mayhew, D., Mitra, R. D., Feschotte, C. and Craig, N. L. (2013). Functional characterization of piggyBat from the bat Myotis lucifugus unveils an active mammalian DNA transposon. *Proceedings of the National Academy of Sciences of the United States of America* 110, 234–239.

Molaro, A., Falciatori, I., Hodges, E., Aravin, A. A., Marran, K., Rafii, S., McCombie, W. R., Smith, A. D. and Hannon, G. J. (2014). Two waves of *de novo* methylation during mouse germ cell development. *Genes & Development* 28, 1544–1549.

Moran, J. V., DeBerardinis, R. J. and Kazazian, H. H., Jr. (1999). Exon shuffling by L1 retrotransposition. *Science* 283, 1530–1534.

Moran, J. V., Holmes, S. E., Naas, T. P., DeBerardinis, R. J., Boeke, J. D. and Kazazian, H. H., Jr. (1996). High frequency retrotransposition in cultured mammalian cells. *Cell* 87, 917–927.

Muotri, A. R., Chu, V. T., Marchetto, M. C., Deng, W., Moran, J. V. and Gage, F. H. (2005). Somatic mosaicism in neuronal precursor cells mediated by L1 retrotransposition. *Nature* 435, 903–910.

Murchison, E. P., Stein, P., Xuan, Z., Pan, H., Zhang, M. Q., Schultz, R. M. and Hannon, G. J. (2007). Critical roles for Dicer in the female germline. *Genes & Development* 21, 682–693.

Ohno, R., Nakayama, M., Naruse, C., Okashita, N., Takano, O., Tachibana, M., Asano, M., Saitou, M. and Seki, Y. (2013). A replication-dependent passive mechanism modulates DNA demethylation in mouse primordial germ cells. *Development* 140, 2892–2903.

Ohshima, K. and Okada, N. (2004). SINEs and LINEs: symbionts of eukaryotic genomes with a common tail. *Cytogenetic and Genome Research* 110, 475–490.

Ollinger, R., Reichmann, J. and Adams, I. R. (2010). Meiosis and retrotransposon silencing during germ cell development in mice. *Differentiation* 79, 147–158.

Ooi, S. K. and Bestor, T. H. (2008). The colorful history of active DNA demethylation. *Cell* 133, 1145–1148.

Ooi, S. K., Qiu, C., Bernstein, E., Li, K., Jia, D., Yang, Z., Erdjument-Bromage, H., Tempst, P., Lin, S. P., Allis, C. D. *et al.* (2007). DNMT3L connects unmethylated lysine 4 of histone H3 to de novo methylation of DNA. *Nature* 448, 714–717.

Orgel, L. E. and Crick, F. H. (1980). Selfish DNA: the ultimate parasite. *Nature* 284, 604–607.

Pace, J. K. and Feschotte, C. (2007). The evolutionary history of human DNA transposons: evidence for intense activity in the primate lineage. *Genome Research* 17, 422–432.

Pardue, M. L., Danilevskaya, O. N., Lowenhaupt, K., Slot, F. and Traverse, K. L. (1996). Drosophila telomeres: new views on chromosome evolution. *Trends in Genetics: TIG* 12, 48–52.

Pastor, W. A., Aravind, L. and Rao, A. (2013). TETonic shift: biological roles of TET proteins in DNA demethylation and transcription. *Nature Reviews Molecular Cell Biology* 14, 341–356.

Peaston, A. E., Evsikov, A. V., Graber, J. H., de Vries, W. N., Holbrook, A. E., Solter, D. and Knowles, B. B. (2004). Retrotransposons regulate host genes in mouse oocytes and preimplantation embryos. *Developmental Cell* 7, 597–606.

Pek, J. W., Anand, A. and Kai, T. (2012). Tudor domain proteins in development. *Development* 139, 2255–2266.

Pepling, M. E. and Spradling, A. C. (1998). Female mouse germ cells form synchronously dividing cysts. *Development* 125, 3323–3328.

Pezic, D., Manakov, S. A., Sachidanandam, R. and Aravin, A. A. (2014). piRNA pathway targets active LINE1 elements to establish the repressive H3K9me3 mark in germ cells. *Genes & Development* 28, 1410–1428.

Piégu, B., Bire, S., Arensburger, P. and Bigot, Y. (2015). A survey of transposable element classification systems — A call for a fundamental update to meet the challenge of their diversity and complexity. *Molecular Phylogenetics and Evolution* 86, 90–109.

Pierre-Étienne, J., Justin, J. and Guillaume, B. (2013). The majority of primate-specific regulatory sequences are derived from transposable elements. *PLoS Genetics* 9, e1003504.

Popp, C., Dean, W., Feng, S., Cokus, S. J., Andrews, S., Pellegrini, M., Jacobsen, S. E. and Reik, W. (2010). Genome-wide erasure of DNA methylation in mouse primordial germ cells is affected by AID deficiency. *Nature* 463, 1101–1105.

Qin, C., Wang, Z., Shang, J., Bekkari, K., Liu, R., Pacchione, S., McNulty, K. A., Ng, A., Barnum, J. E. and Storer, R. D. (2010). Intracisternal A particle genes: Distribution in the mouse genome, active subtypes, and potential roles as species-specific mediators of susceptibility to cancer. *Molecular Carcinogenesis* 49, 54–67.

Rangan, P., Malone, C. D., Navarro, C., Newbold, S. P., Hayes, P. S., Sachidanandam, R., Hannon, G. J. and Lehmann, R. (2011). piRNA production requires heterochromatin formation in Drosophila. *Current Biology* 21, 1373–1379.

Ray, D. A., Feschotte, C., Pagan, H. J., Smith, J. D., Pritham, E. J., Arensburger, P., Atkinson, P. W. and Craig, N. L. (2008). Multiple waves of recent DNA transposon activity in the bat, Myotis lucifugus. *Genome Research* 18, 717–728.

Ribet, D., Harper, F., Dupressoir, A., Dewannieux, M., Pierron, G. and Heidmann, T. (2008). An infectious progenitor for the murine IAP retrotransposon: emergence of an intracellular genetic parasite from an ancient retrovirus. *Genome Research* 18, 597–609.

Rodic, N., Sharma, R., Sharma, R., Zampella, J., Dai, L., Taylor, M. S., Hruban, R. H., Iacobuzio-Donahue, C. A., Maitra, A., Torbenson, M. S. *et al.* (2014). Long interspersed element-1 protein expression is a hallmark of many human cancers. *American Journal of Pathology* 184, 1280–1286.

Roovers, E. F., Rosenkranz, D., Mahdipour, M., Han, C. T., He, N., Chuva de Sousa Lopes, S. M., van der Westerlaken, L. A., Zischler, H., Butter, F., Roelen, B. A. *et al.* (2015). Piwi Proteins and piRNAs in Mammalian Oocytes and Early Embryos. *Cell Reports* 10, 2069–2082.

Rory, J. and Roderic, G. (2014). The RIDL hypothesis: transposable elements as functional domains of long noncoding RNAs. *RNA* 20, 959–976.

Rubin, G. M. and Spradling, A. C. (1983). Vectors for P element-mediated gene transfer in Drosophila. *Nucleic Acids Research* 11, 6341–6351.

Sandmeyer, S. B., Hansen, L. J. and Chalker, D. L. (1989). Integration specificity of retrotransposons and retroviruses. *Annual Review of Genetics* 24, 491–518.

Sassaman, D. M., Dombroski, B. A., Moran, J. V., Kimberland, M. L., Naas, T. P., DeBerardinis, R. J., Gabriel, A., Swergold, G. D. and Kazazian, H. H., Jr. (1997). Many human L1 elements are capable of retrotransposition. *Nature Genetics* 16, 37–43.

Sato, K., Nishida, K. M., Shibuya, A., Siomi, M. C. and Siomi, H. (2011). Maelstrom coordinates microtubule organization during Drosophila oogenesis through interaction with components of the MTOC. *Genes & Development* 25, 2361–2373.

Savage, A. L., Bubb, V. J., Breen, G. and Quinn, J. P. (2013). Characterisation of the potential function of SVA retrotransposons to modulate gene expression patterns. *BMC Evolutionary Biology* 13, 101.

Seisenberger, S., Andrews, S., Krueger, F., Arand, J., Walter, J., Santos, F., Popp, C., Thienpont, B., Dean, W. and Reik, W. (2012). The dynamics of genome-wide DNA methylation reprogramming in mouse primordial germ cells. *Molecular Cell* 48, 849–862.

Seisenberger, S., Peat, J. R., Hore, T. A., Santos, F., Dean, W. and Reik, W. (2013). Reprogramming DNA methylation in the mammalian life cycle: building and breaking epigenetic barriers. *Philosophical Transactions of the Royal Society of London. Series B, Biological Sciences* 368, 20110330.

Seki, Y., Yamaji, M., Yabuta, Y., Sano, M., Shigeta, M., Matsui, Y., Saga, Y., Tachibana, M., Shinkai, Y. and Saitou, M. (2007). Cellular dynamics associated with the genome-wide epigenetic reprogramming in migrating primordial germ cells in mice. *Development* 134, 2627–2638.

Shirane, K., Toh, H., Kobayashi, H., Miura, F., Chiba, H., Ito, T., Kono, T. and Sasaki, H. (2013). Mouse oocyte methylomes at base resolution reveal genome-wide accumulation of non-CpG methylation and role of DNA methyltransferases. *PLoS Genetics* 9, e1003439.

Simons, C., Makunin, I. V., Pheasant, M. and Mattick, J. S. (2006a). Maintenance of transposon-free regions throughout vertebrate evolution. *BMC Genomics* 8, 470.

Simons, C., Pheasant, M., Makunin, I. V. and Mattick, J. S. (2006b). Transposon-free regions in mammalian genomes. *Genome Research* 16, 164–172.

Smallwood, S. A. and Kelsey, G. (2011). *De novo* DNA methylation: a germ cell perspective. *Trends in Genetics: TIG* 28, 33–42.

Smallwood, S. A., Tomizawa, S.-I., Krueger, F., Ruf, N., Carli, N., Segonds-Pichon, A., Sato, S., Hata, K., Andrews, S. R. and Kelsey, G. (2011). Dynamic CpG island methylation landscape in oocytes and preimplantation embryos. *Nature Genetics* 43, 811–814.

Smit, A. F. (1996). The origin of interspersed repeats in the human genome. *Current Opinion in Genetics & Development* 6, 743–748.

Smit, A. F. and Riggs, A. D. (1996). Tiggers and DNA transposon fossils in the human genome. *Proceedings of the National Academy of Sciences of the United States of America* 93, 1443–1448.

Soper, S. F., van der Heijden, G. W., Hardiman, T. C., Goodheart, M., Martin, S. L., de Boer, P. and Bortvin, A. (2008). Mouse maelstrom, a component of nuage, is essential for spermatogenesis and transposon repression in meiosis. *Developmental Cell* 15, 285–297.

Spradling, A. C. (1994). Transposable elements and the evolution of heterochromatin. *Society of General Physiologists Series* 49, 69–83.

Stein, P., Rozhkov, N. V., Li, F., Cárdenas, F. L., Davydenk, O., Vandivier, L. E., Gregory, B. D., Hannon, G. J. and Schultz, R. M. (2015). Essential Role for endogenous siRNAs during meiosis in mouse oocytes. *PLoS Genetics* 11.

Stencel, A. and Crespi, B. (2013). What is a genome? *Molecular Ecology* 22, 3437–3443.

Sundaram, V., Cheng, Y., Ma, Z., Li, D., Xing, X., Edge, P., Snyder, M. P. and Wang, T. (2014). Widespread contribution of transposable elements to the innovation of gene regulatory networks. *Genome Research* 24, 1963–1976.

Tam, O. H., Aravin, A. A., Stein, P., Girard, A., Murchison, E. P., Cheloufi, S., Hodges, E., Anger, M., Sachidanandam, R., Schultz, R. M. *et al.* (2008). Pseudogene-derived small interfering RNAs regulate gene expression in mouse oocytes. *Nature* 453, 534–538.

Tubio, J. M. C., Li, Y., Ju, Y. S., Martincorena, I., Cooke, S. L., Tojo, M., Gundem, G., Pipinikas, C. P., Zamora, J., Raine, K. *et al.* (2014). Extensive transduction of nonrepetitive DNA mediated by L1 retrotransposition in cancer genomes. *Science* 345, 1251343.

Vagin, V. V., Wohlschlegel, J., Qu, J., Jonsson, Z., Huang, X., Chuma, S., Girard, A., Sachidanandam, R., Hannon, G. J. and Aravin, A. A. (2009). Proteomic analysis of murine Piwi proteins reveals a role for arginine methylation in specifying interaction with Tudor family members. *Genes & Development* 23, 1749–1762.

van den Hurk, J. A., Meij, I. C., Seleme, M. C., Kano, H., Nikopoulos, K., Hoefsloot, L. H., Sistermans, E. A., de Wijs, I. J., Mukhopadhyay, A., Plomp, A. S. *et al.* (2007). L1 retrotransposition can occur early in human embryonic development. *Human Molecular Genetics* 16, 1587–1592.

van der Heijden, G. W., Castaneda, J. and Bortvin, A. (2010). Bodies of evidence — compartmentalization of the piRNA pathway in mouse fetal prospermatogonia. *Current Opinion in Cell Biology* 22, 752–757.

Varagona, M. J., Purugganan, M. and Wessler, S. R. (1992). Alternative splicing induced by insertion of retrotransposons into the maize waxy gene. *The Plant Cell* 4, 811–820.

Vourekas, A., Zheng, K., Fu, Q., Maragkakis, M., Alexiou, P., Ma, J., Pillai, R. S., Mourelatos, Z. and Wang, P. J. (2015). The RNA helicase MOV10L1 binds piRNA precursors to initiate piRNA processing. *Genes & Development* 29, 617–629.

Vourekas, A., Zheng, Q., Alexiou, P., Maragkakis, M., Kirino, Y., Gregory, B. D. and Mourelatos, Z. (2012). Mili and Miwi target RNA repertoire reveals piRNA biogenesis and function of Miwi in spermiogenesis. *Nature Structural & Molecular Biology* 19, 773–781.

Wallace, N., Wagstaff, B. J., Deininger, P. L. and Roy-Engel, A. M. (2008). LINE-1 ORF1 protein enhances Alu SINE retrotransposition. *Gene* 419, 1–6.

Walsh, C. P., Chaillet, J. R. and Bestor, T. H. (1998). Transcription of IAP endogenous retroviruses is constrained by cytosine methylation. *Nature Genetics* 20, 116–117.

Wang, P. J., Page, D. C. and McCarrey, J. R. (2005). Differential expression of sex-linked and autosomal germ-cell-specific genes during spermatogenesis in the mouse. *Human Molecular Genetics* 14, 2911–2918.

Wang, W., Yoshikawa, M., Han, B. W., Izumi, N., Tomari, Y., Weng, Z. and Zamore, P. D. (2014). The initial uridine of primary piRNAs does not create the tenth adenine that Is the hallmark of secondary piRNAs. *Molecular Cell* 56, 708–716.

Watanabe, T., Tomizawa, S., Mitsuya, K. and Totoki, Y. (2011). Role for piRNAs and non-coding RNA in *de novo* DNA methylation of the imprinted mouse Rasgrf1 locus. *Science* 332, 848–852.

Watanabe, T., Totoki, Y., Toyoda, A., Kaneda, M., Kuramochi-Miyagawa, S., Obata, Y., Chiba, H., Kohara, Y., Kono, T., Nakano, T. *et al.* (2008). Endogenous siRNAs from naturally formed dsRNAs regulate transcripts in mouse oocytes. *Nature* 453, 539–543.

Waterston, R. H., Lindblad-Toh, K., Birney, E., Rogers, J., Abril, J. F., Agarwal, P., Agarwala, R., Ainscough, R., Alexandersson, M., An, P. *et al.* (2002). Initial sequencing and comparative analysis of the mouse genome. *Nature* 420, 520–562.

Webster, K. E., O'Bryan, M. K., Fletcher, S., Crewther, P. E., Aapola, U., Craig, J., Harrison, D. K., Aung, H., Phutikanit, N., Lyle, R. *et al.* (2005). Meiotic and epigenetic defects in Dnmt3L-knockout mouse spermatogenesis. *Proceedings of the National Academy of Sciences of the United States of America* 102, 4068–4073.

Weick, E. M. and Miska, E. A. (2014). piRNAs: from biogenesis to function. *Development* 141, 3458–3471.

Weiner, A. M., Deininger, P. L. and Efstratiadis, A. (1986). Nonviral retroposons: genes, pseudogenes, and transposable elements generated by the reverse flow of genetic information. *Annual Review of Biochemistry* 55, 631–661.

Weiss, R. A. (2005). The discovery of endogenous retroviruses. *Retrovirology* 3, 67.

Wichman, H. A., Van den Bussche, R. A., Hamilton, M. J. and Baker, R. J. (1992). Transposable elements and the evolution of genome organization in mammals. *Genetica* 86, 287–293.

Wicker, T., Sabot, F., Hua-Van, A., Bennetzen, J. L., Capy, P., Chalhoub, B., Flavell, A., Leroy, P., Morgante, M., Panaud, O. *et al.* (2007). A unified classification system for eukaryotic transposable elements. *Nature Reviews Genetics* 8, 973–982.

Xing, J., Wang, H., Zhang, Y., Ray, D. A., Tosi, A. J., Disotell, T. R. and Batzer, M. A. (2007). A mobile element-based evolutionary history of guenons (tribe Cercopithecini). *BMC Biology* 5, 5.

Xing, J., Zhang, Y., Han, K., Salem, A. H., Sen, S. K., Huff, C. D., Zhou, Q., Kirkness, E. F., Levy, S., Batzer, M. A. *et al.* (2009). Mobile elements create structural variation: analysis of a complete human genome. *Genome Research* 19, 1516–1526.

Yamaguchi, S., Hong, K., Liu, R., Shen, L., Inoue, A., Diep, D., Zhang, K. and Zhang, Y. (2012). Tet1 controls meiosis by regulating meiotic gene expression. *Nature* 492, 443–447.

Yamaguchi, S., Shen, L., Liu, Y., Sendler, D. and Zhang, Y. (2013). Role of Tet1 in erasure of genomic imprinting. *Nature* 504, 460–464.

Yoder, J. A., Walsh, C. P. and Bestor, T. H. (1997). Cytosine methylation and the ecology of intragenomic parasites. *Trends in Genetics: TIG* 13, 335–340.

Yuan, Y.-W. and Wessler, S. R. (2011). The catalytic domain of all eukaryotic cut-and-paste transposase superfamilies. *Proceedings of the National Academy of Sciences* 108, 7884–7889.

Zhao, S., Gou, L. T., Zhang, M., Zu, L. D., Hua, M. M., Hua, Y., Shi, H. J., Li, Y., Li, J., Li, D. *et al.* (2013). piRNA-triggered MIWI ubiquitination and removal by APC/C in late spermatogenesis. *Developmental Cell* 24, 13–25.

Zody, M. C., Garber, M., Adams, D. J., Sharpe, T., Harrow, J., Lupski, J. R., Nicholson, C., Searle, S. M., Wilming, L., Young, S. K. *et al.* (2006). DNA sequence of human chromosome 17 and analysis of rearrangement in the human lineage. *Nature* 440, 1045–1049.

2

Epigenetic Modifications during Mammalian Oogenesis: Emerging Roles of Chromatin Structure during Oocyte Growth and Meiotic Maturation

Rabindranath De La Fuente[*,†], *Claudia Baumann*[*] *and Maria M. Viveiros*[*,†]

[*]*Department of Physiology and Pharmacology,*
College of Veterinary Medicine, University of Georgia,
501 D.W. Brooks Drive, Athens, GA, 30602, USA
[†]*Regenerative Biosciences Center,*
University of Georgia, 501 D.W. Brooks Drive, Athens, GA, 30602, USA

Chromatin modifications during oogenesis are essential for the control of gene expression, the establishment of maternal-specific DNA methylation marks and chromosome stability at critical stages of oocyte growth and meiotic maturation. Compelling evidence indicates that establishment of the correct epigenetic landscape during oocyte growth is indispensable to ensure proper chromosome segregation, for silencing of repetitive elements and potentially deleterious transposons and for meiotic centromere stability. During the critical window of post-natal oocyte growth, chromatin modifications are required for the establishment of maternal specific genomic imprinting and global methylation patterns in the oocyte genome. Importantly, large-scale chromatin remodeling in pre-ovulatory oocytes is essential for pericentric heterochromatin formation and the epigenetic control of centromere function in preparation for meiotic chromosome segregation. Functional differentiation of chromatin structure during oogenesis or 'epigenetic maturation' is regulated by multiple and hierarchical chromatin modifications established at critical developmental transitions that are regulated at the local (single locus) or global (entire chromosomal domains) levels by an increasing list of chromatin remodeling proteins and histone post-translational modifications. Here, we focus on recent advances in our understanding of epigenetic maturation of the mammalian oocyte genome and the emerging links between chromatin remodeling and transcription in the establishment of maternal DNA methylation and maintenance of chromosome stability during oogenesis.

2.1 Introduction

Functional differentiation of chromatin structure during post-natal oocyte growth is essential for the control of transcription, nuclear architecture, establishment of gamete specific DNA methylation patterns and maintenance of chromosome stability during meiosis (De La Fuente 2006). The mammalian oocyte nucleus or germinal vesicle (GV) is subject to dynamic chromatin modifications at several developmental transitions in preparation for meiotic divisions. This process of epigenetic maturation is strictly required for the acquisition of both meiotic and developmental potential. During the critical window of post-natal development, the oocyte genome is subject to multiple levels or regulation. At the local level, targeted chromatin modifications at regulatory DNA sequences of specific loci establish gamete-specific epigenetic marks or genomic imprinting. Importantly, superimposed on these local chromatin modifications lays an important non-targeted strategy for the control of global transcription and large-scale chromatin remodeling (De La Fuente 2006). Multiple and hierarchical large-scale chromatin modifications induced by chromatin remodeling proteins and histone post-translational modifications regulate heterochromatin formation and the epigenetic control of centromere function (De La Fuente *et al.* 2012). Notably, in order to cope with the unique functional requirements of meiosis-specific chromosome dynamics, specialized mechanisms are set in place for the control of large-scale chromatin remodeling in the female germ line (De La Fuente 2006, 2014, Ivanovska and Orr-Weaver 2006, Kimmins and Sassone-Corsi 2005, Revenkova *et al.* 2004).

The mechanisms regulating the 'meiotic histone code' are beginning to be unraveled through the generation of genetic mouse models and the use of genome-wide transcriptome and DNA methylome analyses. These are powerful strategies that will accelerate our understanding of how the meiotic histone code, or changes in large-scale chromatin structure regulated by ATP-dependent chromatin remodeling complexes may regulate female meiosis and maternal inheritance of epigenetic states. This will be critical to gain a mechanistic understanding of the epigenetic origins of chromosome instability in the germ line, genomic reprogramming and the control of gene expression during development and differentiation.

2.2 Functional Differentiation of Chromatin Structure during Post-Natal Oocyte Growth

Differential methylation patterns established during gametogenesis confer allele-specific parental identity through a process of genomic imprinting, as these marks are established either during oogenesis or spermatogenesis (Hackett and Surani 2013, Weaver and Bartolomei 2014). The paternal or maternal-specific expression of imprinted genes is under the control of regulatory sequences called imprinted control regions (ICRs). All known ICRs are germ line differentially methylated regions (DMRs), which are generally located within 5' flanking regions of imprinted genes (Weaver and Bartolomei 2014). Following global genome reprograming in germ cells, maternal-specific DNA methylation patterns are reestablished during a critical window of post-natal oocyte growth. Beginning with day five of post-natal development and coincident with the transition from primary to secondary follicles, maternal methylation imprints at CpG islands begin to be established, progressively on a locus-by-locus basis, a process that is mostly completed by the germinal vesicle stage before global transcriptional repression in pre-ovulatory oocytes (Lucifero *et al.* 2004, Obata and Kono 2002, Smallwood *et al.* 2011).

De novo methylation occurs during oocyte growth around the time when the oocyte reaches a 50 μm diameter and different DMRs acquire methylation in different oocyte size populations. The ontogeny of DNA methylation at specific loci during oogenesis has been previously described (Lucifero *et al.* 2004, Smallwood *et al.* 2011). However, less is known about the mechanisms regulating the establishment and/or maintenance of DNA methylation at DMRs of imprinted genes during oogenesis. A *de novo*, DNA methyltransferase enzyme DNMT3A and its catalytically inactive regulatory co-factor DNMT3L regulate the establishment of DNA methylation at DMRs in mouse oocytes (Bourc'his *et al.* 2001, Hata *et al.* 2002, Kaneda *et al.* 2004). In addition, KDM1B, a histone H3 lysine 4 demethylase, also known as AOF1 or LSD2 is required for DNA methylation of several but not all DMRs of maternally imprinted genes during oocyte growth (Ciccone *et al.* 2009). The mechanisms of KDM1B function remain to be determined. However, it has been suggested that KDM1B may induce de-methylation of H3K4 potentially inducing an open chromatin configuration to recruit the *de novo* DNA methylation machinery to selected loci (Ciccone *et al.* 2009). The mechanisms conferring specificity to select loci remain to be determined. Similarly, the KRAB-zinc finger protein Zfp57 is required for methylation of a DMR in the SNURPN locus through a potential interaction with KAP1 (also known as TRIM28, Tif1b or Krip-1), which in turn interacts with the histone methyltransferase SETDB1 and with DNMT3A (Li *et al.* 2008, Messerschmidt *et al.* 2012, Quenneville *et al.* 2011).

Reduced representation bisulfate sequencing (RRBS) studies identified approximately, 1,600 CpG islands that undergo DNMT3A/Dnmt3L-dependent methylation in mature oocytes. Interestingly, these CpG islands were found to be preferentially located within active transcription units, lacking histone H3 tri-methylated at lysine 4 (H3K4me3), a histone mark associated with transcriptional repression and found enriched at un-methylated CpG islands in embryonic stem cells, supporting previous reports indicating that establishment of DNA methylation at imprinted loci during oocyte growth may be linked with active transcription across the differentially methylated regions of these loci (Chotalia *et al.* 2009, Smallwood *et al.* 2011). The role of transcription in the establishment of DNA methylation at imprinted loci during oogenesis is not fully understood. However, it has been suggested that transcription may be an important mechanism to maintain an open chromatin configuration and allow the recruitment of *de novo* methylases to regulatory sequences (Chotalia *et al.* 2009). An important observation emerging from genome wide studies is that methylation patterns during oocyte growth are very dynamic revealing the presence of more than 85 CpG islands that do not seem to complete the establishment of DNA methylation until the final stages of oocyte meiotic maturation at metaphase II (Smallwood *et al.* 2011), a stage known to be highly susceptible to both hormonal and/or environmental fluctuations (De La Fuente 2006) and, thus, with important clinical implications for assisted reproductive technologies.

Notably genome wide analysis of DNA methylation during oogenesis has also recently revealed the presence of CpG methylation at a subset of non-imprinted genes in mammalian oocytes (Rutledge *et al.* 2014). Results obtained from several independent laboratories indicate that CpG island methylation during post-natal oocyte growth is not exclusively associated with genomic imprinting but it determines a significant proportion of the DNA methylation status inherited by the pre-implantation embryo (Rutledge *et al.* 2014; Smallwood *et al.* 2011). Amongst the 1,600 methylated CpG islands (also called gametic differentially methylated regions; gDMRs) found in the oocyte genome, some are associated with imprinted genes, some seem to play a role in maintaining transcriptional repression of a subset of testis-specific genes and a third functional type is required to repress brain-specific genes during embryonic development (Rutledge *et al.* 2014). Importantly, these gDMRs are evolutionarily conserved between mouse and human and may be important to regulate phenotypic changes induced by pre-natal and post-natal environmental factors that affect DNA methylation (Rutledge *et al.* 2014). Notably, methylome analysis using genome-wide bisulfite sequencing (DNASeq) revealed that the oocyte genome also exhibits significant DNA methylation at non-CpG islands. Indeed, methylation in oocytes seems to occur predominantly over gene bodies (Kobayashi *et al.* 2012, Shirane *et al.* 2013). Genomic non-CpG methylation has been described in pluripotent stem cells, brain and growing oocytes (Lister *et al.* 2009, Shirane *et al.* 2013). Remarkably, non-CpG methylation has been recently shown to comprise up to 25% of all cytosine methylation in embryonic stem cells adding a new layer of regulation to the mammalian genome (Lister *et al.* 2009). Shirane *et al.* constructed methylome maps of non growing oocytes (Day 5), fully grown oocytes at the germinal vesicle stage and oocytes mutant for DNMT3A, DNMT3B and DNMT3L to elucidate the developmental timing of non-CpG methylation and the enzymes responsible for its regulation. Nearly two thirds of all methyl cytosines in the oocyte genome occur in a non-CpG context and showed clear enrichment in gene bodies. Importantly, non-CpG methylation accumulates during oocyte growth (non-growing oocytes were four times less methylated at non CpG sites than fully grown oocytes). Non-CpG methylation depends on the Dnmt3L complex but not on Dnmt3b (Shirane *et al.* 2013).

The biological significance of non-CpG methylation in the oocyte genome is not clear at present. However, both the recent discovery of different types of DNA methylation during oogenesis and the mechanistic insight obtained from genome wide sequencing of the oocyte methylome have now shifted the paradigm from initially considering DNA sequence motifs or structural properties as the potential primary mark for the establishment of specific methylation patterns, to that of the presence of critical histone modifications and transcriptional status at the chromatin template that may initiate the recruitment of the DNMT3A/DNMT3L complex. Exposure of intragenic regions to DNA methyltransferases during transcription as well as other epigenetic marks that induce an open chromatin structure seems critical in this new model. Thus, both transcriptional and epigenetic modifications may be required for Dnmt3L mediated intragenic methylation (Chotalia *et al.* 2009, Smallwood *et al.* 2011, Weaver and Bartolomei 2014). Changes in transcription during oocyte growth may thus coordinate the progressive establishment of methylation marks at different DMRs and may be subject to signaling from granulosa cells at different stages of follicular growth and differentiation as well as hormonal or environmental fluctuations.

2.3 Large-Scale Chromatin Remodeling in Pre-ovulatory Oocytes

Oocyte meiotic maturation entails two consecutive reductional divisions to decrease the number of chromosomes to a haploid complement. Meiotic chromosomes exhibit unique structural and functional characteristics to coordinate the initial segregation of homologous chromosomes (meiosis-I) with the subsequent separation of sister chromatids during meiosis-II. Establishment and timely resolution of centromeric cohesion during this process is under stringent molecular regulation and specialized spindle microtubule attachments are necessary to ensure accurate chromosome segregation (Page and Hawley 2003, Petronczki *et al.* 2003).

In preparation for these meiosis-specific events, pre-ovulatory oocytes are subject to large-scale chromatin remodeling events that are essential for timely meiotic progression and chromosome segregation. Several lines of evidence indicate that chromatin remodeling is critical for the acquisition of both meiotic and developmental competence in mammalian oocytes (Baumann *et al.* 2010, De La Fuente 2006, De La Fuente *et al.* 2004, Yang *et al.* 2009). For example, growing oocytes are characterized by highly decondensed chromatin that encompasses the entire GV, a configuration termed "non-surrounded nucleolus" (NSN) (Debey *et al.* 1993, Mattson and Albertini 1990). At this stage oocytes show high levels of transcriptional activity, required for the synthesis and storage of maternal transcripts essential to support meiotic resumption and the first cleavage divisions until embryonic genome activation ensues in the fertilized embryo. During the final stages of oocyte growth, however, changes in global chromatin structure lead to a progressive DNA condensation and the formation of a perinucleolar heterochromatin rim, termed "surrounded nucleolus" (SN) or karyosphere (De La Fuente 2006, Debey *et al.* 1993, Parfenov *et al.* 1989, Zuccotti 1995). Large-scale chromatin remodeling into the SN configuration is associated with global transcriptional quiescence. This process is critical to confer full developmental potential and to ensure timely progression of meiotic maturation (De La Fuente 2006, Debey *et al.* 1993, Schramm *et al.* 1993, Wickramasinghe *et al.* 1991) (Zuccotti *et al.* 1998). However, the mechanisms involved in this process remain poorly understood.

Evidence suggests that signals originating from companion granulosa cells modulate chromatin remodeling and transcriptional activity in pre-ovulatory oocytes (Chen *et al.* 2013, De La Fuente and Eppig 2001, Liu and Aoki 2002). However, pharmacological manipulation of chromatin structure and several transgenic mouse models indicate that large-scale chromatin remodeling and global transcriptional repression can be experimentally dissociated and are thus regulated through different pathways (Abe *et al.* 2010, De La Fuente *et al.* 2004, Andreu-Vieyra *et al.* 2010, Pattabiraman *et al.* 2015). For example, in pre-ovulatory oocytes from nucleoplasmin 2 (Npm2) knockout mice chromatin remains in a decondensed, NSN-like status following gonadotropin stimulation, yet global transcriptional silencing ensues in a timely manner (De La Fuente *et al.* 2004). Conversely, oocytes lacking the histone H3K4 methyltransferase MLL2 or the bromo-domain protein BRWD1 fail to undergo global transcriptional repression despite large-scale chromatin remodeling from the NSN to the SN configuration (Andreu-Vieyra *et al.* 2010, Pattabiraman *et al.* 2015). Additional insight is derived from studies investigating the role of the vaccinia related kinase (VRK1) in mammalian oocytes. Loss of function analysis revealed a critical role for this conserved histone H2A kinase in nuclear chromatin remodeling in preparation for meiotic resumption (Schober *et al.* 2011). Depletion of VRK1 causes striking epigenetic changes, such as loss of histone H2A phosphorylation (H2AT119ph), decreased histone H3 (H3K14ac) and histone H4 (H4K5ac) acetylation. Notably, loss of VRK1 in fully-grown oocytes leads to abnormal chromatin remodeling and a reduced proportion of gametes that exhibit the SN chromatin configuration. Taken together, these studies provide compelling evidence indicating that large-scale chromatin remodeling and global transcriptional silencing in fully-grown oocytes are concurrent developmental processes under the control of discrete cellular pathways. Notably, chromatin configuration and sub-cellular position of the germinal vesicle in fully-grown human oocytes has been suggested to provide a non-invasive marker for the gamete's meiotic and developmental potential (Bellone *et al.* 2009, Brunet and Maro 2007, Inoue *et al.* 2008). These findings suggest that global chromatin remodeling and the resulting epigenetic landscape play critical roles in the establishment of full meiotic and developmental potential in mammalian oocytes of several species (Ivanovska and Orr-Weaver 2006).

The precise orchestration of establishment, maintenance and/or erasure of covalent histone modifications confers structural and functional identity to nuclear domains (Dillon and Festenstein 2002, Dundr M 2001) and is an essential prerequisite for the control of meiotic chromosome segregation. For instance, while a subset of epigenetic marks, such as histone lysine methylation, remains stably associated with specific chromatin domains of meiotic chromosomes throughout meiosis (Arney *et al.* 2002, Liu *et al.* 2004, Meglicki *et al.* 2008, Ooga *et al.* 2008, Swain *et al.* 2007, Wang *et al.* 2006), other modifications, such as histone acetylation, exhibit highly dynamic patterns during meiotic resumption (Adenot *et al.* 1997, De La Fuente *et al.* 2004, Kim *et al.* 2003, Sarmento *et al.* 2004). High levels of acetylation of histone H3 and H4 at most lysine residues are detectable in fully-grown, pre-ovulatory oocytes. However, concomitant with germinal vesicle breakdown, a wave of global histone deacetylation occurs in the oocyte genome that is particularly extensive at the lysine 12 residue of histone H4 (H4K12ac) (Kim *et al.* 2003) as well as at lysine 5 of histone H4 (H4K5ac) (De La Fuente *et al.* 2004). Although, little is known concerning the enzymatic activities that are involved in this process, recent evidence suggests histone deacetylase 2 (HDAC2) as the major enzyme regulating global histone acetylation levels during oocyte development, in particular H4K16ac (Abidi *et al.* 2004).

Timely coordination of developmental transitions in the global epigenetic landscape during meiotic maturation is of outmost importance for the maintenance of chromosome stability in the female gamete. For example, global histone deacetylation during meiotic resumption is required for the localization and function of centromeric proteins such as ATRX (De La Fuente *et al.* 2004), the chromosome passenger complex (Balboula, 2014) and interference with this process using the HDAC inhibitor trichostatin A (TSA) results in the formation of hyperacetylated, decondensed chromosomes, highly abnormal meiotic figures including misaligned chromosomes and chromosome lagging in a large proportion of oocytes (De La Fuente *et al.* 2004, Yang 2012). TSA exposure leads to severe oocyte aneuploidy and subfertility (Akiyama *et al.* 2006). Notably, several HDAC inhibitors (HDACi) have recently entered phase I and II clinical trials for the treatment of several types of human cancers (Martinet and Bertrand 2011). Oocytes exhibit an extreme susceptibility to HDACi-induced chromatin changes even at nanomolar concentrations leading to rapid euchromatin decondensation and severe chromosome instability (Yang 2012). Similarly, functional ablation of HDAC2 in oocytes interferes with deacetylation of H4K16ac and significantly increases the proportion of oocytes with chromosome condensation and segregation errors (Ma and Schultz 2013). Global histone deacetylation in oocytes is critical for centromere function and binding of chromatin remodeling proteins such as ATRX to centromeric domains as hyperacetylation of chromosomes is sufficient to induce loss of centromeric ATRX (De La Fuente *et al.* 2004) and interferes with the deposition of the kinetochore-specific histone variant CENP-A (Ma and Schultz 2013, Wiens and Sorger 1998).

Reproductive aging is now widely recognized as a major risk factor for oocyte aneuploidy. Notably, oocytes obtained from older mice commonly exhibit high levels of histone acetylation at the metaphase-II stage, in particular of lysine 8 and lysine 12 in histone H4 (Akiyama *et al.* 2006, Suo *et al.* 2010) and a similar correlation between advanced maternal age, defective deacetylation of the lysine 12 residue in histone H4 and chromosome misalignment has been reported in human oocytes (van den Berg *et al.* 2011). In addition, reduced global tri-methylation of histone H3 (H3K9me3) as well as di-methylation of histone H4 (H4K20me2) and histone H3 (H3K36me2 and HeK79me2) is associated with highly abnormal chromatin configurations, such as condensed, irregular or clumped chromatin at the GV stage in aged mouse oocytes (Manosalva and Gonzalez 2010).

Advanced maternal age is also associated with meiotic spindle defects (Hassold and Hunt 2001, Hunt and Hassold 2008, Vialard *et al.* 2006). These findings are supported by evidence obtained by microarray analyses of oocytes from young and old mice suggesting deregulation of mechanisms controlling the spindle assembly checkpoint, kinetochore function, a deterioration of cohesion complexes as well as abnormal chromatin remodeling processes during reproductive aging (Chiang *et al.* 2010, Hodges *et al.* 2005, Lister *et al.* 2010, Pan *et al.* 2008, Revenkova *et al.* 2004, 2010) that collectively contribute to age-related sub-fertility and impaired embryonic development (Baird *et al.* 2005, Pellestor *et al.* 2003). The epigenetic maturation of centromeric heterochromatin domains may be particularly susceptible to disturbances leading to detrimental effects on oocyte and embryo quality.

2.4 Epigenetic Control of Centromere Function and Chromosome Stability during the Oocyte to Embryo Transition

Centromeric heterochromatin in mammals forms the interface between condensed chromosomes and spindle microtubules (Guenatri *et al.* 2004, Sullivan and Karpen 2004). It consists of several megabases of repetitive DNA sequences that are essential, albeit not sufficient, for the formation of a functional centromere, indicating that centromere formation is under strict epigenetic control (Dillon and Festenstein 2002, Karpen and Allshire 1997). Several epigenetic marks, such as tri-methylation of H3K9 (H3K9me3) and H4K20 (H4K20me3), have been implicated in centromeric heterochromatin formation and provide docking sites for chromatin remodeling proteins such as heterochromatin protein 1 (HP1) and ATRX (Bannister *et al.* 2001, Kourmouli *et al.* 2005, Lachner *et al.* 2001, Rea *et al.* 2000, Schotta *et al.* 2004, De La Fuente *et al.* 2004, McDowell *et al.* 1999). ATRX is a member of the SWI/SNF2 family of ATP-dependent chromatin remodeling proteins and is found at pericentric heterochromatin in mouse and human somatic cells (De La Fuente *et al.* 2004, McDowell *et al.* 1999). In mouse oocytes, ATRX localizes to pericentric heterochromatin in GV stage oocytes and acquires a strictly centromeric localization during metaphase-I and metaphase-II of meiosis. Importantly, loss of maternal ATRX function results in chromosome misalignment at the metaphase-II spindle and severe centromere instability as indicated by a high incidence of DNA breaks in zygote stage embryos. Pre-implantation embryos deficient for maternal ATRX exhibit rampant aneuploidy and embryo demise resulting in severe sub-fertility (Baumann *et al.* 2010, De La Fuente *et al.* 2004). Loss of ATRX interferes with recruitment of the transcriptional regulator DAXX to pericentric heterochromatin in GV stage oocytes and zygote stage embryos suggesting that ATRX regulates the molecular composition of pericentric heterochromatin in mammalian oocytes (Baumann *et al.* 2010, De La Fuente *et al.* 2015). ATRX mRNA and protein expression are significantly reduced in oocytes of older mice (Pan *et al.* 2008). Notably, ATRX-deficient oocytes exhibit the same type of chromosome segregation defects (i.e. premature centromere separation and chromosome lagging) detected in aneuploid oocytes obtained from women of advanced reproductive age (Baumann *et al.* 2010, Vialard *et al.* 2006). Thus, transgenic mice with an oocyte-conditional deletion of ATRX constitute an important animal model to determine the mechanisms leading to the transmission of aneuploidy during reproductive senescence and the epigenetic origins of chromosome instability during the oocyte to embryo transition.

For example, recent studies conducted in cleavage stage human embryos indicate that a striking 70–90% of embryos exhibit one or more blastomeres carrying an aneuploid chromosome complement including deletions and/or translocations (Robberecht *et al.* 2010, Vanneste *et al.* 2009, Voet *et al.* 2011). The mechanisms responsible for such high levels of chromosome instability are not known. However, our ATRX model provides novel evidence indicating that centromeric instability is a major mechanism leading to the formation of large-scale chromosomal rearrangements in the cleavage stage embryo (De La Fuente *et al.* 2015). Our results indicate that the exclusive maternal inheritance of ATRX contributes to the striking epigenetic asymmetry of maternal and paternal centromeres during the transition to the first mitosis (Figure 2.1). Notably, ATRX is essential for transcriptional silencing of major satellite repeats in the maternal genome and required to prevent deleterious centromeric mitotic recombination (De La Fuente *et al.* 2015). Loss of ATRX during this critical developmental window leads to centromeric breaks, sister chromatid exchanges and large-scale chromosomal rearrangements in a striking proportion of cleavage stage embryos (De La Fuente *et al.* 2015). These findings suggest that ATRX-deficiency is associated with the formation of mitotic recombination breakpoints and the activation of DNA repair pathways providing a unique model to determine the epigenetic origins of cleavage stage chromosome instability (De La Fuente *et al.* 2015).

Severe chromosome instability during female meiosis has recently also been observed in mice lacking BRWD1 (bromo- and WD-containing protein 1) (Pattabiraman *et al.* 2015). Bromodomains play critical roles in a variety of cellular processes, such as chromatin remodeling and transcriptional regulation, through

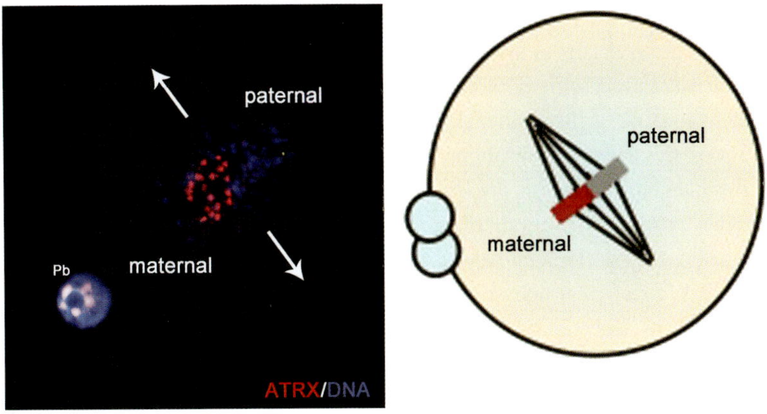

Figure 2.1. Maternal inheritance of ATRX confers epigenetic asymmetry to centromeres during the first mitosis in the zygote stage embryo. The chromatin remodeling protein ATRX (red) is inherited through the maternal germ line and is present at centromeres of maternal chromosomes and the polar body (Pb). Notably, ATRX is absent from the paternal chromosome complement during the transition to the first mitosis in the fertilized zygote. This results in a unique arrangement of maternal and paternal chromosomes based on distinct epigenetic marks. Arrows indicate the direction of the mitotic spindle poles (see diagram). DNA in chromosomes was counterstained with DAPI (blue).

the recognition of acetyl-lysine residues (Bottomley 2004, Hudson *et al.* 2000, Zeng and Zhou 2002). Loss of BRWD1 function impairs global transcriptional silencing in pre-ovulatory mouse oocytes, as indicated by nascent RNA transcripts in transcription run-on-assays, while acquisition of the SN chromatin configuration appeared unaffected (Pattabiraman *et al.* 2015). Further investigation of the protein coding and non-coding transcriptome by micro-array, pathway-specific PCR arrays as well as qPCR revealed surprisingly similar expression patterns of protein coding transcripts in oocytes from mutants and controls. However, loss of BRWD1 is associated with overexpression of mixed lineage leukemia protein 5 (MLL5) as well as long interspersed nuclear elements (LINE's), indicating that continued run-on expression in mutant oocytes might be due to persistent expression of repetitive elements. Most importantly, BRWD1-deficient ova present highly disrupted meiotic spindle configurations and show a severely limited potential to mature to the metaphase-II stage resulting in complete infertility. Prominent defects in axial chromatid condensation and abnormal chromosome-microtubule interactions lead to major defects in chromosome congression and homologous chromosome separation during anaphase-I and result in severe chromatid elongation with loss of structural integrity and chromatin clumping in this model. Extensive chromosomal breaks and general chromosome instability in BRDW1 mutant oocytes are, however, not associated with defects in histone deacetylation, mono-methylation of H4K20 or loading of the condensin protein Smc4 (Pattabiraman *et al.* 2015). While aberrant expression levels or abnormal chromosomal localization of specific epigenetic markers have, thus far, not been identified in this model, it is conceivable that large-scale chromatin remodeling processes are disrupted as a consequence of MLL5 and LINE-1 element overexpression leading to severe chromosome instability in BRDW1-deficient oocytes.

Notably, it is becoming increasingly clear that the establishment of a proper epigenetic landscape and the epigenetic control of transposon expression during oogenesis are essential mechanisms to ensure normal progression of meiotic transitions and the preservation of genome integrity in the female germ line. For example, overexpression of intracisternal A particle (IAP) transposons in MLL2 knockout oocytes is associated with severe meiotic defects and oocyte DNA damage resulting in premature ovarian failure and female sterility (Andreu-Vieyra *et al.* 2010). Abnormal expression of the mouse transposon (MTs) in DICER knockout oocytes and mutants of the recently discovered oocyte specific isoform of DICER (DICERo) exhibit severe meiotic defects with lack of proper chromosome alignment to the meiotic spindle in maturing oocytes (Flemr *et al.* 2013, Murchison *et al.* 2007). In addition, targeted deletion of MARF1 (meiotic arrest female 1) during oocyte growth results in overexpression of both IAP transposons and LINE-1 elements resulting in widespread DNA damage and meiotic arrest (Su *et al.* 2012a, 2012b). Finally, experimental overexpression of LINE-1 elements in mouse oocytes results in oocyte aneuploidy and embryonic lethality (Malki *et al.* 2014). Elucidating the precise epigenetic mechanisms responsible for predisposing female gametes to aneuploidy and CIN is an important first step towards the identification of reliable early detection markers of chromosome stability, oocyte quality and developmental potential.

2.4.1 *Chromosome-microtubule interactions in the oocyte*

The accuracy of chromosome segregation during mitotic and meiotic division is critically dependent on the establishment of correctly oriented and stable chromosome-microtubule attachments. Failure to form the appropriate interactions between the chromosomes and spindle microtubules during meiosis can result in deleterious chromosome segregation errors, leading to aneuploidy and genomic instability in gametes (Nagaoka *et al.* 2012). Spindle microtubules bind to the chromosomes at the kinetochore, a specialized region of the centromere

(Cheeseman and Desai 2008). During mitosis, kinetochores from each sister chromatid normally attach to microtubules from opposing spindle poles, referred to as bi-oriented or amphitelic attachment. Disruptions in this process can lead to different types of attachment errors, including (i) syntelic errors in which kinetochores from sister chromatids both attach to the same spindle pole, or (ii) merotelic errors, whereby a single kinetochore attaches to both apposing spindle poles simultaneously. Similar attachment errors also occur during meiotic division in mammalian oocytes. Notably, chromosome-microtubule interactions are more complex during meiosis-I since it entails the separation of homologous chromosomes. The sister chromatids of each homologue must attach and segregate to the same pole (co-orientation), necessitating that the sister kinetochores function as a unit. These unique aspects of meiotic division likely involve distinctive regulatory processes, yet our knowledge of the underlying mechanisms that regulate kinetochore-microtubule interactions during meiotic division is based mainly on functional comparisons with mitosis.

The kinetochore is a large multi-protein complex that only assembles at the centromere when a cell enters M-phase and then disassembles upon mitotic exit (Oegema *et al.* 2001). It is crucial that a kinetochore assemble only at a single location on each chromosome, and specific aspects of the centromere mark it as the exclusive site for kinetochore formation. Key centromere features include a combination of epigenetic marks such as nucleosomes that contain the histone H3 variant Centromere Protein A (CENP-A), which is required for kinetochore assembly, as well as histone modifications that affect chromatin structure (Black *et al.* 2007, Howman *et al.* 2000, Oegema *et al.* 2001). However, satellite DNA sequences normally associated with the centromere are not required for kinetochore assembly (Amor and Choo 2002, Bassett *et al.* 2010). The assembled kinetochore has a tri-layer structure composed of an inner kinetochore plate that contacts centromeric heterochromatin, as well as an outer kinetochore plate and corona layer that contain proteins required for microtubule interaction. Recognized proteins with microtubule binding activity include the dimeric kinesin CENP-E (Duesbery *et al.* 1997), Ska1 complex (Welburn *et al.* 2009) as well as the KMN network composed of KNL1, Mis12 and Ndc80 (Cheeseman *et al.* 2006). Studies in somatic cells demonstrate that the NDC80 microtubule interacting complex directly binds to microtubules and is essential for the formation of kinetochore-microtubule attachments, as depletion of NDC80 complex proteins leads to severe microtubule attachment defects during mitosis (Cheeseman *et al.* 2006, Cheeseman and Desai 2008). While it is not known if these mechanisms are fully conserved in gametes, studies have shown that mouse oocytes express CENP-E at the kinetochores and inhibition of CENP-E function lead to MI-arrest and unstable kinetochore–microtubule interactions (Duesbery *et al.* 1997, Gui and Homer 2012). CENP-C is also expressed at kinetochores in both maternal and paternal chromosomes during the first cleavage division of the one-cell embryo (De La Fuente *et al.* 2015). Notably, during the first two divisions in the early embryo kinetochore domains are significantly larger in the maternal chromosomes, demonstrating a gradual and unique conversion from meiotic to mitotic kinetochores during the oocyte to embryo transition (De La Fuente *et al.* 2015).

Key proteins also assemble at the outer kinetochore plate that function in the detection and correction of inappropriate chromosome–microtubule interactions, whereby appropriate bi-oriented microtubule attachments at the kinetochore are stabilized, while incorrect attachments are eliminated. Notably, seminal experiments by Bruce Nicklas established that tension across the centromere is a key factor for stabilizing chromosome–microtubule interactions (Nicklas and Ward 1994, Nicklas *et al.* 2001). In contrast, phosphorylation of kinetochore substrates by the conserved Ipl1/Aurora B kinase selectively eliminates incorrect attachments by promoting the turnover of microtubule attachments in the absence of tension. In mitotic cells, Aurora B reportedly controls kinetochore microtubule attachments by modulating the microtubule binding activity of key kinetochore proteins such as Hec1/Ndc8 (Cheeseman *et al.* 2002, DeLuca *et al.* 2006, Welburn *et al.* 2010). Intriguingly, Aurora C kinase, and not Aurora B, is primarily expressed in germ cells and plays an essential role in regulating chromosome-microtubule interactions during meiotic division (Yang *et al.* 2010). Inhibition of Aurora C activity in mouse oocytes promotes abnormal kinetochore–microtubule attachments as well as premature chromosome segregation and cytokinesis failure during meiosis-I (Yang *et al.* 2010). Analysis of the knockout mice (Kimmins *et al.* 2007) reveal that Aurora C null females exhibit both meiotic defects in oocytes as well as cell division errors during early embryonic development, leading to subfertility with approximately 30% fewer pups per litter (Schindler *et al.* 2012). These studies underscore an important function for Aurora C, but also indicate some degree of compensatory function by Aurora B. Intriguingly, Aurora B transcripts are significantly up-regulated in ATRX-deficient embryos characterized by centromeric instability in the early embryo, and potentially functions to correct inappropriate kinetochore–microtubule interactions (De La Fuente *et al.* 2015).

Chromosome–microtubule attachment errors activate the spindle assembly checkpoint (SAC) that inhibits anaphase promoting complex APC/C^{cdc20} activity to prevent anaphase onset and allow for error correction mechanisms to promote bi-orientation. Key SAC proteins including Mad2, BubR1 and Bub3, function as an important surveillance mechanism and localize to the outer kinetochore until stable chromosome-microtubule interactions are established. Studies support the expression and function of key SAC proteins in mammalian oocytes (Homer *et al.* 2005, Leland *et al.* 2009, McGuinness *et al.* 2009, Niault *et al.* 2007). High-resolution live cell imaging analyses also indicate that homologous chromosome co-orientation during meiosis-I is potentially error prone in mammalian oocytes, as the majority of chromosomes undergo error correction of kinetochore–microtubule attachments prior to anaphase onset and completion of the first meiotic division (Kitajima *et al.* 2011). Yet, high rates of aneuploidy in gametes and mammalian embryos (Nagaoka *et al.* 2012) suggest that some attachment errors potentially elude detection and/or correction by SAC. Studies with *Sycp3*$^{-/-}$ mice reveal that during meiosis-I univalents are bi-oriented and show loss of Mad2 at the kinetochores indicating that the requirements for SAC are satisfied (Kouznetsova *et al.* 2007). Moreover, single chromatids in *Sycp3*$^{-/-}$ oocytes, which separated prematurely during metaphase-II, establish merotelic (bi-directional) kinetochore-microtubule attachments (Kouznetsova *et al.* 2014), leading to equal tension that potentially elude SAC-mediated arrest. Such merotelic attachments can promote aneuploidy, and suggest that in mammalian oocytes, kinetochores are potentially organized into multiple domains that enable bi-directional attachment by chromatids (Kouznetsova *et al.* 2014). Nevertheless, other studies with *Mlh1* mutant mice demonstrate that metaphase alignment and bi-directional attachment of all chromosomes is not required for anaphase onset in oocytes (Nagaoka *et al.* 2011). It is proposed that in contrast to mitosis, anaphase onset during meiotic division in oocytes likely requires bipolar attachments of a 'critical mass', but not all, chromosomes (Nagaoka *et al.* 2011). This premise is also supported by studies in which meiotic spindle organization and stability is disrupted. Depletion of a key microtubule organizing center (MTOC)-associated protein, NEDD1, leads to significant defects in microtubule organization and spindle stability such that NEDD1-depleted oocytes exhibit a high incidence of misaligned and lagging chromosomes at metaphase-I (Ma *et al.* 2010). Yet, despite SAC activation, more than 50% of the NEDD1-depleted oocytes that progressed to metaphase-II were aneuploid (Ma *et al.* 2010). Thus, confirming that SAC was satisfied and anaphase onset was initiated in oocytes with inappropriate kinetochore–microtubule attachments. Why SAC is potentially less sensitive in mammalian oocytes to chromosome attachment errors is not known. Nevertheless, it poses a significant risk in contributing to aneuploidy in gametes and embryos.

References

Abe, K., Inoue, A., Suzuki, M. G. and Aoki, F. (2010). Global gene silencing is caused by the dissociation of RNA polymerase II from DNA in mouse oocytes. *The Journal of Reproduction and Development* 56, 502–507.

Abidi, F. E., Cardoso, C., Lossi, A.-M., Lowry, R. B., Depetris, D., Mattei, M.-G., Lubs, H. A., Stevenson, R. E., Fontes, M., Chudley, A. E. *et al.* (2004). Mutation in the 5[prime] alternatively spliced region of the XNP//ATR-X gene causes Chudley-Lowry syndrome. *European Journal of Human Genetics* 13, 176–183.

Adenot, P. G., Mercier, Y., Renard, J. P. and Thompson, E. M. (1997). Differential H4 acetylation of paternal and maternal chromatin precedes DNA replication and differential transcriptional activity in pronuclei of 1-cell mouse embryos. *Development* 124, 4615–4625.

Akiyama, T., Nagata, M. and Aoki, F. (2006). Inadequate histone deacetylation during oocyte meiosis causes aneuploidy and embryo death in mice. *Proceedings of the National Academy of Sciences U S A* 103, 7339–7344.

Amor, D. J. and Choo, K. H. A. (2002). Neocentromeres: role in human disease, evolution, and centromere study. *The American Journal of Human Genetics* 71, 695–714.

Andreu-Vieyra, C. V., Chen, R., Agno, J. E., Glaser, S., Anastassiadis, K., Stewart, A. F. and Matzuk, M. M. (2010). MLL2 is required in oocytes for bulk histone 3 lysine 4 trimethylation and transcriptional silencing. *PLoS Biology* 8, e1000453.

Arney, K. L., Bao, S., Bannister, A. J., Kouzarides, T. and Surani, M. A. (2002). Histone methylation defines epigenetic asymmetry in the mouse zygote. *The International Journal of Developmental Biology* 46, 317–320.

Baird, D. T., Collins, J., Egozcue, J., Evers, L. H., Gianaroli, L., Leridon, H., Sunde, A., Templeton, A., Van Steirteghem, A., Cohen, J. *et al.* (2005). Fertility and ageing. *Human Reproduction Update* 11, 261–276.

Balboula, A. Z., Stein, P., Schultz, R. M. and Schindler, K. (2014) Knockdown of RBBP7 unveils a requirement of histone deacetylation for CPC function in mouse oocytes. *Cell Cycle* 13, 600–611.

Bannister, A. J., Zegerman, P., Partridge, J. F., Miska, E. A., Thomas, J. O., Allshire, R. C. and Kouzarides, T. (2001). Selective recognition of methylated lysine 9 on histone H3 by the HP1 chromo domain. *Nature* 410, 120–124.

Bassett, E. A., Wood, S., Salimian, K. J., Ajith, S., Foltz, D. R. and Black, B. E. (2010). Epigenetic centromere specification directs aurora B accumulation but is insufficient to efficiently correct mitotic errors. *The Journal of Cell Biology* 190, 177–185.

Baumann, C., Viveiros, M. M. and De La Fuente, R. (2010). Loss of maternal ATRX results in centromere instability and aneuploidy in the mammalian oocyte and pre-implantation embryo. *PLoS Genetics* 6, e1001137.

Bellone, M., Zuccotti, M., Redi, C. A. and Garagna, S. (2009). The position of the germinal vesicle and the chromatin organization together provide a marker of the developmental competence of mouse antral oocytes. *Reproduction* 138, 639–643.

Black, B. E., Brock, M. A., Bedard, S., Woods, V. L. and Cleveland, D. W. (2007). An epigenetic mark generated by the incorporation of CENP-A into centromeric nucleosomes. *Proceedings of the National Academy of Sciences USA* 104, 5008–5013.

Bottomley, M. J. (2004). Structures of protein domains that create or recognize histone modifications. *EMBO Reports* 5, 464–469.

Bourc'his, D., Xu, G. L., Lin, C. S., Bollman, B. and Bestor, T. H. (2001). Dnmt3L and the establishment of maternal genomic imprints. *Science* 294, 2536–2539.

Brunet, S. and Maro, B. (2007). Germinal vesicle position and meiotic maturation in mouse oocyte. *Reproduction* 133, 1069–1072.

Cheeseman, I. M., Anderson, S., Jwa, M., Green, E. M., Kang, J.-S., Yates Iii, J. R., Chan, C. S. M., Drubin, D. G. and Barnes, G. (2002). Phospho-regulation of kinetochore-microtubule attachments by the Aurora Kinase Ipl1p. *Cell* 111, 163–172.

Cheeseman, I. M., Chappie, J. S., Wilson-Kubalek, E. M. and Desai, A. (2006). The conserved KMN network constitutes the core microtubule-binding site of the kinetochore. *Cell* 127, 983–997.

Cheeseman, I. M. and Desai, A. (2008). Molecular architecture of the kinetochore-microtubule interface. *Nature Reviews Molecular Cell Biology* 9, 33–46.

Chen, J., Torcia, S., Xie, F., Lin, C.-J., Cakmak, H., Franciosi, F., Horner, K., Onodera, C., Song, J. S., Cedars, M. I. *et al.* (2013). Somatic cells regulate maternal mRNA translation and developmental competence of mouse oocytes. *Nature Cell Biology* 15, 1415–1423.

Chiang, T., Duncan, F. E., Schindler, K., Schultz, R. M. and Lampson, M. A. (2010). Evidence that weakened centromere cohesion is a leading cause of age-related aneuploidy in oocytes. *Current Biology* 20, 1522–1528.

Chotalia, M., Smallwood, S. A., Ruf, N., Dawson, C., Lucifero, D., Frontera, M., James, K., Dean, W. and Kelsey, G. (2009). Transcription is required for establishment of germline methylation marks at imprinted genes. *Genes & Development* 23, 105–117.

Ciccone, D. N., Su, H., Hevi, S., Gay, F., Lei, H., Bajko, J., Xu, G., Li, E. and Chen, T. (2009). KDM1B is a histone H3K4 demethylase required to establish maternal genomic imprints. *Nature* 461, 415–418.

De La Fuente, R. (2006). Chromatin modifications in the germinal vesicle (GV) of mammalian oocytes. *Developmental Biology* 292, 1–12.

De La Fuente, R. (2014). Histone deacetylation: Establishing a meiotic histone code. *Cell Cycle* 13, 879–880.

De La Fuente, R., Baumann, C. and Viveiros, M. (2012). Chromatin structure and ATRX function in mouse oocytes. *Results and Problems in Cell Differentiation* 55, 45–68.

De La Fuente, R., Baumann, C. and Viveiros, M. (2015). ATRX contributes to epigenetic asymmetry and silencing of major satellite transcripts in the maternal genome of the mouse embryo. *Development* 142(10), 1806–1817.

De La Fuente, R. and Eppig, J. J. (2001). Transcriptional activity of the mouse oocyte genome: Companion granulosa cells modulate transcription and chromatin remodeling. *Developmental Biology* 229, 224–236.

De La Fuente, R., Viveiros, M., Wigglesworth, K. and Eppig, J. (2004). ATRX, a member of the SNF2 family of helicase /ATPases, is required for chromosome alignment and meiotic spindle organization in metaphase II stage mouse oocytes. *Developmental Biology* 272, 1–14.

De La Fuente, R., Viveiros, M., Burns, K., Adashi, E., Matzuk, M. and Eppig, J. (2004). Major chromatin remodeling in the germinal vesicle (GV) of mammalian oocytes is dispensable for global transcriptional silencing but required for centromeric heterochromatin function. *Developmental Biology* 275, 447–458.

Debey, P., Szollosi, M. S., Szollosi, D., Vautier, D., Girousse, A. and Besombes, D. (1993). Competent mouse oocytes isolated from antral follicles exhibit different chromatin organization and follow different maturation dynamics. *Molecular Reproduction Development* 36, 59–74.

DeLuca, J. G., Gall, W. E., Ciferri, C., Cimini, D., Musacchio, A. and Salmon, E. D. (2006). Kinetochore microtubule dynamics and attachment stability are regulated by Hec1. *Cell* 127, 969–982.

Dillon, N. and Festenstein, R. (2002). Unravelling heterochromatin: competition between positive and negative factors regulates accessibility. *Trends in Genetics* 18, 252–258.

Duesbery, N. S., Choi, T., Brown, K. D., Wood, K. W., Resau, J., Fukasawa, K., Cleveland, D. W. and Vande Woude, G. F. (1997). CENP-E is an essential kinetochore motor in maturing oocytes and is masked during mos-dependent, cell cycle arrest at metaphase II. *Proceedings of the National Academy of Sciences USA* 94, 9165–9170.

Dundr, M. and Misteli, T. (2001). Functional architecture in the cell nucleus. *Biochemical Journal* 356, 297–310.

Flemr, M., Malik, R., Franke, V., Nejepinska, J., Sedlacek, R., Vlahovicek, K. and Svoboda, P. (2013). A retrotransposon-driven dicer isoform directs endogenous small interfering RNA production in mouse oocytes. *Cell* 155, 807–816.

Guenatri, M., Bailly, D., Maison, C. and Almouzni, G. (2004). Mouse centric and pericentric satellite repeats form distinct functional heterochromatin. *Journal of Cell Biology* 166, 493–505.

Gui, L. and Homer, H. (2012). Spindle assembly checkpoint signalling is uncoupled from chromosomal position in mouse oocytes. *Development* 139, 1941–1946.

Hackett, J. A. and Surani, M. A. (2013). DNA methylation dynamics during the mammalian life cycle. *Philosophical Transactions of the Royal Society B: Biological Sciences* 368, 20110328.

Hassold, T. and Hunt, P. (2001). To err (meiotically) is human: The genesis of human aneuploidy. *Nature Reviews Genetics* 2, 280–291.

Hata, K., Okano, M., Lei, H. and Li, E. (2002). Dnmt3L cooperates with the Dnmt3 family of *de novo* DNA methyltransferases to establish maternal imprints in mice. *Development* 129, 1983–1993.

Hodges, C. A., Revenkova, E., Jessberger, R., Hassold, T. J. and Hunt, P. A. (2005). SMC1beta-deficient female mice provide evidence that cohesins are a missing link in age-related nondisjunction. *Nature Genetics* 37, 1351–1355.

Homer, H. A., McDougall, A., Levasseur, M. and Herbert, M. (2005). Restaging the spindle assembly checkpoint in female mammalian meiosis I. *Cell Cycle* 4, 650–653.

Howman, E. V., Fowler, K. J., Newson, A. J., Redward, S., MacDonald, A. C., Kalitsis, P. and Choo, K. H. A. (2000). Early disruption of centromeric chromatin organization in centromere protein A (Cenpa) null mice. *Proceedings of the National Academy of Sciences USA* 97, 1148–1153.

Hudson, B. P., Martinez-Yamout, M. A., Dyson, H. J. and Wright, P. E. (2000). Solution structure and acetyl-lysine binding activity of the GCN5 bromodomain. *Journal of Molecular Biology* 304, 355–370.

Hunt, P. A. and Hassold, T. J. (2008). Human female meiosis: What makes a good egg go bad? *Trends in Genetics* 24, 86–93.

Inoue, A., Nakajima, R., Nagata, M. and Aoki, F. (2008). Contribution of the oocyte nucleus and cytoplasm to the determination of meiotic and developmental competence in mice. *Human Reproduction* 23, 1377–1384.

Ivanovska, I. and Orr-Weaver, T. L. (2006). Histone modifications and the chromatin scaffold for meiotic chromosome architecture. *Cell Cycle* 5, 2064–2071.

Kaneda, M., Okano, M., Hata, K., Sado, T., Tsujimoto, N., Li, E. and Sasaki, H. (2004). Essential role for *de novo* DNA methyltransferase Dnmt3a in paternal and maternal imprinting. *Nature* 429, 900–903.

Karpen, G. H. and Allshire, R. C. (1997). The case for epigenetic effects on centromere identity and function. *Trends in Genetics* 13, 489–496.

Kim, J., Liu, H., Tazaki, M., Nagata, M. and Aoki, F. (2003). Changes in histone acetylation during mouse oocyte meiosis. *Journal of Cell Biology* 162, 37–46.

Kimmins, S., Crosio, C., Kotaja, N., Hirayama, J., Monaco, L., Höög, C., van Duin, M., Gossen, J. A. and Sassone-Corsi, P. (2007). Differential functions of the Aurora-B and Aurora-C kinases in mammalian spermatogenesis. *Molecular Endocrinology* 21, 726–739.

Kimmins, S. and Sassone-Corsi, P. (2005). Chromatin remodelling and epigenetic features of germ cells. *Nature* 434, 583–589.

Kitajima, T. S., Ohsugi, M. and Ellenberg, J. (2011). Complete kinetochore tracking reveals error-prone homologous chromosome biorientation in mammalian oocytes. *Cell* 146, 568–581.

Kobayashi, H., Sakurai, T., Imai, M., Takahashi, N., Fukuda, A., Yayoi, O., Sato, S., Nakabayashi, K., Hata, K., Sotomaru, Y. *et al.* (2012). Contribution of intragenic DNA methylation in mouse gametic DNA methylomes to establish oocyte-specific heritable marks. *PLoS Genetics* 8, e1002440.

Kourmouli, N., Sun, Y. M., van der Sar, S., Singh, P. B. and Brown, J. P. (2005). Epigenetic regulation of mammalian pericentric heterochromatin *in vivo* by HP1. *Biochemical Biophysical Research and Communications* 337, 901–907.

Kouznetsova, A., Hernández-Hernández, A. and Höög, C. (2014). Merotelic attachments allow alignment and stabilization of chromatids in meiosis II oocytes. *Nature Communications* 5, 4409.

Kouznetsova, A., Lister, L., Nordenskjold, M., Herbert, M. and Hoog, C. (2007). Bi-orientation of achiasmatic chromosomes in meiosis I oocytes contributes to aneuploidy in mice. *Nature Genetics* 39, 966–968.

Lachner, M., O'Carroll, D., Rea, S., Mechtler, K. and Jenuwein, T. (2001). Methylation of histone H3 lysine 9 creates a binding site for HP1 proteins. *Nature* 410, 116–120.

Leland, S., Nagarajan, P., Polyzos, A., Thomas, S., Samaan, G., Donnell, R., Marchetti, F. and Venkatachalam, S. (2009). Heterozygosity for a Bub1 mutation causes female-specific germ cell aneuploidy in mice. *Proceedings of the Natational Academy of Sciences USA* 106, 12776–12781.

Li, X., Ito, M., Zhou, F., Youngson, N., Zuo, X., Leder, P. and Ferguson-Smith, A. C. (2008). A maternal-zygotic effect gene, Zfp57, maintains both maternal and paternal imprints. *Developmental Cell* 15, 547–557.

Lister, L. M., Kouznetsova, A., Hyslop, L. A., Kalleas, D., Pace, S. L., Barel, J. C., Nathan, A., Floros, V., Adelfalk, C., Watanabe, Y. *et al.* (2010). Age-related meiotic segregation errors in mammalian oocytes are preceded by depletion of cohesin and Sgo2. *Current Biology* 20, 1511–1521.

Lister, R., Pelizzola, M., Dowen, R. H., Hawkins, R. D., Hon, G., Tonti-Filippini, J., Nery, J. R., Lee, L., Ye, Z., Ngo, Q. M. *et al.* (2009). Human DNA methylomes at base resolution show widespread epigenomic differences. *Nature* 462, 315–322.

Liu, H. and Aoki, F. (2002). Transcriptional activity associated with meiotic competence in fully grown mouse GV oocytes. *Zygote* 10, 327–332.

Liu, H., Kim, J. and Aoki, F. (2004). Regulation of histone H3 lysine 9 methylation in oocytes and early pre-implantation embryos. *Development* 131, 2269–2280.

Lucifero, D., Mann, M. R., Bartolomei, M. S. and Trasler, J. M. (2004). Gene-specific timing and epigenetic memory in oocyte imprinting. *Human Molecular Genetics* 13, 839–849.

Ma, P. and Schultz, R. M. (2013). Histone deacetylase 2 (HDAC2) regulates chromosome segregation and kinetochore function via H4K16 deacetylation during oocyte maturation in mouse. *PLoS Genetics* 9, e1003377.

Ma, W., Baumann, C. and Viveiros, M. M. (2010). NEDD1 is crucial for meiotic spindle stability and accurate chromosome segregation in mammalian oocytes. *Developmental Biology* 339, 439–450.

Malki, S., van der Heijden, Godfried W., O'Donnell, Kathryn A., Martin, Sandra L. and Bortvin, A. (2014). A role for retrotransposon LINE-1 in fetal oocyte attrition in mice. *Developmental Cell* 29, 521–533.

Manosalva, I. and Gonzalez, A. (2010). Aging changes the chromatin configuration and histone methylation of mouse oocytes at germinal vesicle stage. *Theriogenology* 74, 1539–1547.

Martinet, N. and Bertrand, P. (2011). Interpreting clinical assays for histone deacetylase inhibitors. *Cancer Management and Research* 3, 117–141.

Mattson, B. A. and Albertini, D. F. (1990). Oogenesis: chromatin and microtubule dynamics during meiotic prophase. *Molecular Reproduction and Development* 25, 374–383.

McDowell, T. L., Gibbons, R. J., Sutherland, H., O'Rourke, D. M., Bickmore, W. A., Pombo, A., Turley, H., Gatter, K., Picketts, D. J. and Buckle, V. J. (1999). Localization of a putative transcriptional regulator (ATRX) at pericentromeric heterochromatin and the short arms of acrocentric chromosomes. *Proceedings of the National Academy of Sciences USA* 96, 13983–13988.

McGuinness, B. E., Anger, M., Kouznetsova, A., Gil-Bernabe, A. M., Helmhart, W., Kudo, N. R., Wuensche, A., Taylor, S., Hoog, C., Novak, B. *et al.* (2009). Regulation of APC/C activity in oocytes by a Bub1-dependent spindle assembly checkpoint. *Current Biology* 19, 369–380.

Meglicki, M., Zientarski, M. and Borsuk, E. (2008). Constitutive heterochromatin during mouse oogenesis: the pattern of histone H3 modifications and localization of HP1alpha and HP1beta proteins. *Molecular Reproduction and Development* 75, 414–428.

Messerschmidt, D. M., de Vries, W., Ito, M., Solter, D., Ferguson-Smith, A. and Knowles, B. B. (2012). Trim28 is required for epigenetic stability during mouse oocyte to embryo transition. *Science* 335, 1499–1502.

Murchison, E. P., Stein, P., Xuan, Z., Pan, H., Zhang, M. Q., Schultz, R. M. and Hannon, G. J. (2007). Critical roles for Dicer in the female germline. *Genes and Development* 21, 682–693.

Nagaoka, S. I., Hassold, T. J. and Hunt, P. A. (2012). Human aneuploidy: mechanisms and new insights into an age-old problem. *Nature Reviews Genetics* 13, 493–504.

Nagaoka, S. I., Hodges, C. A., Albertini, D. F. and Hunt, P. A. (2011). Oocyte-specific differences in cell-cycle control create an innate susceptibility to meiotic errors. *Current Biology* 21, 651–657.

Niault, T., Hached, K., Sotillo, R., Sorger, P. K., Maro, B., Benezra, R. and Wassmann, K. (2007). Changing mad2 levels affects chromosome segregation and spindle assembly checkpoint control in female mouse meiosis I. *PLoS ONE* 2, e1165.

Nicklas, R. and Ward, S. (1994). Elements of error correction in mitosis: microtubule capture, release, and tension. *The Journal of Cell Biology* 126, 1241–1253.

Nicklas, R. B., Waters, J. C., Salmon, E. D. and Ward, S. C. (2001). Checkpoint signals in grasshopper meiosis are sensitive to microtubule attachment, but tension is still essential. *Journal of Cell Science* 114, 4173–4183.

Obata, Y. and Kono, T. (2002). Maternal primary imprinting is established at a specific time for each gene throughout oocyte growth. *Journal of Biological Chemistry* 277, 5285–5289.

Oegema, K., Desai, A., Rybina, S., Kirkham, M. and Hyman, A. A. (2001). Functional analysis of kinetochore assembly in Caenorhabditis elegans. *The Journal of Cell Biology* 153, 1209–1226.

Ooga, M., Inoue, A., Kageyama, S., Akiyama, T., Nagata, M. and Aoki, F. (2008). Changes in H3K79 methylation during preimplantation development in mice. *Biology of Reproduction* 78, 413–424.

Page, S. and Hawley, R. (2003). Chromosome choreography: the meiotic ballet. *Science* 301, 785–789.

Pan, H., Ma, P., Zhu, W. and Schultz, R. M. (2008). Age-associated increase in aneuploidy and changes in gene expression in mouse eggs. *Developmental Biology* 316, 397–407.

Parfenov, V., Potchukalina, G., Dudina, L., Kostyuchek, D. and Gruzova, M. (1989). Human antral follicles: oocyte nucleus and the karyosphere formation (electron microscopic and autoradiographic data). *Gamete Research* 22, 219–231.

Pattabiraman, S., Baumann, C., Guisado, D., Eppig, J. J., Schimenti, J. C. and De La Fuente, R. (2015). Mouse BRWD1 is critical for spermatid postmeiotic transcription and female meiotic chromosome stability. *The Journal of Cell Biology* 208, 53–69.

Pellestor, F., Andreo, B., Arnal, F., Humeau, C. and Demaille, J. (2003). Maternal aging and chromosomal abnormalities: new data drawn from *in vitro* unfertilized human oocytes. *Human Genetics* 112, 195–203.

Petronczki, M., Siomos, M. and Nasmyth, K. (2003). Un menage a quatre: the molecular biology of chromosome segregation in meiosis. *Cell* 112, 423–440.

Quenneville, S., Verde, G., Corsinotti, A., Kapopoulou, A., Jakobsson, J., Offner, S., Baglivo, I., Pedone, Paolo V., Grimaldi, G., Riccio, A. *et al.* (2011). In embryonic stem cells, ZFP57/KAP1 recognize a methylated hexanucleotide to affect chromatin and DNA methylation of imprinting control regions. *Molecular Cell* 44, 361–372.

Rea, S., Eisenhaber, F., O'Carroll, D., Strahl, B. D., Sun, Z. W., Schmid, M., Opravil, S., Mechtler, K., Ponting, C. P., Allis, C. D. *et al.* (2000). Regulation of chromatin structure by site-specific histone H3 methyltransferases. *Nature* 406, 593–599.

Revenkova, E., Eijpe, M., Heyting, C., Hodges, C., Hunt, P., Liebe, B., Scherthan, H. and Jessberger, R. (2004). Cohesin SMC1 beta is required for meiotic chromosome dynamics, sister chromatid cohesion and DNA recombination. *Nature Cell Biology* 6, 555–562.

Revenkova, E., Herrmann, K., Adelfalk, C. and Jessberger, R. (2010). Oocyte cohesin expression restricted to predictyate stages provides full fertility and prevents aneuploidy. *Current Biology* 20, 1529–1533.

Robberecht, C., Vanneste, E., Pexsters, A., D'Hooghe, T., Voet, T. and Vermeesch, J. R. (2010). Somatic genomic variations in early human prenatal development. *Current Genomics* 11, 397–401.

Rutledge, C. E., Thakur, A., O'Neill, K. M., Irwin, R. E., Sato, S., Hata, K. and Walsh, C. P. (2014). Ontogeny, conservation and functional significance of maternally inherited DNA methylation at two classes of non-imprinted genes. *Development* 141, 1313–1323.

Sarmento, O. F., Digilio, L. C., Wang, Y., Perlin, J., Herr, J. C., Allis, C. D. and Coonrod, S. A. (2004). Dynamic alterations of specific histone modifications during early murine development. *Journal of Cell Science* 117, 4449–4459.

Schindler, K., Davydenko, O., Fram, B., Lampson, M. A. and Schultz, R. M. (2012). Maternally recruited Aurora C kinase is more stable than Aurora B to support mouse oocyte maturation and early development. *Proceedings of the National Academy of Sciences USA* 109, E2215–2222.

Schotta, G., Lachner, M., Sarma, K., Ebert, A., Sengupta, R., Reuter, G., Reinberg, D. and Jenuwein, T. (2004). A silencing pathway to induce H3-K9 and H4-K20 trimethylation at constitutive heterochromatin. *Genes & Development* 18, 1251–1262.

Schramm, R. D., Tennier, M. T., Boatman, D. E. and Bavister, B. D. (1993). Chromatin configurations and meiotic competence of oocytes are related to follicular diameter in nonstimulated rhesus monkeys. *Biology of Reproduction* 48, 349–356.

Shirane, K., Toh, H., Kobayashi, H., Miura, F., Chiba, H., Ito, T., Kono, T. and Sasaki, H. (2013). Mouse Oocyte Methylomes at Base Resolution Reveal Genome-Wide Accumulation of Non-CpG Methylation and Role of DNA Methyltransferases. *PLoS Genetics* 9, e1003439.

Smallwood, S. A., Tomizawa, S., Krueger, F., Ruf, N., Carli, N., Segonds-Pichon, A., Sato, S., Hata, K., Andrews, S. R. and Kelsey, G. (2011). Dynamic CpG island methylation landscape in oocytes and preimplantation embryos. *Nation Genetics* 43, 811–814.

Su, Y.-Q., Sugiura, K., Sun, F., Pendola, J. K., Cox, G. A., Handel, M. A., Schimenti, J. C. and Eppig, J. J. (2012a). MARF1 Regulates Essential Oogenic Processes in Mice. *Science* 335, 1496–1499.

Su, Y.-Q., Sun, F., Handel, M. A., Schimenti, J. C. and Eppig, J. J. (2012b). Meiosis arrest female 1 (MARF1) has nuage-like function in mammalian oocytes. *Proceedings of the National Academy of Sciences USA* 109, 18653–18660.

Sullivan, B. A. and Karpen, G. H. (2004). Centromeric chromatin exhibits a histone modification pattern that is distinct from both euchromatin and heterochromatin. *Nature Structural & Molecular Biology* 11, 1076–1083.

Suo, L., Meng, Q.-G., Pei, Y., Yan, C.-L., Fu, X.-W., Bunch, T. D. and Zhu, S.-E. (2010). Changes in acetylation on lysine 12 of histone H4 (acH4K12) of murine oocytes during maternal aging may affect fertilization and subsequent embryo development. *Fertility and Sterility* 93, 945–951.

Swain, J. E., Ding, J., Brautigan, D. L., Villa-Moruzzi, E. and Smith, G. D. (2007). Proper chromatin condensation and maintenance of histone H3 phosphorylation during mouse oocyte meiosis requires protein phosphatase activity. *Biology of Reproduction* 76, 628–638.

van den Berg, I. M., Eleveld, C., van der Hoeven, M., Birnie, E., Steegers, E. A., Galjaard, R. J., Laven, J. S. and van Doorninck, J. H. (2011). Defective deacetylation of histone 4 K12 in human oocytes is associated with advanced maternal age and chromosome misalignment. *Human Reproducion* 26, 1181–1190.

Vanneste, E., Voet, T., Le Caignec, C., Ampe, M., Konings, P., Melotte, C., Debrock, S., Amyere, M., Vikkula, M., Schuit, F., *et al.* (2009). Chromosome instability is common in human cleavage-stage embryos. *Nature Medicine* 15, 577–583.

Vialard, F., Petit, C., Bergere, M., Gomes, D. M., Martel-Petit, V., Lombroso, R., Ville, Y., Gerard, H. and Selva, J. (2006). Evidence of a high proportion of premature unbalanced separation of sister chromatids in the first polar bodies of women of advanced age. *Human Reproduction* 21, 1172–1178.

Voet, T., Vanneste, E., Van der Aa, N., Melotte, C., Jackmaert, S., Vandendael, T., Declercq, M., Debrock, S., Fryns, J.-P., Moreau, Y. *et al.* (2011). Breakage–fusion–bridge cycles leading to inv dup del occur in human cleavage stage embryos. *Human Mutation* 32, 783–793.

Wang, Q., Wang, C. M., Ai, J. S., Xiong, B., Yin, S., Hou, Y., Chen, D. Y., Schatten, H. and Sun, Q. Y. (2006). Histone phosphorylation and pericentromeric histone modifications in oocyte meiosis. *Cell Cycle* 5, 1974–1982.

Weaver, J. R. and Bartolomei, M. S. (2014). Chromatin regulators of genomic imprinting. *Biochimica et Biophysica Acta (BBA) — Gene Regulatory Mechanisms* 1839, 169–177.

Welburn, J. P. I., Grishchuk, E. L., Backer, C. B., Wilson-Kubalek, E. M., Yates, J. R. and Cheeseman, I. M. (2009). The human kinetochore Ska1 complex facilitates microtubule depolymerization-coupled motility. *Developmental Cell* 16, 374–385.

Welburn, J. P. I., Vleugel, M., Liu, D., Yates, J. R., Lampson, M. A., Fukagawa, T. and Cheeseman, I. M. (2010). Aurora B phosphorylates spatially distinct targets to differentially regulate the kinetochore-microtubule interface. *Molecular Cell* 38, 383–392.

Wickramasinghe, D., Ebert, K. M. and Albertini, D. F. (1991). Meiotic competence acquisition is associated with the appearance of M-phase characteristics in growing mouse oocytes. *Developmental Biology* 143, 162–172.

Wiens, G. R. and Sorger, P. K. (1998). Centromeric chromatin and epigenetic effects in kinetochore assembly. *Cell* 93, 313–316.

Yang, F., Baumann, C. and De La Fuente, R. (2009). Persistence of histone H2AX phosphorylation after meiotic chromosome synapsis and abnormal centromere cohesion in poly (ADP-ribose) polymerase (Parp-1) null oocytes. *Developmental Biology* 331, 326–338.

Yang, F., Baumann C., Viveiros M. and De La Fuente, R. (2012). Histone hyperacetylation during meiosis interferes with large-scale chromatin remodeling, axial chromatid condensation and sister chromatid separation in the mammalian oocyte. *International Journal of Developmental Biology* 56, 889–899.

Yang, K. T., Li, S. K., Chang, C. C., Tang, C. J. C., Lin, Y. N., Lee, S. C. and Tang, T. K. (2010). Aurora-C kinase deficiency causes cytokinesis failure in meiosis I and production of large polyploid oocytes in mice. *Molecular Biology of the Cell* 21, 2371–2383.

Zeng, L. and Zhou, M. M. (2002). Bromodomain: an acetyl-lysine binding domain. *FEBS Letters* 513, 124–128.

Zuccotti, M., Giorgi Rossi, P., Martinez, A., Garagna, S., Forabosco, A. and Redi, C. A. (1998). Meiotic and developmental competence of mouse antral oocytes. *Biology of Reproduction* 58, 700–704.

Zuccotti, M., Piccinelli, A., Rossi P. G., Garagna, S. and Redi, C. A. (1995). Chromatin organization during mouse oocyte growth. *Molecular Reproduction and Development* 4, 479–485.

Meiotic Recombination in the Human Germ Line

3

Fatima Smagulova

Irset Inserm, U1085, 9 Avenue du Professeur Leon Bernard, 35000, Rennes, France

Homologous recombination (HR) is the exchange of genetic information between similar deoxyribonucleic acid (DNA) molecules. In mammalian meiosis, HR is essential for correct chromosome segregation during gamete formation. Errors of HR lead to aneuploidy, formation of gametes with incorrect number of chromosomes and genetic disorders. Understanding the mechanisms of recombination is important for identifying disease-associated genomic regions and for the study of human genome evolution. With the development of new whole-genome technologies, knowledge of meiotic recombination in germ cells have advanced significantly. The recent discovery of the role of the PR domain containing 9 (PRDM9) protein in the recombination process opens new perspectives in HR research. In this review, I describe the basics of HR; introduce the pedigree, sperm typing and chromatin immunoprecipitation Sequencing (ChIP-seq) approaches that are currently used to study recombination in human germ cells and summarize the properties of recombination "hotspots", discovered as a result of the understanding of the process involved in double-strand break (DSB) formation. Finally, I discuss the implications of the recombination events for genetic disorders in humans.

3.1 Basics

Homologous recombination (HR) is a process, which normally occurs during cell division in eukaryotes. During mitosis, HR is involved in repair of deoxyribonucleic acid (DNA) double-strand breaks (DSB) caused by radiation or DNA-damaging chemicals (reviewed in (Moynahan and Jasin 2010)), it also plays a role in repair of replication fork stalling, the transient pausing of the replication (Nagaraju and Scully 2007, Hastings *et al.* 2009, Ciccia and Elledge 2010). In germ cells, HR is required for the correct formation of haploid gametes. Recombination is the main driving force in producing genetic diversity within populations, as it shuffles homologous parts of parental chromosomes. Meiosis is highly conservative in eukaryotes. Most of our knowledge of HR mechanisms comes from the studies in budding yeast *Saccharomyces cerevisiae* (reviewed in (Keeney 2007)). Meiotic recombination initiates with DSB introduction by SPO11 (SPO11 is a meiotic protein covalently bound to DSBs), a topoisomerase-like enzyme, which exist in yeast and mammals (Figure 3.1) (Keeney *et al.* 1997, Romanienko and Camerini-Otero 2000, Neale and Keeney 2006). Immediately after formation of DSBs, the MRE11-RAD50-NBN (MRN); MRE11 homolog A, double strand break repair (DSBR) nuclease-RAD50 homolog, DSBR protein-Nibrin) complex binds to DNA on each side of the break (Ohta *et al.* 1998, van den Bosch *et al.* 2003, Cannavo and Cejka 2014). The MRN complex stabilizes the broken chromosome, serves as a sensor of lesions and also participates in the assembly of large macromolecular complexes at DSBs (reviewed in (van den Bosch *et al.* 2003)). One essential step in DSB formation is the generation of extended single-stranded DNA (ssDNA) regions at the break sites. Human exonucleases EXO1 (Exonuclease 1), BLM (Bloom syndrome, RecQ helicase-like) and DNA replication helicase/nuclease 2 (DNA2L) are required for 5′-3′ resection (Mimitou and Symington 2009, Bolderson *et al.* 2010). These exonucleases create 3′ overhangs of ssDNA (Mimitou and Symington 2009) that serve to bind replication protein A (RPA) to the 3′ overhangs (Wold 1997, Wyka *et al.* 2003). Interestingly, the binding properties of human and yeast RPA/Rpa1 are similar. Both proteins bind preferentially to the pyrimidine-rich strand (Kim *et al.* 1992). The RPA protein interacts with RAD52 (Recombinase RAD52) to facilitate the formation of a filament of the RAD51 (RAD51 homolog) (Gibb *et al.* 2014). The strand-exchange proteins RAD51 (Aboussekhra *et al.* 1992, Basile *et al.* 1992, Shinohara *et al.* 1992) and DNA meiotic recombinase 1 (DMC1) (Bishop *et al.* 1992, Story *et al.* 1993) are the central players in HR. These proteins bind to 3′ overhangs to form DNA-protein filaments and mediate the homologous search process (Sung *et al.* 2003). The single-stranded nucleoprotein filaments invade the recipient DNA duplex from non-sister

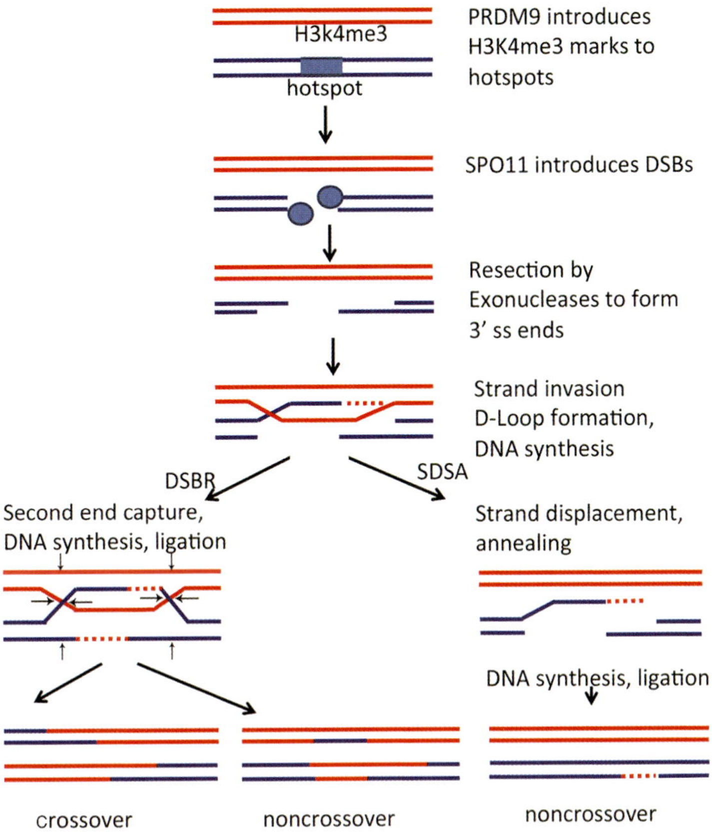

Figure 3.1. Pathways of homologous recombination. DSBs can be repaired by DSBR and SDSA pathways. DSB ends can form an intermediate two Holliday junctions, accompanied by DNA synthesis and ligation. The resolution of two Holliday junctions by endonucleases can result in either non-crossover or crossover products. In the SDSA pathway, the D loop does not form a Holliday junctions and the ssDNA strand anneals with the complementary ssDNA strand that is associated with the other DSB end. The reaction is completed by DNA synthesis and ligation. This pathway forms only noncrossover products.

chromatids in a process termed strand invasion. A displacement loop (D-loop) is formed during strand invasion between the invading 3′ overhang strand and the homologous chromosome. In humans, it was proposed that DNA polymerase delta (POLD) extends the end of the invading 3′ strand by synthesizing new complementary DNA (Sneeden *et al.* 2013). This changes the D-loop to a structure known as a Holliday junction. Two pathways were proposed for HR, the

DSBR and the synthesis dependent strand annealing (SDSA) (Sung and Klein 2006). In the DSBR pathway, the remaining 3′ overhang also forms a Holliday junction with the homologous chromosome. The Holliday junctions may then be resolved to yield either crossover (CO) or non-crossover (NCO) outcome. Only 10% of all DSBs are repaired via formation of a CO in mammals (Andersen and Sekelsky 2010). The SDSA pathway begins with DNA strand invasion and D-loop formation similar to a DSBR pathway. In this pathway, the invading 3′ strand is extended along the recipient DNA duplex. The Holliday junction between the donor and recipient DNA molecules is resolved in a process called *branch migration*. The newly synthesized 3′ end anneals to the other 3′ overhang in the recipient chromosome through complementary base pairing. During meiosis, COs are formed by resolution of double Holliday junctions via the DSBR mechanism, while NCOs are primarily produced via the SDSA mechanism (Allers and Lichten 2001, Hunter and Kleckner 2001).

Finally, first meiotic division will accomplish the production of two cells with diploid number of chromosomes. During the second meiotic division each diploid cell will form two haploid cells.

3.2 Methodologies

It is now well established that recombination initiation DSBs are not randomly distributed along the genome (Lindahl 1991, Clark *et al.* 2010, Paigen and Petkov 2012). Many studies have demonstrated the existence of localized genomic regions in eukaryotes where DSBs preferentially occur, referred as hotspots of recombination (Shiroishi *et al.* 1991, Lichten and Goldman 1995, Yip *et al.* 1999, Jeffreys *et al.* 2000, Kauppi *et al.* 2003, Kong *et al.* 2010). There are several approaches, which are currently being used to study HR (reviewed in (Arnheim *et al.* 2003, Kauppi *et al.* 2004, Arnheim *et al.* 2007, Paigen and Petkov 2012)). The classic tool is pedigree approach, based on the parent-to-offspring transmission of alleles and on the analysis of several polymorphic loci combination, so-called haplotype blocks. The first human recombination hotspots, such as hotspots within major histocompatibility complex (MHC) genes (Bouissou *et al.* 1995), Duchenne muscular dystrophy (Grimm *et al.* 1989), insulin and hotspots proximal to beta globin genes (Lebo *et al.* 1983, Chakravarti *et al.* 1984), were identified or detected by this approach. However, this method requires a large number of parent–offspring samples to produce accurate estimates of the recombination and applies to known hotspots because this method detects COs, not all hotspots. Another approach is a direct sperm typing (Tiemann-Boege *et al.* 2006). It does not require parent–offspring samples, recombination can be studied at the level of an individual sample and allows to search for new hotspots (reviewed in (Buard and de Massy 2007)). The sperm-typing approach is very often combined with *linkage disequilibrium analysis*. Typically, the single sperm-typing approach is based on fluorescence-activated cell sorting (FACS) of sperm or serial dilutions of DNA followed by whole genome amplification and polymerase chain reaction (PCR) of hotspot regions (Arnheim *et al.* 2007). The advantage of this method is precision in hotspot length estimation and single cell resolution. For example, analysis of sorted cells allowed the localization of a hotspot to an 11 kb region in the human beta globin locus (Schneider *et al.* 2002). The major limitation of this method is the requirement to analyze a large number (10^4 to 10^6) of single cells to achieve statistical significance. For example, 1 kb long hotspot, with recombination rate 20 times higher than the genome averaged rate, requires analysis of 4×10^4 single cells (Arnheim *et al.* 2003). To overcome the need to examine large numbers of individual sperm, the pooled-sperm DNA genotyping was developed. This method is based on allele-specific PCR analysis. Allele-specific PCR is designed to detect single nucleotide polymorphism (SNP) at some genomic loci, the detection of allele-specific PCR fragment have a risk of false-positive signal. Several techniques were developed to avoid false-positive rate of allele-specific PCR that include: DNA sequencing, fluorescence resonance energy transfer (FRET), luminescence, absorbance and melting temperature analysis (reviewed in (Wang *et al.* 2004)). Many human recombination hotspots were analyzed by pooled-sperm typing analysis (reviewed in (Arnheim *et al.* 2007)). For example, sperm typing was used to characterize more than 30 hotspots including hotspot in transporter 2 (*TAP2*), ATP-binding cassette, sub-family B (MDR/TAP) (Jeffreys *et al.* 2000), MHC, (Jeffreys *et al.* 2001) and pseudoautosomal region (PAR) (May *et al.* 2002), while valuable, sperm typing studies are aimed at the analysis of only a few individual hotspots. To better understand the mechanisms of HR and build a genome-wide recombination map, comprehensive analysis of hotspots is required. Progress in recombination hotspot mapping at genome-wide level was accelerated by using computer-inferred approaches and statistical estimates. Since 1990, many human SNPs were identified (Botstein *et al.* 1980). A large number of SNPs were deposited at several databases, including Hapmap database (http://hapmap.ncbi.nlm.nih.gov), Seattle (http://pga.gs.washington.edu) and Perlegen (Peacock and Whiteley 2005). At Hapmap database phase I, four populations were included: 30 adult-and-both-parents trios of Yoruba from Ibadan, Nigeria (YRI), 30 trios of Northern and Western European ancestry (CEU), 44 unrelated Japanese individuals and 45 unrelated Han Chinese individuals. In phase III, 11 global ancestry groups have been assembled (Altshuler *et al.* 2010). Using LDhat, a computer approach for coalescent analysis of patterns of linkage disequilibrium and estimation of the population recombination rate Myers *et al.* generated the first fine-scale human recombination map (Myers *et al.* 2005). In this work, coalescent model was used, under this model, the expected time between successive coalescence events, by which two gene copies arise from a single ancestral copy, increases almost exponentially back in time (Kingman 2000). The first fine-scale recombination map showed a good correlation for some of the previously published sperm-typing data. A common 13-bp degenerate motif CCNCCNTNNCCNC was identified in 41% of all human hotspots (Myers *et al.* 2008). Myers *et al.* showed an enrichment of retrotransposon elements in hotspots, including THE1A and THE1B retrotransposons (Table 3.1), and 80% of recombination hotspots were found to occur in 10–20% of sequences. LDhat method, however, require high SNP density, necessary to perform fine-scale molec-

Table 3.1. Properties of human recombination initiation double strand hotspots.

Property	Recombination initiation hotspots	Reference
Number of hotspots per genome	25000–50000	(Myers *et al.* 2005, Pratto *et al.* 2014)
Averaged hotspots width/ variation	1.2Kb/ 0.50–10 Kb	(Genomes Project, Abecasis *et al.* 2010)
Common motifs,		
PRDM9 variant A	CCNCCNTNNCCNC	(Myers *et al.* 2005)
Variant C	CCNCNNTNNNCNTNNC	(Kong *et al.* 2010)
Overrepresented repeats	THE1A, THE1B, GA/rich, CT/rich, l2	(Myers *et al.* 2008)
Underrepresented repeats	L1PA4, L1PA4, LPA7L1P, L1PA5	(Myers *et al.* 2008)
Hotspots activity is high in region	Subtelomere, PAR	(Kong *et al.* 2002, Coop *et al.* 2008, Khil and Camerini-Otero 2010, Wang *et al.* 2012, Kirkness *et al.* 2013)
Hotspots activity is depleted in regions	Centromere	(Nachman 2002)
Other feature	Higher GC content at hotspot center	(Arnheim *et al.* 2007, Buard *et al.* 2009, Paigen and Petkov 2012)

ular analysis as it is based on identification of COs, which represent just 10% of DSBs. The breakthrough in recombination analysis only became possible with advent of direct genome-wide sequencing of hotspots, specifically methods based on chromatin immunoprecipitation (ChIP) techniques, followed by high throughput Illumina sequencing technologies (ChIP-Seq) (Smagulova *et al.* 2011). The method allows detecting hotspots, which are not COs. A kinetic approach to library preparation and computational identification of single stranded ends increased the power of hotspots detection (Khil *et al.* 2012). This combined approach allowed building a fine high-resolution hotspot map in five individuals (Pratto *et al.* 2014). The direct sequencing approach significantly advanced detailed studies and helped to reveal the genomic features of human recombination hotspots (Pratto *et al.* 2014). The high-resolution map of recombination in human males revealed that DSBs, like COs occur more frequently at subtelomeric regions of chromosomes and the CO rate is proportional to the DSBs frequency. The authors show that DSBs affect genome sequence at the level of structural rearrangements (Pratto *et al.* 2014).

3.3 Regulation of Recombination at the Hotspots

The hotspot studies that involved crossing different strains of mice determined that the region on mouse chromosome 17 determines the activity of recombination hotspots. Detailed analysis of this region revealed that the PR domain containing 9 (PRDM9) gene encoding histone H3 methyltransferase, is a factor that determines the hotspot activity in mice (Baudat *et al.* 2010, Myers *et al.* 2010, Parvanov *et al.* 2010). In mice, PRDM9 protein expression is limited to germ line cells and deletion (knockout) of PRDM9 leads to sterility in both sexes (Hayashi and Matsui 2006). The involvement of PRDM9 activity in genome-wide hotspots localization in mice and human was shown by several groups (Baudat *et al.* 2010, Myers *et al.* 2010, Parvanov *et al.* 2010). In mice and human, PRDM9 protein contains N-terminal KRAB domain, methyltransferase SET domain, which catalyzes histone H3 Lys4 trimethylation (H3K4me3) at hotspot loci and contains highly variable zinc finger array (Figure 3.2). It was shown that H3K4me3 marks are highly enriched in hotspots of yeast and mice (Borde *et al.* 2009, Buard *et al.* 2009). However, H3K4me3 marks in mice are not specific for recombination hotspots; they are also enriched at active chromatin sites, where they act as a platform for the binding of multiple histone acetyltransferase, histone demethylases and nucleosome remodeling complexes (Kouzarides 2007, Vermeulen *et al.* 2010). Histone H3 trimethylation marks at hotspots of mice are distinct from trimethylation found at gene promoters as they are established by different methyltransferases

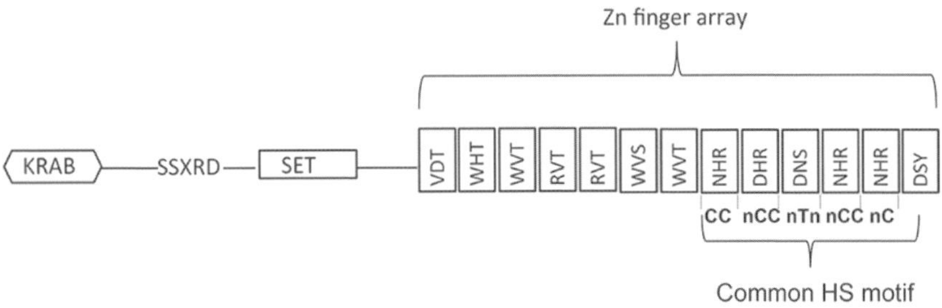

Figure 3.2. Schematic of PRDM9 protein. PRDM9 protein contains kruppel-associated domain-A box (KRAB) domain, SSXRD motif, and SET methyltransferase domain and Zinc finger array. The amino acids of PRDM9 binding sites to hotspot motif are shown in boxes.

(Brick *et al.* 2012). In mice, the major function of PRDM9 is to introduce H3K4me3 at hotspots and to direct the formation of recombination hotspots away from promoters of the genes (Brick *et al.* 2012). Mice lacking the PRDM9 gene have hotspots at promoters (Brick *et al.* 2012) and meiosis arrest at pachytene stage (Mihola *et al.* 2009). Since the discovery of function for PRDM9 as a critical determinant of hotspot location in 2010, dozens of allelic variants were found in many species including mice (Parvanov *et al.* 2010, Buard *et al.* 2014), chimpanzee (Groeneveld *et al.* 2012) and human (Baudat *et al.* 2010, Berg *et al.* 2010, Berg *et al.* 2011). Up to date, more than 20 alleles of human PRDM9 were identified within various human populations (Oliver *et al.* 2009) (Baudat *et al.* 2010, Berg *et al.* 2010, Kong *et al.* 2010, Parvanov *et al.* 2010). The number of human alleles of PRDM9 is constantly increasing as more individual samples are sequenced. Human individuals of European ancestry have less variation in PRDM9: the majority have allele A, (defined by PRDM9 Zn finger sequence, which is CCnCCnTnnCCnC) with the frequency of ~ 84–90% (Baudat *et al.* 2010, Berg *et al.* 2010). For example, in Central European population (CEPH) the allele A occurrs with a frequency of 90%, in Hutterite (small ethnic group living in Western Canada and US) allele A, B, and I, were present at frequencies of 94%, 4% and 2%, respectively (Baudat *et al.* 2010). The individuals of African origin have a much greater diversity in PRDM9 allele sequences, so more variants of PRDM9 were identified (Berg *et al.* 2011), (Hinch *et al.* 2011). For example, in African Americans, the alleles A and C have a frequency of 50% and 15%, respectively (Berg *et al.* 2011). Allelic variants of PRDM9 protein define the different sets of hotspots in mice and human (Baudat *et al.* 2010). Binding capacity of PRDM9/Prdm9 is determined by matching its zinc finger array to a motif located in the middle of hotspots (Baudat *et al.* 2010, Smagulova *et al.* 2011, Pratto *et al.* 2014). It was established that the common European variant of human PRDM9 (coded by allele A) binds to a common CCNCCNTNNCCNC motif *in vivo* and *in vitro* (Baudat *et al.* 2010). However, for individuals with different PRDM9 alleles, for example with allele I, the CCNCCNTNNCCNC motif is not enriched in hotspots (Baudat *et al.* 2010). PRDM9 variant C recognizes motif CCNCNNTNNNCNTNNC and this motif may contribute to difference in LD profiles between European and African populations (Kong *et al.* 2010). As a result of recognizing different DNA motifs, the allelic variants of PRDM9 define different sets of hotspots in human (Baudat *et al.* 2010, Berg *et al.* 2011, Kong *et al.* 2014, Pratto *et al.* 2014). For example, non-A allele individuals showed ~30 times lower recombination frequencies at common hotspots found in AA individuals (Berg *et al.* 2011). The dominance of a particular PRDM9 allele in heterozygous individuals depends on combination of alleles of PRDM9 (Baudat *et al.* 2010). Heterozygous individuals with AI alleles differ significantly from AA homozygotes in their use of recombination hotspots (Baudat *et al.* 2010), with allele I contributing more to hotspots usage than allele A. PRDM9 allele B does not specify a distinct set of hotspots and individuals with alleles AB do not differ significantly from AA individuals in their sets of hotspots (Pratto *et al.* 2014). In contrast, in individuals with allele AC a distinct set of hotspots is present compared to AA individuals (Pratto *et al.* 2014). Dominance of the PRDM9 variant C on hotspot determination over the A allele could be (1) due to a stronger association of allele C PRDM9 to its cognate hotspots than that of allele A (Pratto *et al.* 2014), (2) the alleles of PRDM9 may differ in methyltransferase activity capacity; (3) other factors can contribute to binding specificity of PRDM9.

3.4 Mutagenic Effects of Recombination

As repair of DSBs is associated with gene conversion (GC), Myers and colleagues (Myers *et al.* 2005) suggested that the recombination hotspots are "self-destructive". Sequencing of individual genomes with AA1 and AA2 alleles showed that sequence variants were strongly enriched around the center of hotspots (Pratto *et al.* 2014). Polymorphism analysis in humans showed biased GC (Birdsell 2002). High GC content appears to be highly correlated with recombination activity and it is enriched in hotspot centers (Arnheim *et al.* 2007, Buard *et al.* 2009, Paigen and Petkov 2012). Meiotic recombination favors the fixation of G and C nucleotides at hotspots and increases the rate of A/T to G/C substitutions. This process is known as GC-biased gene conversion (gBGC) (Spencer *et al.* 2006). gBGC plays a role in self-destructing of the recombination hotspots. The study of a large number of human COs by a single sperm analysis showed that COs carry more *de novo* mutations than non-recombinant molecules, when analyzed in the same donor (Arbeithuber *et al.* 2015). The observed mutations were CG to TA transitions with the highest transition at CpG than non-CpG sites (Arbeithuber *et al.* 2015). This transition is the opposite of gBGC. The authors suggested that gBGC could be an adaptation to counteract the mutational load of recombination (Arbeithuber *et al.* 2015). In recent work, the authors showed that DSB hotspots are enriched at the breakpoints of copy number variants that arise via homology-mediated mechanisms. The activity of DSB hotspots during recombination could increase the mutation rate at these regions due to possible rearrangements (duplication, deletion, inversion) caused by structural features of regions (Pratto *et al.* 2014).

3.5 Evolution of PRDM9

PRDM9 is a rapidly evolving gene. It differs significantly between species and within the same species. The fast evolution of PRDM9 was suggested to be a result of non-allelic GC, which has contributed to the mutational mechanism in human and mouse PRDM9/Prdm9 (Oliver *et al.* 2009). It was suggested that the rapid evolution of hotspots could eventually create a selective pressure to generate and recognize new target hotspot sequences and favoring new PRDM9 variants (Coop and Myers 2007).

In dogs (*Canis familiaris*) (Axelsson *et al.* 2012) and their close relatives wolfs (*Canis lupus)* (Munoz-Fuentes *et al.* 2011), PRDM9 is not functional due to frame shift mutations, however, these species have normal meiosis. Recombination appears to be similar to other mammalian species. However, fine-scale estimates show that in canines highly elevated recombination rates are observed in the vicinity of CpG-rich regions including gene promoter regions (Auton *et al.* 2013), so mechanisms of meiosis are different from what was observed in mice and human. It was suggested that in dogs other mechanisms were developed to compensate for the inactive PRDM9 function (Auton *et al.* 2013). PRDM9 is lacking in chicken (*Gallus Gallus*), frog (*Xenopus tropicalis*) and marsupial (*Monodelphis domestica*) (Oliver *et al.* 2009). In chimpanzee (*Pan troglodytes verus*), PRDM9 does not play a major role in specifying hotspot location but the protein itself is functional (Auton *et al.* 2012). The recombination rate in chimpanzee is similar to the human (Auton *et al.* 2012), so other mechanisms must contribute to the determination of hotspots positions and frequency.

How old are the modern human hotspots? Comparison of DSBs hotspots of Denisovan human (the extinct relative of Neanderthals, which fossils were found in Denisova Cave, Russia) and those of modern humans showed the substantial difference in hotspot usage (Lesecque *et al.* 2014). Although the PRDM9 sequence in the Denisovan human is not complete, sequence homology of certain regions resembles PRDM9 alleles in African population (Schwartz *et al.* 2014). According to estimates, the modern human hotspots are relatively young (Lesecque *et al.* 2014). The combined analysis of modern and Denisovan humans demonstrates that the life expectancy of hotspots is estimated between 70.000–130.000 years, which is only 10% of time when human split from chimpanzee (700.000–130000 years ago) (Lesecque *et al.* 2014). Rapid evolution of PRDM9 and GC at hotspots have made a strong contribution in shaping recombination profiles in different populations of modern humans (Lesecque *et al.* 2014).

3.6 Properties of Recombination Hotspots: Rate, Width, Number, Genomic Distribution and Sex-Differences

Recombination rate. Initially, the recombination rate on average was estimated to be 1.1 cM/Mb, with some regions in the genome having a higher recombination rate. For example, the pseudoautosomal region 1 (PAR1) on the human X chromosome, has a 30-fold higher recombination rate than the genome average (Rouyer *et al.* 1986, Weissenbach *et al.* 1987, Henke *et al.* 1993). Later studies of recombination in human populations showed a substantial difference in recombination rates across the genome (Genomes Project, Abecasis *et al.* 2010).

Hotspots width. Sperm-typing data estimates hotspots length to be from several hundred to several thousand bps (Buard *et al.* 2000, Jeffreys *et al.* 2001, Guillon and de Massy 2002, Kauppi *et al.* 2004, Holloway *et al.* 2006, Tiemann-Boege *et al.* 2006). The 1,000 genomes study found the average hotspot width to be 2.3 kb (Genomes Project, Abecasis *et al.* 2010).

Genomic distribution. The earliest human genetic recombination studies already suggested that recombination was not uniform (Botstein *et al.* 1980, Donis-Keller *et al.* 1987, Buetow *et al.* 1994, Broman *et al.* 1998, Nachman 2002). So why is distribution of HR not even across the chromosome? One possibility is that a cell should balance two opposing needs; while recombination is necessary to ensure proper chromosome segregation, the breaking up should be minimized to ensure genome stability. It was discovered that recombination is suppressed near centromeres (Nachman 2002), perhaps it would otherwise interfere with meiotic chromosome segregation (Lamb *et al.* 2005). Recombination is associated with increased risk of chromosome missegregation (Rockmill *et al.* 2006). For example, in humans non-disjunction has been found to be associated with CO that occurs too close to the centromere (Hassold and Hunt 2001). It has been established that COs in human males are enriched in subtelomeric regions (Kong *et al.* 2002, Coop *et al.* 2008, Khil and Camerini-Otero 2010, Wang *et al.* 2012, Kirkness *et al.* 2013). DSB hotspots are also stronger and densely placed at subtelomeric regions (Linardopoulou *et al.* 2005, Pratto *et al.* 2014). Telomere clustering in meiotic cells (Bass 2003) may favor exchange of chromosome ends during DSB repair and it was suggested that it may lead to high recombination activity (Linardopoulou *et al.* 2005).

Sex-differences in HR. Studies of recombination in humans showed that there are substantial differences in female and male recombination. Analysis of MLH1 foci (protein which can be detected at COs) showed large variations in the number of COs among oocytes of the same female (Lenzi *et al.* 2005). Recombination rate is higher in females; all human chromosomes have longer genetic maps in females than in males, by 1.6-fold on average (Kong *et al.* 2002, Kong *et al.* 2010). Although male recombination results in more exon shuffling within genes, female recombination generates more new combinations of nearby genes (Kong *et al.* 2010). Around 15% of female hotspots are sex-specific (Kong *et al.* 2010). Sex-specific hotspots tend to occur in clusters (Kong *et al.* 2010). Despite the fact that spermatocytes have fewer number of COs that are required to keep chromosomes together during formation of the meiotic spindle, they have a lower rate of aneuploidy than oocytes (Hassold and Hunt 2001). The explanation lies in the differences of the meiotic pathway in males and females. The mammalian meiosis proceeds differently during development in males and females. In human and mouse spermatocytes, meiosis is usually arrested if synaptic or pairing errors are present (Hunt and Hassold 2002). Meiotic recombination occurs in males throughout life while female meiosis begins in the fetal ovary, is arrested after COs are formed and is completed after fertilization. A prominent feature of aneuploidies is that they frequently affect sex chromosomes; HR in sex chromosomes differs from other chromosomes. DNA polymorphism analysis demonstrated that most sex chromosomes aneuploidies occur in the paternal germ line (Hall *et al.* 2006). The common reason of paternal-driven X chromosome aneuploidies is a recombination failure at the PAR in male meiosis. Recombination at the PAR differs significantly between sexes: in the male it has the highest recombination rate (Hall *et al.* 2006), while in the female is has the average rate of recombination (Page *et al.* 1987, Henke *et al.* 1993, Flaquer *et al.* 2009). The PAR region has short homologous sequences both in X- and Y-chromosomes, where obligatory CO formation occurs during meiosis. Reduced recombination activity at PAR may cause X chromosome monosomy (Turner syndrome) or Kleinfelter syndrome in the offspring or lead to infertility ((Hassold *et al.* 1991, Shi *et al.* 2001), reviewed in (Heard and Turner

2011)). A striking difference between mice and humans is that the mouse PRDM9 is not involved in recombination in PAR. In humans, PRDM9 localizes all peaks of recombination and among human population the recombination activity in PAR significantly varies (Hinch *et al.* 2014).

3.7 Implications of the Recombination Hotspot Variation in Genetic Disorders in Humans

In mice, PRDM9 is a hybrid sterility gene and particular combinations of crosses between different mouse strains lead to infertility (Mihola *et al.* 2009). In a human, DSB sites in certain genomic contexts are predicted to be risk loci for disease-causing genomic rearrangements in individuals carrying certain PRDM9 alleles (Hinch *et al.* 2011). The activity of PRDM9 is a major contributor to Non-Allelic Homologous Recombination (NAHR), which results in recurrent rearrangements that include the same genomic intervals occurring in unrelated individuals. The activity of some human PRDM9 allele variants cause the instability of mini-satellite repeats (Berg *et al.* 2010). In human, due to recombination events, the region in chromosome 17 undergoes recurrent deletion/duplication in Charcot–Marie–Tooth type 1A (CMT1A) and hereditary neuropathy with liability to pressure palsies (HNPP) genetic diseases (Berg *et al.* 2010). Activity of PRDM9 variant A leads to unequal meiotic exchanges between 1.5-Mb separated CMT1A-REP genomic regions. The individuals with no A alleles have nearly 25 times reduced chance to have these disorders (Berg *et al.* 2010). Study using twins showed that NAHR deletion rate at CMT1A is heritable and variable within populations (MacArthur *et al.* 2014). In a population-based study (Borel *et al.* 2012), the significant increase in frequency of PRDM9 non-A variants in parents who transmitted *de novo* 7q11.23 deletions to their offspring was shown. These datas suggest that non-A PRDM9 alleles are associated with an increased susceptibility to recurrent 7q11.23 microdeletions that cause Williams–Beuren syndrome. Rare allelic forms of PRDM9 are associated with childhood leukemogenesis (Hussin *et al.* 2013). Different PRDM9 alleles are associated with the development of preleukemic clones in B-acute lymphoblastic leukemia (ALL) patients (Hussin *et al.* 2013). The authors reported an excess of alleles containing *k-finger* repeat (k finger alleles are PRDM9 alleles C and D, which characterized by presence of additional zinc finger k) in the parents of children with ALL, with no preferential carriage by either parent (Hussin *et al.* 2013). Another group confirmed these results and detected an excess of PRDM9 *k-finger* in a cohort of parents of children with ALL and their children compared with a control cohort of individuals of similar descent (Woodward *et al.* 2014). The model was proposed, in which the role of PRDM9 histone H3K4 methyltransferase activity in the parental germline is to use alternative sites of genome for trimethylation that could lead to the genomic instability associated with childhood ALL (Hussin *et al.* 2013). Specific mutations were identified in the PRDM9 allele in patients with azoospermia (Irie *et al.* 2009). Neither variant was identified in fertile subjects. Thus, some mutations in PRDM9 gene might cause idiopathic infertility in human males (Irie *et al.* 2009). It was determined that the recurrence in NAHR is mediated by a common genomic structure in which the rearranged interval is flanked by paralogous sequences or low copy repeats (LCRs) (reviewed in (Lupski and Stankiewicz 2005)). If a CO occurs between LCRs in non-homologous chromosomes, recurrent translocation may be produced. NAHR-mediated events may result in deletions, duplications or inversions (Table 3.2). Structural alterations, the copy number of dosage-sensitive genes changes lead to genomic disorders (reviewed in (Zhang *et al.* 2009)). In a human genome, the existence of 1,143 LCRs was predicted (Coutton *et al.* 2013). Eighty-nine regions were identified as NAHR-prone regions based on LCR data. For example, deletion of the DPY19L2 locus was discovered in 75% of patients with globozoospermia associated with NAHR between two 28kb low copies number repeats (Coutton *et al.* 2013). The breakpoints of NAHR rearrangements are coincident with meiotic hotspots (Coutton *et al.* 2013). About 14 out of 27 known human disease-associated breakpoints coincide with PRDM9 allele A-defined hotspots while individuals with PRDM9 allele C are not risk alleles for these diseases (Pratto *et al.* 2014). The 726 hotspots allele defined by PRDM9-allele A are located in the regions of LCR and these regions could be targets for future studies to identify genomic regions that cause genetic disorders in humans (Pratto *et al.* 2014).

3.8 Effects of Environmental Factors and Age on HR

A number of experimental studies have examined the effect of bisphenol A (BPA) on reproduction. A positive correlation between BPA treatment during embryonic development and decreased COs in progeny of mice was demonstrated (Vrooman *et al.* 2015). BPA affects the recombination processes and oocyte survival in mice (Brieno-Enriquez *et al.* 2011). In experiments with DDT treatment, a significant correlation between exposure to DDT and increased CO numbers in fly species was observed (Flexon and Rodell 1982).

There are only a few studies of the impacts of environmental factors on human recombination. Cigarette smoking delays the meiotic maturation of human oocytes and causes abnormalities of sperm chromosome condensation (Zenzes *et al.* 1990, Zenzes *et al.* 1995, Zenzes *et al.* 1997). There is a significant decrease of recombination rates with advancing maternal age in humans (Hussin *et al.* 2011). The effect appears

Table 3.2. Genetic diseases are likely associated with NAHR.

Disorders	Chr:NAHR event/gene	Phenotypes	Ref.
Bartter syndrome type 3	chr1: 11 kb deletion on 1p/ CLCNKB	Defective renal function, retardation of growth and mental development	(Simon *et al*. 1997)
Gaucher disease	chr1: 16 kb deletion on 1q/ GBA	Hepatosplenomegaly, pancytopenia, bone marrow infiltration defects	(Zimran *et al*. 1990)
Sotos syndrome	chr5: 1.9 Mb deletion on 5q/ NSD1	Facial dysmorphism, nonprogressive cerebral disorder with mental retardation	(Kurotaki *et al*. 2005)
Congenital adrenal hyperplasia	chr6:30 kb deletion or gene conversion on 6p/ CYP21L2	Abnormal sexual development and growth defects	(White *et al*. 1988, Tusie-Luna and White 1995)
Williams–Beuren syndrome	chr7: 1.5 Mb deletion on 7q/28 genes are deleted	Supravalvular aortic stenosis (SVAS), mental retardation, and distinctive facial features.	(Urban *et al*. 1996)
Glucocorticoid-remediable aldosteronism	chr8: 45 kb duplication on 8q/CYP11B1	Hypertension, low plasma renin activity and increased aldosterone secretion	(Lifton *et al*. 1992)
Prader–Willi syndrome	chr15: Paternal 4 Mb deletion on 15q/ SNRPNa and NDN	Muscular hypotonia, mental retardation, short stature, hypogonadism	(Carrozzo *et al*. 1997)
Angelman syndrome	chr15: Maternal 4 Mb deletion on 15q/UBE3a, MECP2	Neurodevelopmental disorder characterized by mental retardation	(Carrozzo *et al*. 1997)

(*Continued*)

Table 3.2. (*Continued*)

Disorders	Chr:NAHR event/gene	Phenotypes	Ref.
15q13.3 deletion syndrome*	chr15: 680 kb, 1.5 Mb or 3.95 Mb deletion on 15q	Mental retardation, epilepsy and variable dysmorphism of the face and digits	(Sharp *et al.* 2008, Shinawi *et al.* 2009)
Smith–Magenis syndrome*	chr15: 3.7 Mb deletion on 17p/ RAI1	mental retardation, hypotonia, speech delay	(Chen *et al.* 1997, Potocki *et al.* 2000)
Charcot–Marie–Tooth disease type 1A*	chr17: 1.4 Mb duplication on17p12/gene/PMP22	Weakness and atrophy of muscles of the lower legs and hands	(Pentao *et al.* 1992)
Hereditary neuropathy with liability to pressure palsies*	chr17: 1.4 Mb deletion on 17p/PMP22/ PMP22	Weakening and atrophy of muscles	(Chance *et al.* 1994)
Potocki–Lupski syndrome*	chr17: 3.7 Mb duplication on 17p	Hypotonia, mental retardation, pervasive developmental disorders	(Potocki *et al.* 2000)
Isolated growth hormone deficiency, type 1A	chr17: 6.7 kb deletion on 17q/GHI	Decreased growth velocity	(Vnencak-Jones *et al.* 1988)
Velocardiofacial syndrome	chr:22 1.5 Mb or 3 Mb deletion on 22q/TBX1	Parathyroid hypoplasia, thymic hypoplasia and outflow tract defects of the heart	(Edelmann *et al.* 1999)
Polycystic kidney disease 1*	chr16: Gene conversion on 16p/PKD1	Impairment of renal functions	(Watnick *et al.* 1998)
Ichthyosis*	chrX: 1.5 Mb deletion on Xp/SHOX	Dry scaly skin	(Van Esch *et al.* 2005)
β-thalassaemia*	chr11: 7 kb deletion on 11p/ HBB	Reduced beta globin, Anemia	(Metzenberg *et al.* 1991)
Haemophilia A	chrX: 500 kb or 600 kb inversion on Xq	Impaired blood coagulation	(Lakich *et al.* 1993)

* DSB hotspots coinciding with NAHR breakpoint from (Pratto *et al.* 2014).

to be localized in the middle section of chromosomal arms and near subtelomeric regions (Hussin *et al.* 2011). The advancing paternal age affects pairing, synapsis and recombination between homologous chromosomes (Vrooman *et al.* 2014).

3.9 Perspectives

Sequencing of human genomes and discovery of recombination hotspots in human populations has greatly advanced the understanding of the mechanisms for HR. The identification of PRDM9 variants and association of some of them with genetic diseases could be a predictive tool in diagnostics of human disorders. As sequencing data from human populations accumulates, it will provide the foundation to mine for the new knowledge of the HR in evolution of human genomes. We still have little understanding of the mechanistic model for PRDM9 action during initiation of the recombination. The impacts of environmental factors, including toxic reagents, smoking and age on recombination are yet poorly understood.

Glossary

HR	Exchange between similar molecules of DNA within or between chromosome
Linkage disequilibrium	A measure of whether alleles at two loci coexist in a population in a non-random fashion
Coalescent model	A model for a genealogy of nominally unrelated individuals in a population sample
Haplotype	A combination of alleles at different loci that is transmitted together from one generation to another
GC	Non-reciprocal exchange of sequence from one chromosome to another
CO	Recombination process that produces a reciprocal exchange of DNA sequences
Genomic disorders	Conditions arising in the germline by chromosomal rearrangements that alter the copy number of dosage-sensitive genes, disrupt genes or generate fusion genes
Telomere	Segment at the end of each chromosome arm that consists of repetitive sequences
Centromere	Segment of chromosome that links sister chromatids

Acknowledgment

I apologize to colleagues whose work could not be cited due to limited space. This work was supported by European University of Brittany for Chair of Excellence and Atip-Avenir programs to FS. I would like to thank Sergei Tevosian for a critical reading of the manuscript.

References

Aboussekhra, A., Chanet, R., Adjiri, A. and Fabre F. (1992). Semidominant suppressors of Srs2 helicase mutations of Saccharomyces cerevisiae map in the RAD51 gene, whose sequence predicts a protein with similarities to procaryotic RecA proteins. *Molecular Cell Biology* 12(7), 3224–3234.

Allers, T. and Lichten M. (2001). Differential timing and control of noncrossover and crossover recombination during meiosis. *Cell* 106(1), 47–57.

Altshuler, D. M., Gibbs, R. A., Peltonen, L., Altshuler, D. M., Gibbs, R. A., Peltonen, L., Dermitzakis, E., Schaffner, S. F., Yu, F., Peltonen, L., Dermitzakis, E., Bonnen, P. E., Altshuler, D. M., Gibbs, R. A., de Bakker, P. I., Deloukas, P., Gabriel, S. B., Gwilliam, R., Hunt, S., Inouye, M., Jia, X., Palotie, A., Parkin, M., Whittaker, P., Yu, F., Chang, K., Hawes, A., Lewis, L. R., Ren, Y., Wheeler, D., Gibbs, R. A., Muzny, D. M., Barnes, C., Darvishi, K., Hurles, M., Korn, J. M., Kristiansson, K., Lee, C., McCarrol, S. A., Nemesh, J., Dermitzakis, E., Keinan, A., Montgomery, S. B., Pollack, S., Price, A. L., Soranzo, N., Bonnen, P. E., Gibbs, R. A., Gonzaga-Jauregui, C., Keinan, A., Price, A. L., Yu, F., Anttila, V., Brodeur, W., Daly, M. J., Leslie, S., McVean, G., Moutsianas, L., Nguyen, H., Schaffner, S. F., Zhang, Q., Ghori, M. J., McGinnis, R., McLaren, W., Pollack, S., Price, A. L., Schaffner, S. F., Takeuchi, F., Grossman, S. R., Shlyakhter, I., Hostetter, E. B., Sabeti, P. C., Adebamowo, C. A., Foster, M. W., Gordon, D. R., Licinio, J., Manca, M. C., Marshall, P. A., Matsuda, I., Ngare, D., Wang, V. O., Reddy, D., Rotimi, C. N., Royal, C. D., Sharp, R. R., Zeng, C., Brooks, L. D. and McEwen, J. E. (2010). Integrating common and rare genetic variation in diverse human populations. *Nature* 467(7311), 52–58.

Andersen, S. L. and Sekelsky J. (2010). Meiotic versus mitotic recombination: two different routes for double-strand break repair: the different functions of meiotic versus mitotic DSB repair are reflected in different pathway usage and different outcomes. *Bioessays* 32(12), 1058–1066.

Arbeithuber, B., Betancourt, A. J., Ebner, T. and Tiemann-Boege, I. (2015). Crossovers are associated with mutation and biased gene conversion at recombination hotspots. *Proceedings of the National Academy of Sciences USA* 112(7), 2109–2114.

Arnheim, N., Calabrese, P. and Nordborg, M. (2003). Hot and cold spots of recombination in the human genome: the reason we should find them and how this can be achieved. *The American Journal of Human Genetics* 73(1), 5–16.

Arnheim, N., Calabrese, P. and Tiemann-Boege, I. (2007). Mammalian meiotic recombination hot spots. *Annual Review of Genetics* 41, 369–399.

Auton, A., Fledel-Alon, A., Pfeifer, S., Venn, O., Segurel, L., Street, T., Leffler, E. M., Bowden, R., Aneas, I., Broxholme, J., Humburg, P., Iqbal, Z., Lunter, G., Maller, J., Hernandez, R. D., Melton, C., Venkat, A., Nobrega, M. A., Bontrop, R., Myers, S., Donnelly, P., Przeworski, M. and McVean, G. (2012). A fine-scale chimpanzee genetic map from population sequencing. *Science* 336(6078), 193–198.

Auton, A., Rui Li, Y., Kidd, J., Oliveira, K., Nadel, J., Holloway, J. K., Hayward, J. J., Cohen, P. E., Greally, J. M., Wang, J., Bustamante, C. D. and Boyko, A. R. (2013). Genetic recombination is targeted towards gene promoter regions in dogs. *PLoS Genetics* 9(12), e1003984.

Axelsson, E., Webster, M. T., Ratnakumar, A., Consortium, L., Ponting, C. P. and Lindblad-Toh, K. (2012). Death of PRDM9 coincides with stabilization of the recombination landscape in the dog genome. *Genome Research* 22(1), 51–63.

Basile, G., Aker, M. and Mortimer, R. K. (1992). Nucleotide sequence and transcriptional regulation of the yeast recombinational repair gene RAD51. *Molecular and Cellular Biology* 12(7), 3235–3246.

Bass, H. W. (2003). Telomere dynamics unique to meiotic prophase: formation and significance of the bouquet. *Cellular and Molecular Life Sciences* 60(11), 2319–2324.

Baudat, F., Buard, J., Grey, C., Fledel-Alon, A., Ober, C., Przeworski, M., Coop, G. and de Massy, B. (2010). PRDM9 is a major determinant of meiotic recombination hotspots in humans and mice. *Science* 327(5967), 836–840.

Berg, I. L., Neumann, R., Lam, K. W., Sarbajna, S., Odenthal-Hesse, L. May, C. A. and Jeffreys, A. J. (2010). PRDM9 variation strongly influences recombination hot-spot activity and meiotic instability in humans. *Nature Genetics* 42(10), 859–863.

Berg, I. L., Neumann, R., Sarbajna, S., Odenthal-Hesse, L. Butler, N. J. and Jeffreys, A. J. (2011). Variants of the protein PRDM9 differentially regulate a set of human meiotic recombination hotspots highly active in African populations. *Proceedings of the National Academy of Sciences USA* 108(30), 12378–12383.

Birdsell, J. A. (2002). Integrating genomics, bioinformatics, and classical genetics to study the effects of recombination on genome evolution. *Molecular Biology Evolution* 19(7), 1181–1197.

Bishop, D. K., Park, D., Xu, L. and Kleckner, N. (1992). DMC1: a meiosis-specific yeast homolog of E. coli recA required for recombination, synaptonemal complex formation, and cell cycle progression. *Cell* 69(3), 439–456.

Bolderson, E., Tomimatsu, N., Richard, D. J., Boucher, D., Kumar, R., Pandita, T. K., Burma, S. and Khanna, K. K. (2010). Phosphorylation of Exo1 modulates homologous recombination repair of DNA double-strand breaks. *Nucleic Acids Research* 38(6), 1821–1831.

Borde, V., Robine, N., Lin, W., Bonfils, S., Geli, V. and Nicolas, A. (2009). Histone H3 lysine 4 trimethylation marks meiotic recombination initiation sites. *The EMBO Journal* 28(2), 99–111.

Borel, C., Cheung, F., Stewart, H., Koolen, D. A., Phillips, C., Thomas, N. S., Jacobs, P. A., Eliez, S. and Sharp, A. J. (2012). Evaluation of PRDM9 variation as a risk factor for recurrent genomic disorders and chromosomal non-disjunction. *Human Genetics* 131(9), 1519–1524.

Botstein, D., White, R. L., Skolnick, M. and Davis, R. W. (1980). Construction of a genetic linkage map in man using restriction fragment length polymorphisms. *The American Journal of Human Genetics* 32(3), 314–331.

Bouissou, C., Pontarotti, P. and Crouau-Roy, B. (1995). A precise meiotic map in the class I region of the human major histocompatibility complex. *Genomics* 30(3), 486–492.

Brick, K., Smagulova, F., Khil, P., Camerini-Otero, R. D. and Petukhova, G. V. (2012). Genetic recombination is directed away from functional genomic elements in mice. *Nature* 485(7400), 642–645.

Brieno-Enriquez, M. A., Robles, P., Camats-Tarruella, N., Garcia-Cruz, R., Roig, I., Cabero, L., Martinez, F. and Caldes, M. G. (2011). Human meiotic progression and recombination are affected by Bisphenol A exposure during in vitro human oocyte development. *Human Reproduction* 26(10), 2807–2818.

Broman, K. W., Murray, J. C., Sheffield, V. C., White, R. L. and Weber, J. L. (1998). Comprehensive human genetic maps: individual and sex-specific variation in recombination. *The American Journal of Human Genetics* 63(3), 861–869.

Buard, J., Barthes, P., Grey, C. and de Massy, B. (2009). Distinct histone modifications define initiation and repair of meiotic recombination in the mouse. *The EMBO Journal* 28(17), 2616–2624.

Buard, J. and de Massy, B. (2007). Playing hide and seek with mammalian meiotic crossover hotspots. *Trends in Genetics* 23(6), 301–309.

Buard, J., Rivals, E., Dunoyer de Segonzac, D., Garres, C., Caminade, P., de Massy, B. and Boursot, P. (2014). Diversity of Prdm9 zinc finger array in wild mice unravels new facets of the evolutionary turnover of this coding minisatellite. *PLoS One* 9(1), e85021.

Buard, J., Shone, A. C. and Jeffreys, A. J. (2000). Meiotic recombination and flanking marker exchange at the highly unstable human minisatellite CEB1 (D2S90). *The American Journal of Human Genetics* 67(2), 333–344.

Buetow, K. H., Weber, J. L., Ludwigsen, S., Scherpbier-Heddema, T., Duyk, G. M., Sheffield, V. C., Wang, Z. and Murray, J. C. (1994). Integrated human genome-wide maps constructed using the CEPH reference panel. *Nature Genetics* 6(4), 391–393.

Cannavo, E. and Cejka, P. (2014). Sae2 promotes dsDNA endonuclease activity within Mre11-Rad50-Xrs2 to resect DNA breaks. *Nature* 514(7520), 122–125.

Chakravarti, A., Buetow, K. H., Antonarakis, S. E., Waber, P. G., Boehm, C. D. and Kazazian, H. H. (1984). Nonuniform recombination within the human beta-globin gene cluster. *The American Journal of Human Genetics* 36(6), 1239–1258.

Ciccia, A. and Elledge, S. J. (2010). The DNA damage response: making it safe to play with knives. *Molecular Cell* 40(2), 179–204.

Clark, A. G., Wang, X. and Matise, T. (2010). Contrasting methods of quantifying fine structure of human recombination. *Annual Review of Genomics Human Genetics* 11, 45–64.

Coop, G. and Myers, S. R. (2007). Live hot, die young: transmission distortion in recombination hotspots. *PLoS Genetics* 3(3), e35.

Coop, G., Wen, X., Ober, C., Pritchard, J. K. and Przeworski, M. (2008). High-resolution mapping of crossovers reveals extensive variation in fine-scale recombination patterns among humans. *Science* 319(5868), 1395–1398.

Coutton, C., Abada, F., Karaouzene, T., Sanlaville, D., Satre, V., Lunardi, J., Jouk, P. S., Arnoult, C., Thierry-Mieg, N. and Ray, P. F. (2013). Fine characterisation of a recombination hotspot at the DPY19L2 locus and resolution of the paradoxical excess of duplications over deletions in the general population. *PLoS Genetics* 9(3), e1003363.

Donis-Keller, H., Green, P., Helms, C., Cartinhour, S., Weiffenbach, B., Stephens, K., Keith, T. P., Bowden, D. W., Smith, D. R., Lander, E. S. and *et al.* (1987). A genetic linkage map of the human genome. *Cell* 51(2), 319–337.

Flaquer, A., Fischer, C. and Wienker, T. F. (2009). A new sex-specific genetic map of the human pseudoautosomal regions (PAR1 and PAR2). *Human Heredity* 68(3), 192–200.

Flexon, P. B. and Rodell, C. F. (1982). Genetic recombination and directional selection for DDT resistance in Drosophila melanogaster. *Nature* 298(5875), 672–674.

Genomes Project, C., Abecasis, G. R., Altshuler, D., Auton, A., Brooks, L. D., Durbin, R. M., Gibbs, R. A., Hurles, M. E. and McVean, G. A. (2010). A map of human genome variation from population-scale sequencing. *Nature* 467(7319), 1061–1073.

Gibb, B., Ye, L. F., Kwon, Y., Niu, H., Sung, P. and Greene, E. C. (2014). Protein dynamics during presynaptic-complex assembly on individual single-stranded DNA molecules. *Nature Structural and Molecular Biology* 21(10), 893–900.

Grimm, T., Muller, B., Dreier, M., Kind, E., Bettecken, T., Meng, G. and Muller, C. R. (1989). Hot spot of recombination within DXS164 in the Duchenne muscular dystrophy gene. *The American Journal of Human Genetics* 45(3), 368–372.

Groeneveld, L. F., Atencia, R., Garriga, R. M. and Vigilant, L. (2012). High diversity at PRDM9 in chimpanzees and bonobos. *PLoS One* 7(7), e39064.

Guillon, H. and de Massy, B. (2002). An initiation site for meiotic crossing-over and gene conversion in the mouse. *Nature Genetics* 32(2), 296–299.

Hall, H., Hunt, P. and Hassold, T. (2006). Meiosis and sex chromosome aneuploidy: how meiotic errors cause aneuploidy; how aneuploidy causes meiotic errors. *Current Opinion in Genetics & Development* 16(3), 323–329.

Hassold, T. and Hunt, P. (2001). To err (meiotically) is human: the genesis of human aneuploidy. *Nature Review Genetics* 2(4), 280–291.

Hassold, T. J., Sherman, S. L., Pettay, D., Page, D. C. and Jacobs, P. A. (1991). XY chromosome nondisjunction in man is associated with diminished recombination in the pseudoautosomal region. *The American Journal of Human Genetics* 49(2), 253–260.

Hastings, P. J., Ira, G. and Lupski, J. R. (2009). A microhomology-mediated break-induced replication model for the origin of human copy number variation. *PLoS Genetics* 5(1), e1000327.

Hayashi, K. and Matsui, Y. (2006). Meisetz, a novel histone tri-methyltransferase, regulates meiosis-specific epigenesis. *Cell Cycle* 5(6), 615–620.

Heard, E. and Turner, J. (2011). Function of the sex chromosomes in mammalian fertility. *Cold Spring Harbor Perspectives Biology* 3(10), a002675.

Henke, A., Fischer, C. and Rappold, G. A. (1993). Genetic map of the human pseudoautosomal region reveals a high rate of recombination in female meiosis at the Xp telomere. *Genomics* 18(3), 478–485.

Hinch, A. G., Altemose, N., Noor, N., Donnelly, P. and Myers, S. R. (2014). Recombination in the human Pseudoautosomal region PAR1. *PLoS Genetics* 10(7), e1004503.

Hinch, A. G., Tandon, A., Patterson, N., Song, Y., Rohland, N., Palmer, C. D., Chen, G. K., Wang, K., Buxbaum, S. G., Akylbekova, E. L., Aldrich, M. C., Ambrosone, C. B., Amos, C., Bandera, E. V., Berndt, S. I., Bernstein, L., Blot, W. J., Bock, C. H., Boerwinkle, E., Cai, Q., Caporaso, N., Casey, G., Cupples, L. A., Deming, S. L., Diver, W. R., Divers, J., Fornage, M., Gillanders, E. M., Glessner, J., Harris, C. C., Hu, J. J., Ingles, S. A., Isaacs, W., John, E. M., Kao, W. H., Keating, B., Kittles, R. A., Kolonel, L. N., Larkin, E., Le Marchand, L., McNeill, L. H., Millikan, R. C., Murphy, A., Musani, S., Neslund-Dudas, C., Nyante, S., Papanicolaou, G. J., Press, M. F., Psaty, B. M., Reiner, A. P., Rich, S. S., Rodriguez-Gil, J. L., Rotter, J. I., Rybicki, B. A., Schwartz, A. G., Signorello, L. B., Spitz, M., Strom, S. S., Thun, M. J., Tucker, M. A., Wang, Z., Wiencke, J. K., Witte, J. S., Wrensch, M., Wu, X., Yamamura, Y., Zanetti, K. A., Zheng, W., Ziegler, R. G., Zhu, X., Redline, S., Hirschhorn, J. N., Henderson, B. E., Taylor, H. A., Jr., Price, A. L., Hakonarson, H., Chanock, S. J., Haiman, C. A., Wilson, J. G., Reich, D. and Myers, S. R. (2011). The landscape of recombination in African Americans. *Nature* 476(7359), 170–175.

Holloway, K., Lawson, V. E. and Jeffreys, A. J. (2006). Allelic recombination and de novo deletions in sperm in the human beta-globin gene region. *Human Molecular Genetics* 15(7), 1099–1111.

Hunt, P. A. and Hassold, T. J. (2002). Sex matters in meiosis. *Science* 296(5576), 2181–2183.

Hunter, N. and Kleckner, N. (2001). The single-end invasion: an asymmetric intermediate at the double-strand break to double-holliday junction transition of meiotic recombination. *Cell* 106(1), 59–70.

Hussin, J., Roy-Gagnon, M. H., Gendron, R., Andelfinger, G. and Awadalla, P. (2011). Age-dependent recombination rates in human pedigrees. *PLoS Genetics* 7(9), e1002251.

Hussin, J., Sinnett, D., Casals, F., Idaghdour, Y., Bruat, V., Saillour, V., Healy, J., Grenier, J. C., de Malliard, T., Busche, S., Spinella, J. F., Lariviere, M., Gibson, G., Andersson, A., Holmfeldt, L., Ma, J., Wei, L., Zhang, J., Andelfinger, G., Downing, J. R., Mullighan, C. G. and Awadalla, P. (2013). Rare allelic forms of PRDM9 associated with childhood leukemogenesis. *Genome Research* 23(3), 419–430.

Irie, S., Tsujimura, A., Miyagawa, Y., Ueda, T., Matsuoka, Y., Matsui, Y., Okuyama, A., Nishimune, Y. and Tanaka, H. (2009). Single-nucleotide polymorphisms of the PRDM9 (MEISETZ) gene in patients with nonobstructive azoospermia. *Journal of Andrology* 30(4), 426–431.

Jeffreys, A. J., Kauppi, L. and Neumann, R. (2001). Intensely punctate meiotic recombination in the class II region of the major histocompatibility complex. *Nature Genetics* 29(2), 217–222.

Jeffreys, A. J., Ritchie, A. and Neumann, R. (2000). High resolution analysis of haplotype diversity and meiotic crossover in the human TAP2 recombination hotspot. *Human Molecular Genetics* 9(5), 725–733.

Kauppi, L., Jeffreys, A. J. and Keeney, S. (2004). Where the crossovers are: recombination distributions in mammals. *Nature Review Genetics* 5(6), 413–424.

Kauppi, L., Sajantila, A. and Jeffreys, A. J. (2003). Recombination hotspots rather than population history dominate linkage disequilibrium in the MHC class II region. *Human Molecular Genetics* 12(1), 33–40.

Keeney, S. (2007). Recombination and Meiosis: Crossing-over and Disjunction. Heidelberg, Springer-Verlag.

Keeney, S., Giroux, C. N. and Kleckner, N. (1997). Meiosis-specific DNA double-strand breaks are catalyzed by Spo11, a member of a widely conserved protein family. *Cell* 88(3), 375–384.

Khil, P. P. and Camerini-Otero, R. D. (2010). Genetic crossovers are predicted accurately by the computed human recombination map. *PLoS Genetics* 6(1), e1000831.

Khil, P. P., Smagulova, F., Brick, K. M., Camerini-Otero, R. D. and Petukhova, G. V. (2012). Sensitive mapping of recombination hotspots using sequencing-based detection of ssDNA. *Genome Research* 22(5), 957–965.

Kim, C., Snyder, R. O. and Wold, M. S. (1992). Binding properties of replication protein A from human and yeast cells. *Molecular and Cellular Biology* 12(7), 3050–3059.

Kingman, J. F. (2000). Origins of the coalescent. 1974–1982. *Genetics* 156(4), 1461–1463.

Kirkness, E. F., Grindberg, R. V., Yee-Greenbaum, J., Marshall, C. R., Scherer, S. W., Lasken, R. S. and Venter, J. C. (2013). Sequencing of isolated sperm cells for direct haplotyping of a human genome. *Genome Research* 23(5), 826–832.

Kong, A., Gudbjartsson, D. F., Sainz, J., Jonsdottir, G. M., Gudjonsson, S. A., Richardsson, B., Sigurdardottir, S., Barnard, J., Hallbeck, B., Masson, G., Shlien, A., Palsson, S. T., Frigge, M. L., Thorgeirsson, T. E., Gulcher, J. R. and Stefansson, K. (2002). A high-resolution recombination map of the human genome. *Nature Genetics* 31(3), 241–247.

Kong, A., Thorleifsson, G., Frigge, M. L., Masson, G., Gudbjartsson, D. F., Villemoes, R., Magnusdottir, E., Olafsdottir, S. B., Thorsteinsdottir, U. and Stefansson, K. (2014). Common and low-frequency variants associated with genome-wide recombination rate. *Nature Genetics* 46(1), 11–16.

Kong, A., Thorleifsson, G., Gudbjartsson, D. F., Masson, G., Sigurdsson, A., Jonasdottir, A., Walters, G. B., Jonasdottir, A., Gylfason, A., Kristinsson, K. T., Gudjonsson, S. A., Frigge, M. L., Helgason, A., Thorsteinsdottir, U. and Stefansson, K. (2010). Fine-scale recombination rate differences between sexes, populations and individuals. *Nature* 467(7319), 1099–1103.

Kouzarides, T. (2007). Chromatin modifications and their function. *Cell* 128(4), 693–705.

Lamb, N. E., Sherman, S. L. and Hassold, T. J. (2005). Effect of meiotic recombination on the production of aneuploid gametes in humans. *Cytogenetic and Genome Research* 111(3–4), 250–255.

Lebo, R. V., Chakravarti, A., Buetow, K. H., Cheung, M. C., Cann, H., Cordell, B. and Goodman, H. (1983). Recombination within and between the human insulin and beta-globin gene loci. *Proceedings of the National Academy Sciences USA* 80(15), 4808–4812.

Lenzi, M. L., Smith, J., Snowden, T., Kim, M., Fishel, R., Poulos, B. K. and Cohen, P. E. (2005). Extreme heterogeneity in the molecular events leading to the establishment of chiasmata during meiosis i in human oocytes. *The American Journal of Human Genetics* 76(1), 112–127.

Lesecque, Y., Glemin, S., Lartillot, N., Mouchiroud, D. and Duret, L. (2014). The red queen model of recombination hotspots evolution in the light of archaic and modern human genomes. *PLoS Genetics* 10(11), e1004790.

Lichten, M. and Goldman, A. S. (1995). Meiotic recombination hotspots. *Annual Review of Genetics* 29, 423–444.

Linardopoulou, E. V., Williams, E. M., Fan, Y., Friedman, C., Young, J. M. and Trask, B. J. (2005). Human subtelomeres are hot spots of interchromosomal recombination and segmental duplication. *Nature* 437(7055), 94–100.

Lindahl, K. F. (1991). His and hers recombinational hotspots. *Trends in Genetics* 7(9), 273–276.

Lupski, J. R. and Stankiewicz, P. (2005). Genomic disorders: molecular mechanisms for rearrangements and conveyed phenotypes. *PLoS Genetics* 1(6), e49.

MacArthur, J. A., Spector, T. D., Lindsay, S. J., Mangino, M., Gill, R., Small, K. S. and Hurles, M. E. (2014). The rate of nonallelic homologous recombination in males is highly variable, correlated between monozygotic twins and independent of age. *PLoS Genetics* 10(3), e1004195.

May, C. A., Shone, A. C., Kalaydjieva, L., Sajantila, A. and Jeffreys, A. J. (2002). Crossover clustering and rapid decay of linkage disequilibrium in the Xp/Yp pseudoautosomal gene SHOX. *Nature Genetics* 31(3), 272–275.

Mihola, O., Trachtulec, Z., Vlcek, C., Schimenti, J. C. and Forejt, J. (2009). A mouse speciation gene encodes a meiotic histone H3 methyltransferase. *Science* 323(5912), 373–375.

Mimitou, E. P. and Symington, L. S. (2009). Nucleases and helicases take center stage in homologous recombination. *Trends in Biochemical Sciences* 34(5), 264–272.

Moynahan, M. E. and Jasin, M. (2010). Mitotic homologous recombination maintains genomic stability and suppresses tumorigenesis. *Nature Review Molecular Cell Biology* 11(3), 196–207.

Munoz-Fuentes, V., Di Rienzo, A. and Vila, C. (2011). Prdm9, a major determinant of meiotic recombination hotspots, is not functional in dogs and their wild relatives, wolves and coyotes. *PLoS One* 6(11), e25498.

Myers, S., Bottolo, L., Freeman, C., McVean, G. and Donnelly, P. (2005). A fine-scale map of recombination rates and hotspots across the human genome. *Science* 310(5746), 321–324.

Myers, S., Bowden, R., Tumian, A., Bontrop, R. E., Freeman, C., MacFie, T. S., McVean, G. and Donnelly, P. (2010). Drive against hotspot motifs in primates implicates the PRDM9 gene in meiotic recombination. *Science* 327(5967), 876–879.

Myers, S., Freeman, C., Auton, A., Donnelly, P. and McVean, G. (2008). A common sequence motif associated with recombination hot spots and genome instability in humans. *Nature Genetics* 40(9), 1124–1129.

Nachman, M. W. (2002). Variation in recombination rate across the genome: evidence and implications. *Current Opinion in Genetics & Development* 12(6), 657–663.

Nagaraju, G. and Scully, R. (2007). Minding the gap: the underground functions of BRCA1 and BRCA2 at stalled replication forks. *DNA Repair (Amst)* 6(7), 1018–1031.

Neale, M. J. and Keeney, S. (2006). Clarifying the mechanics of DNA strand exchange in meiotic recombination. *Nature* 442(7099), 153–158.

Ohta, K., Nicolas, A., Furuse, M., Nabetani, A., Ogawa, H. and Shibata, T. (1998). Mutations in the MRE11, RAD50, XRS2, and MRE2 genes alter chromatin configuration at meiotic DNA double-stranded break sites in premeiotic and meiotic cells. *Proceedings of the National Academy of Sciences USA* 95(2), 646–651.

Oliver, P. L., Goodstadt, L., Bayes, J. J., Birtle, Z., Roach, K. C., Phadnis, N., Beatson, S. A., Lunter, G., Malik, H. S. and Ponting, C. P. (2009). Accelerated evolution of the Prdm9 speciation gene across diverse metazoan taxa. *PLoS Genetics* 5(12), e1000753.

Page, D. C., Bieker, K., Brown, L. G., Hinton, S., Leppert, M., Lalouel, J. M., Lathrop, M., Nystrom-Lahti, M., de la Chapelle, A. and White, R. (1987). Linkage, physical mapping, and DNA sequence analysis of pseudoautosomal loci on the human X and Y chromosomes. *Genomics* 1(3), 243–256.

Paigen, K. and Petkov, P. (2012). Meiotic DSBs and the control of mammalian recombination. *Cell Research* 22(12), 1624–1626.

Parvanov, E. D., Petkov, P. M. and Paigen, K. (2010). Prdm9 controls activation of mammalian recombination hotspots. *Science* 327(5967), 835.

Peacock, E. and Whiteley, P. (2005). Perlegen sciences, inc. *Pharmacogenomics* 6(4), 439–442.

Pratto, F., Brick, K., Khil, P., Smagulova, F., Petukhova, G. V. and Camerini-Otero, R. D. (2014). DNA recombination. Recombination initiation maps of individual human genomes. *Science* 346(6211), 1256442.

Rockmill, B., K. Voelkel-Meiman and Roeder, G. S. (2006). Centromere-proximal crossovers are associated with precocious separation of sister chromatids during meiosis in Saccharomyces cerevisiae. *Genetics* 174(4), 1745–1754.

Romanienko, P. J. and Camerini-Otero, R. D. (2000). The mouse Spo11 gene is required for meiotic chromosome synapsis. *Molecular Cell* 6(5), 975–987.

Rouyer, F., Simmler, M. C., Johnsson, C., Vergnaud, G., Cooke, H. J. and Weissenbach, J. (1986). A gradient of sex linkage in the pseudoautosomal region of the human sex chromosomes. *Nature* 319(6051), 291–295.

Schneider, J. A., Peto, T. E., Boone, R. A., Boyce, A. J. and Clegg, J. B. (2002). Direct measurement of the male recombination fraction in the human beta-globin hot spot. *Human Molecular Genetics* 11(3), 207–215.

Schwartz, J. J., Roach, D. J., Thomas, J. H. and Shendure, J. (2014). Primate evolution of the recombination regulator PRDM9. *Nature Communications* 5, 4370.

Shi, Q., Spriggs, E., Field, L. L., Ko, E., Barclay, L. and Martin, R. H. (2001). Single sperm typing demonstrates that reduced recombination is associated with the production of aneuploid 24,XY human sperm. *American Journal of Medical Genetics* 99(1), 34–38.

Shinohara, A., Ogawa, H. and Ogawa, T. (1992). Rad51 protein involved in repair and recombination in S. cerevisiae is a RecA-like protein. *Cell* 69(3), 457–470.

Shiroishi, T., Sagai, T., Hanzawa, N., Gotoh, H. and Moriwaki, K. (1991). Genetic control of sex-dependent meiotic recombination in the major histocompatibility complex of the mouse. *The EMBO Journal* 10(3), 681–686.

Smagulova, F., Gregoretti, I. V., Brick, K., Khil, P., Camerini-Otero, R. D. and Petukhova, G. V. (2011). Genome-wide analysis reveals novel molecular features of mouse recombination hotspots. *Nature* 472(7343), 375–378.

Sneeden, J. L., Grossi, S. M., Tappin, I., Hurwitz, J. and Heyer, W. D. (2013). Reconstitution of recombination-associated DNA synthesis with human proteins. *Nucleic Acids Research* 41(9), 4913–4925.

Spencer, C. C., Deloukas, P., Hunt, S., Mullikin, J., Myers, S., Silverman, B., Donnelly, P., Bentley, D. and McVean, G. (2006). The influence of recombination on human genetic diversity. *PLoS Genetics* 2(9), e148.

Story, R. M., Bishop, D. K., Kleckner, N. and Steitz, T. A. (1993). Structural relationship of bacterial RecA proteins to recombination proteins from bacteriophage T4 and yeast. *Science* 259(5103), 1892–1896.

Sung, P. and Klein, H. (2006). Mechanism of homologous recombination: mediators and helicases take on regulatory functions. *Nature Review Molecular Cell Biology* 7(10), 739–750.

Sung, P., Krejci, L., Van Komen, S. and Sehorn, M. G. (2003). Rad51 recombinase and recombination mediators. *The Journal of Biological Chemistry* 278(44), 42729–42732.

Tiemann-Boege, I., Calabrese, P., Cochran, D. M., Sokol, R. and Arnheim, N. (2006). High-resolution recombination patterns in a region of human chromosome 21 measured by sperm typing. *PLoS Genetics* 2(5), e70.

van den Bosch, M., Bree, R. T. and Lowndes, N. F. (2003). The MRN complex: coordinating and mediating the response to broken chromosomes. *EMBO Reports* 4(9), 844–849.

Vermeulen, M., Eberl, H. C., Matarese, F., Marks, H., Denissov, S., Butter, F., Lee, K. K., Olsen, J. V., Hyman, A. A., Stunnenberg, H. G. and Mann, M. (2010). Quantitative interaction proteomics and genome-wide profiling of epigenetic histone marks and their readers. *Cell* 142(6), 967–980.

Vrooman, L. A., Nagaoka, S. I., Hassold, T. J. and Hunt, P. A. (2014). Evidence for paternal age-related alterations in meiotic chromosome dynamics in the mouse. *Genetics* 196(2), 385–396.

Vrooman, L. A., Oatley, J. M., Griswold, J. E., Hassold, T. J. and Hunt, P. A. (2015). Estrogenic exposure alters the spermatogonial stem cells in the developing testis, permanently reducing crossover levels in the adult. *PLoS Genetics* 11(1), e1004949.

Wang, J., Fan, H. C., Behr, B. and Quake, S. R. (2012). Genome-wide single-cell analysis of recombination activity and *de novo* mutation rates in human sperm. *Cell* 150(2), 402–412.

Wang, X., Ira, G., Tercero, J. A., Holmes, A. M., Diffley, J. F. and Haber, J. E. (2004). Role of DNA replication proteins in double-strand break-induced recombination in Saccharomyces cerevisiae. *Molecular and Cell Biology* 24(16), 6891–6899.

Weissenbach, J., Levilliers, J., Petit, C., Rouyer, F. and Simmler, M. C. (1987). Normal and abnormal interchanges between the human X and Y chromosomes. *Development* 101, 67–74.

Wold, M. S. (1997). Replication protein A: a heterotrimeric, single-stranded DNA-binding protein required for eukaryotic DNA metabolism. *Annual Review of Biochemistry* 66, 61–92.

Woodward, E. L., Olsson, M. L., Johansson, B. and Paulsson, K. (2014). Allelic variants of PRDM9 associated with high hyperdiploid childhood acute lymphoblastic leukaemia. *British Journal of Haematology* 166(6), 947–949.

Wyka, I. M., Dhar, K., Binz, S. K. and Wold, M. S. (2003). Replication protein A interactions with DNA: differential binding of the core domains and analysis of the DNA interaction surface. *Biochemistry* 42(44), 12909–12918.

Yip, S. P., Lovegrove, J. U., Rana, N. A., Hopkinson, D. A. and Whitehouse, D. B. (1999). Mapping recombination hotspots in human phosphoglucomutase (PGM1). *Human Molecular Genetics* 8(9), 1699–1706.

Zenzes, M. T., de Geyter, C., Bordt, J., Schneider, H. P. and Nieschlag, E. (1990). Abnormalities of sperm chromosome condensation in the cytoplasm of immature human oocytes. *Human Reproduction* 5(7), 842–846.

Zenzes, M. T., Reed, T. E. and Casper, R. F. (1997). Effects of cigarette smoking and age on the maturation of human oocytes. *Human Reproduction* 12(8), 1736–1741.

Zenzes, M. T., Wang, P. and Casper, R. F. (1995). Cigarette smoking may affect meiotic maturation of human oocytes. *Human Reproduction* 10(12), 3213–3217.

Zhang, F., Gu, W., Hurles, M. E. and Lupski, J. R. (2009). Copy number variation in human health, disease, and evolution. *Annual Review of Genomics and Human Genetics* 10, 451–481.

Epigenetics of Male Infertility 4

Debbie Montjean, Antoine D. Rolland[†] and Celia Ravel[‡]*

**Service de Médecine et Biologie de la Reproduction, Hopital Saint-Joseph, Marseille, France*
[†]INSERM U1085-Irset, Rennes, France
[‡]Service de Biologie de la Reproduction, CECOS, Universite de Rennes, CHU de Rennes, Université de Rennes 1, France

The contribution of the male gamete to embryogenesis is more complex than a simple DNA transmission. The sperm nucleus has a very heterogeneous structure containing a haploid genome packaged by basic nucleoproteins that carries various epigenetic information, including DNA methylation, histone modifications, protamines and a large population of transcripts, of which their exact role is not yet well known. This epigenetic landscape may be altered in the sperm of a significant proportion of infertile men. This chapter will discuss what we now know about the relationship between male infertility and epigenetic factors, with a special emphasis on the sperm epigenome.

4.1 Introduction

Infertility is the failure of a couple to achieve pregnancy after over a year of unprotected sexual intercourse (Cooper *et al.* 2010). It is estimated that 15% of couples worldwide are infertile (Boivin *et al.* 2007). While male factor infertility accounts for approximately 50% of the cases (Reviewed in Poongothai *et al.* 2009), it can be attributed to numerous factors and it includes a wide variety of disorders

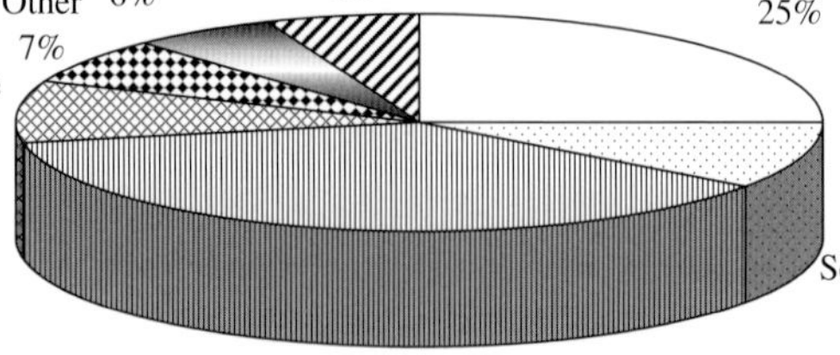

Figure 4.1. Distribution of male infertility disorders.

Table 4.1. Overview of standard values for semen analysis according to the 1999 and 2010 WHO criteria.

	1999 WHO criteria	2010 WHO criteria
Volume (ml)	≥2	≥1,5
pH	7, 2-8, 0	Not precised
Sperm concentration (M/ml)	≥20	≥15
Total sperm count (M/ml ejaculate)	≥40	≥22,5
Morphology (%Normal)*	≥30	≥15
Vitality (% Live)	≥75	≥60
White blood cells (M/ml)	<1,0	<1,0

* Assessment according to Kruger and Menkfeld criteria.

(Figure 4.1). This encompasses anomalies of development of the urogenital system (e.g. cryptorchidism), genetic defects of the endocrine system leading to testicular failure and defects in gametogenesis, erection, ejaculation, gamete function and fertilization. Testicular failure and semen disorders have been linked with genetic, environmental abnormalities and epigenetic disorders. Thanks to the large development of non invasive tools and the easy access of biological material, semen disorders became the most investigated male infertility factor (Benkhalifa *et al.* 2014). A progressive worldwide decline of semen quality was described requiring the revision of normal parameters for semen analysis in 2010 (Cooper *et al.* 2010). Normal sperm parameters and World Health Organization (WHO) nomenclature for semen variables are summarized in Tables 4.1 and 4.2.

Table 4.2. WHO nomenclature for semen variables. Factors associated with male infertility.

Pathology	Description
Azoospermia	Complete absence of sperm.
Aspermia	Ejaculation does not emit any semen.
Oligozoospermia	≤ 15 million sperm/ml of semen.
Asthenozoospermia	> 40% of sperm have low motility.
Teratozoospermia	> 96% of sperm with abnormal morphology*.
Necrospermia	Non-viable/dead sperm.
Hematospermia	Red blood cells present in semen.
Pyospermia	White blood cells present in semen.
Polyzoospermia	> 200 million sperm/ml of semen.

* Assessment according to Kruger and Menkfeld criteria.

The underlying cause for increasing disruptive spermatogenesis remains unknown. Although several genetic alterations were described in male infertility, these abnormalities only explain a fraction of the cases (Tiepolo *et al.* 1976, Gekas *et al.* 2001, Dohle *et al.* 2002, Reviewed in Maduro *et al.* 2003, Skaletsky *et al.* 2003, Reviewed in Matzuk and Lamb 2008, Reviewed in Hamada *et al.* 2013). The decline in male fertility may be attributed to environmental hazard perhaps through alteration of sperm epigenome. Indeed male reproductive system is particularly vulnerable to the effects of chemical (e.g. endocrine disruptors, phtalates, fungicides, pesticides) and physical agents (e.g. electromagnetic waves emitted by cell phones) introduced and spread by human industrial, pharmaceutical, agricultural and recreational activities (reviewed in Guerrero-Bosagna and Skinner 2009, reviewed in Jurewicz *et al.* 2009, Reviewed in Bonde 2010). Such epigenetic alterations during male germ cell development may disrupt spermatogenesis, leading to the production of a reduced number of spermatozoa and/or their poor quality (morphology, mobility, fertilization capacity). Additionally, more discrete epigenetic defects of apparently normal sperm may also alter early development of the fertilized egg. The contribution of male gamete to embryogenesis is indeed more complex than a simple genetic material transmission. Sperm nucleus has a very heterogeneous structure containing a haploid genome packaged by basic nucleoproteins (protamine) in chromatin that carries various epigenetic information including DNA methylation, histone modifications, protamines and a large population of transcripts whose exact role is not yet well known.

The present chapter will discuss the knowledge on the relationship between male infertility and epigenetics factors, with a special emphasis on the sperm epigenome.

4.2 The Role of Epigenetics in Male Infertility

The term "epi (over)genetics" comprises all the modifications that regulate gene expression by mechanisms other than changes in the underlying DNA sequence and refers to the regulation of various genomic functions that are controlled by partially stable modifications of DNA and chromatin proteins. Epigenetic modifications in mature spermatozoa include a range of marks that can be applied directly on DNA or on histones bound to DNA. Protamines and small non-coding RNA are also considered as epigenetic factors as they impact gene expression while DNA sequence remains unchanged. Among these marks, DNA methylation has been widely studied in the context of male infertility because it is technically simple to investigate through both biochemistry and genetics.

4.2.1 *DNA methylation*

4.2.1.1 *DNA methylation, an essential epigenetic mark*

In the mammalian genome, methylation occurs at the five-position of cytosine residues mainly at CpG dinucleotides, which are typically grouped in clusters termed CpG islands that are present in the 5′ regulatory regions of many genes. 60–90% of all CpGs over the entire genome are methylated (Antequara and Bird 1993). Along with other epigenetic modifications, DNA methylation establishes and maintains gene expression status: Methylation on CpG dinucleotides is associated with repressed chromatin and gene expression inhibition, while demethylated CpG islands are associated with transcriptionally active genes (Reviewed in Bird and Wolffe 1999). Importantly, DNA methylation pattern is maintained and transmitted from mother to daughter cells during cell division by copying pre-existing methylation pattern onto newly synthesized DNA strand (Reviewed in Reik *et al.* 2001, Reviewed in Klose and Bird 2006).

4.2.1.2 *Genomic imprinting*

Imprinted genes express only one of their alleles in a parent-of-origin specific manner (Reviewed in Reik *et al.* 2001). This is due to a differential methylation of CpGs within differentially methylated regions (DMRs) that are key regulatory sequences controlling the parent-of-origin expression of the imprinted genes. Here

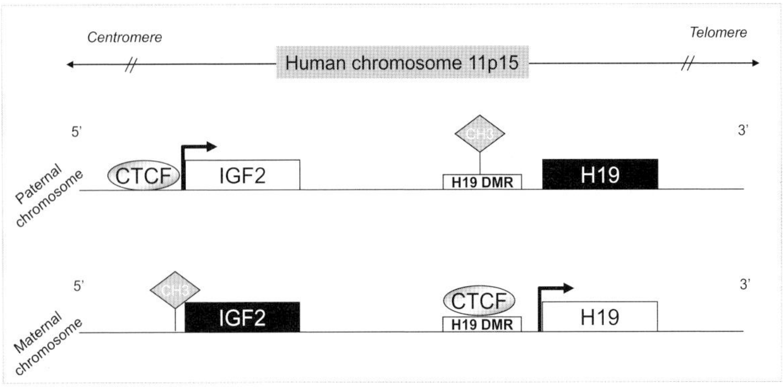

Figure 4.2. Imprinting center region on chromosome 11p15. White boxes represent transcriptionally active genes. Black arrows indicate transcription orientation. Grey tetragones indicate methylation at DMR (Differential Methylated Regions). CTCF (CCTC-Binding Factor) is a transcription factor that binds to repeats of non-methylated core sequence CCCTC and induces transcription. The presence of methylation inhibits CTCF binding and subsequently gene transcription.

again, methylation of DMRs may inhibit transcription, whereas hypomethylation of DMRs is associated with gene expression (Reviewed in Bird and Wolffe 1999) (Figure 4.2). Approximately 0.2% of CpG islands over the entire genome show parent-of-origin specific differential methylation profile and at least 80 imprinted genes are known in humans (Reviewed in Isles and Holland 2005, reviewed in Morison *et al.* 2005, reviewed in Wood and Oakey 2006).

Imprinted genes control important aspects of cell proliferation, embryonic development and cognitive functions (many imprinted genes being indeed expressed only in the brain). Interestingly, many paternally expressed genes (methylated maternal allele) favor cell proliferation or placental growth, whereas maternally expressed genes (methylated paternal allele) tend to do the opposite. This may allow the establishment of a balance for normal embryonal development and a successful pregnancy (reviewed in Isles and Holland 2005, reviewed in Wood and Oakey 2006). Genes subjected to parental imprint are clustered in large regions where they are regulated by the same Imprinted Center (IC), e.g. chromosome 11p15 (Figure 4.2), where genetic or epigenetic errors are responsible for various growth and behavioural syndromes such as Beckwith–Wiedmann syndrome (BWS) (Reik *et al.* 2000, Reviewed in Maher and Reik 2000).

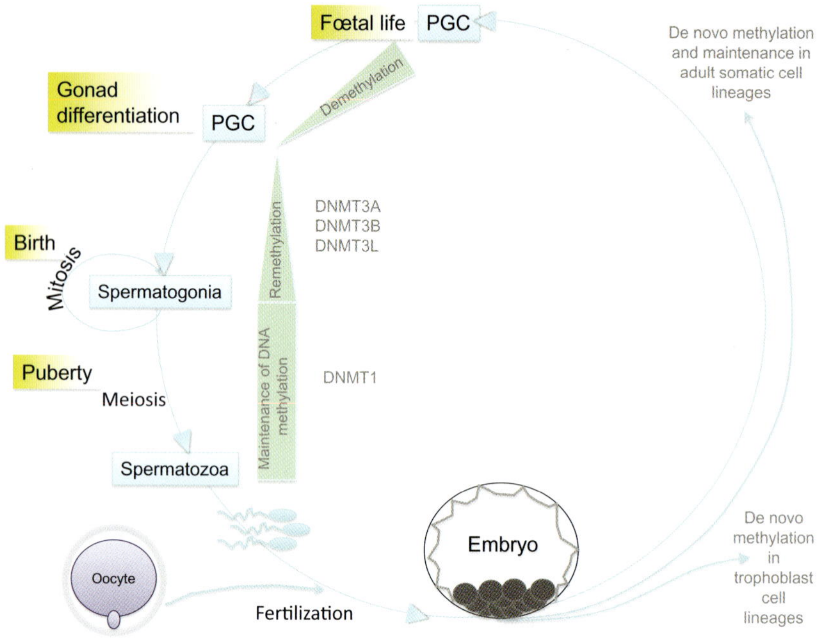

Figure 4.3. Methylation dynamics.

4.2.1.3 *DNA methylation in germ cells*

Appropriate DNA methylation in spermatozoa is crucial for proper sperm genome transmission and subsequent embryo development. The setting of methylation marks in developing germ cells is a complex process that starts during foetal life (Figure 4.3). Primordial germ cells (PGC) undergo extensive epigenetic reprogramming including complete erasure of imprint and reestablishment of a sex-specific methylation pattern during their migration to the genital ridge (Reviewed in Reik *et al.* 2001, Hajkova *et al.* 2002, reviewed in O'Doherty and Gettigan 2014). Consequently, the DNA methylation pattern of male germ cells is significantly different from that of somatic cells and is maintained throughout spermatogenesis and in mature spermatozoa (Figure 4.3). After fertilization, DNA methylation patterns of imprinted genes remain unchanged and are clonally transmitted to all the cell types of the individual while non-imprinting genes undergo *de novo* methylation in somatic cell lineages in a cell type specific manner (Figure 4.3).

DNA methylation is catalyzed by a family of enzymes called DNA (cytosine-5) methyltransferases (DNMTs). In mammals, three active DNMTs have been characterized: DNMT1, DNMT3A and DNMT3B. DNMT3L (DNMT3-Like) is structurally closely related to DNMT3A and DNMT3B but is catalytically inactive on its own and is known to potentiate DNA methylation by interacting with DNMT3A and DNMT3B (Reviewed in Goll and Bestor. 2005). DNMT transcripts are sequentially expressed in pre- and postnatal developing germ cells (La Salle *et al.* 2004, La Salle and Trasler 2006). As depicted in Figure 4.3, these enzymes are required in the germline for the maintenance of methylation, but also for *de novo* methylation (Davis *et al.* 1999, Davis *et al.* 2000, Oakes *et al.* 2007).

DNMT1 is a maintenance methyltransferase that is crucial for the preservation of methylation patterns during DNA replication (Figure 4.3) whereas DNMT3A and DNMT3B play a key role in *de novo* methylation of DNA (Okano *et al.* 1999). In germ cells, DNMT3A and DNMT3B are responsible for establishment of a sex-specific DNA methylation profile. In theory, paternally imprinted genes are fully methylated in mature spermatozoa while maternally imprinted ones are expected to be left unmethylated.

The role of DNA methylation during spermatogenesis has been investigated mainly using gene-targeting (DNMTs knock-out animal models) (Kaneda *et al.* 2004, Bourc'his *et al.* 2001, Webster *et al.* 2005, Kato *et al.* 2007, La Salle *et al.* 2007) or chemical treatments using DNA methylation inhibitors (5-azacytidine and 5-aza-2'-deoxycytidine) (Doerksen *et al.* 1996 2000, Kelly *et al.* 2003, Oakes *et al.* 2007). Specific conditional knockout for *Dnmt3a* and *Dnmt3b* in germ cells showed that male mice display DNA methylation alterations (including imprinted genes e.g. *H19*, *Gtl2/Dlk1*) and suffer from infertility (Bourc'his *et al.* 2004, Webster *et al.* 2005, La Salle *et al.* 2007). Absence of DNMT1 induces apoptosis of germ line stem cells as well as severe defects during embryogenesis (Takashima *et al.* 2009). Although catalytically inactive, DNMT3L was demonstrated to be essential since knock-out mice display germ line defects (Webster *et al.* 2005). Overall, these studies show that DNMTs, and thus methylation process, are essential for the development and function of the germ cells.

Drugs that inhibit DNA methylation also provided insights into the effect of genome-wide hypomethylation on spermatogenesis. 5-azacytidine and 5-aza-2′-deoxycytidine can incorporate in the replicating DNA and inhibit the activity of DNMTs, resulting in a genome-wide DNA demethylation (Reviewed in Egger *et al.* 2004). Rodent models have revealed that exposure to 5-azacytidine triggers DNA hypomethylation in sperm accompanied by a dose-dependent drop in sperm counts (Doerksen *et al.* 1996). Rats treated with 5-aza-2′-deoxycytidine show impaired testicular histology, reduced sperm production and infertility and/or reduced ability to support preimplantation development (Kelly *et al.* 2003).

It has also been shown that environmental factors may induce transgenerationally inheritable DNA methylation changes in the germline (Anway *et al.* 2005, Chang *et al.* 2006, Anway and Skinner 2008, Guerrero-Bosagna 2012, Reviewed in Wei *et al.* 2015). Transient exposure of a gestating female rat during the period of gonadal sex determination to endocrine disruptors decreases sperm counts and sperm vitality and consequently increases the incidence of male infertility accompanied by an alteration of DNA methylation in germline. These aberrant imprints and consequential effects can be inherited through the male germline for up to four generations. Testes from males of the fourth generation still show more than a two-fold increase in spermatogenic cell apoptosis resulting in a 20% reduction of sperm numbers (Anway *et al.* 2005, Anway and Skinner 2006). These studies have thus demonstrated the functional importance of appropriate DNA methylation for normal male germ cell development in rodents.

4.2.1.4 *Methylation changes and male infertility*

While several clinical studies have reported an increased incidence of rare imprinting diseases in children conceived with Assisted Reproduction Technology (ART) (Cox *et al.* 2002, DeBaun *et al.* 2003, Gicquel *et al.* 2003, Maher *et al.* 2003, Orstavik *et al.* 2003), other studies showed a transgenerational inheritance of methylation changes possibly from sperm to the offspring (Anway *et al.* 2005, Chang *et al.* 2006, Anway and Skinner 2008, Kobayashi *et al.* 2009, Guerrero-Bosagna 2012, Reviewed in Wei *et al.* 2015). This hypothesis may be supported by the identification of methylation defects in sperm. Interestingly, observations made in rodents turned out to have a parallel in humans since methylation changes in human sperm DNA have also been identified in disruptive spermatogenesis and more recently in idiopathic infertility cases (Marques *et al.* 2004, Houshdaran *et al.* 2007, Kobayashi *et al.* 2007, Marques *et al.* 2008, Boissonnas *et al.* 2010, Montjean *et al.* 2013, Kuhtz *et al.* 2014, Lauretino *et al.* 2015, Montjean *et al.* 2015, Urdinquio *et al.* 2015). Analysis of DNA methylation pattern in DMRs of two oppositely imprinted genes, *H19* and *PEG1/MEST*, by bisulfite sequencing in spermatozoa of oligozoospermic men showed an association between reduced sperm counts and defective methylation in *H19*-DMR but not on maternal imprint (Marques *et al.* 2004). Subsequently, several studies have analyzed DMRs of a larger number of genes including *IGF2*, *PEG3*, *ZAC*, *LIT1*, *SNRPN* as well as non imprinted repetitive elements, such as long interspersed nucleotide elements (LINE1) and Alu elements, as surrogates for global DNA methylation changes (Kobayashi *et al.*

2007, Boissonnas *et al.* 2010). These studies, however, observed an increased methylation in DMRs of maternally imprinted genes in addition to the aberrant methylation of the paternally imprinted genes. Furthermore, the results of the analysis of LINE1 and Alu elements suggested that methylation defects were specific to imprinted genes (Kobayashi *et al.* 2007, Boissonnas *et al.* 2010). However, other reports suggest that methylation changes associated with male infertility are prone to be genome-wide (Houshdaran *et al.* 2007, Montjean *et al.* 2015) but a controversy remains regarding the trend of global methylation changes in DNA from poor quality sperm as compared to normospermic controls.

The contradictory results in these studies call for further investigations to improve characterization of DNA methylation errors associated with abnormal semen parameter and to elucidate the biological and the clinical consequences of these epimutations. Furthermore, whether these DNA methylation alterations represent the cause(s) or consequence(s) of associated infertility remains unclear.

4.2.2 *Chromatin organisation in sperm*

Chromatin is subjected to tremendous remodeling during spermiogenesis. The majority of somatic histones are gradually removed and replaced by transition proteins which are themselves replaced by small basic proteins called protamines. This mechanism is essential for the protection of sperm genome against environmental hazard during its journey in the female genital tract to reach the oocyte. Additionally, the existence of specific chromosome territories within the sperm nucleus together with the non-random distribution of some chromatin marks suggest that this specific chromatin organization of the male gamete is likely to also have functional importance probably for early fertilization event and/or embryo development.

4.2.2.1 *The nucleo-protamine domain*

Protamines are sperm-specific basic nuclear proteins which bind to 85% of the sperm DNA. In mammals, two protamines PRM1 and PRM2 are described in mature spermatozoon. Unlike PRM1 which is synthesized directly in mature protein, PRM2 results from the proteolysis of a large precursor. Protamines mediate strong DNA binding owing to a high level of arginine residues, while cysteine residues facilitate formation of disulfide bonds associated to zinc bridges, both of them being essential for a highly compacted chromatin, facilitating male haploid genome packaging (Reviewed in Björndahl *et al.* 2010). The structural organization of DNA-protamine complex is significantly different from that of nucleosomes

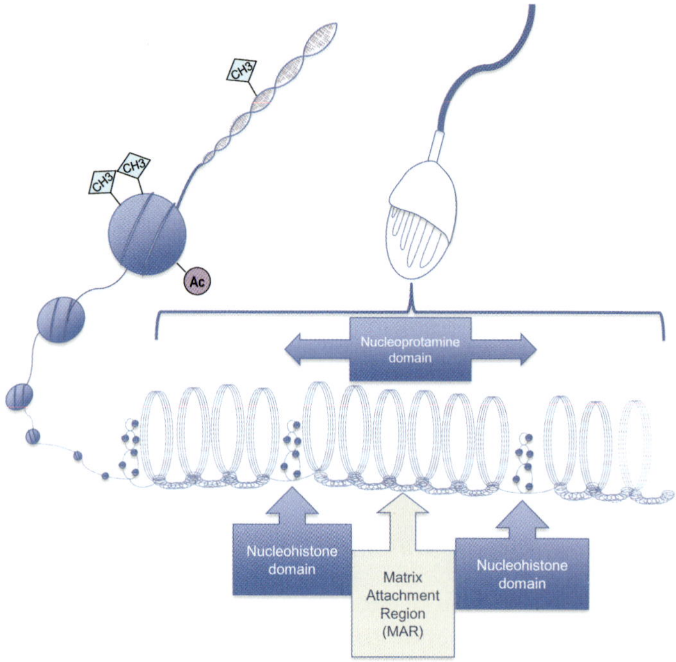

Ac: Acetyl group, CH3: methyl group

Figure 4.4. Sperm DNA organisation.

as depicted in Figure 4.4. Sperm DNA is compacted by protamines that form toroids containing about 25–50 kb of DNA. These toroids are linked by matrix attachment regions (MARs), which are sensitive to nuclease, and bind them to the nuclear matrix (Ribas-Maynou *et al.* 2013). Structural and functional components of sperm nuclear matrix are crucial as they are directly involved in early oocyte activation (Fujimoto *et al.* 2004).

4.2.2.2 *The nucleo-histone domain*

Mature sperm nucleus still contains some chromatin domains containing a typical nucleosome organization. Chromatin of human sperm contains about 15% of somatic histones in addition to protamine and other basic proteins.

During spermatogenesis, histone methylation is regulated by H3-K4 and H3-K9 methyltransferase family proteins or enzymes facilitating gene repression through mono, di or tri-methylation of lysine or arginine. The number and location of methyl groups on sperm-specific histones is part of the control parameters of spermatogenesis. During spermatogenesis, a "wave" of histone acetylation occurs which is essential for the reorganization and subsequent compaction of the sperm genome. The majority of histones exist primarily in a hypoacetylated state until spermiogenesis, when global hyperacetylation occurs within the elongating spermatid as a signal for replacement of the majority of the histones by protamines. As the spermatids condense, the acetylated histones are removed while the pericentromeric regions remain bound to histones (Rousseaux *et al.* 2009). This binding is mediated by a testis specific double bromodomain containing factor BRDT. The mechanism of BRDT action seems to entail recognition of acetylated chromatin via its first bromodomain in elongating spermatids, which then would facilitate histone eviction (Dhar *et al.* 2012, Gaucher *et al.* 2012, Tropberger *et al.* 2013). Additionally, H3K79 methylation temporally overlaps with this histone acetylation (Dottermush-Heidel *et al.* 2014).

Histones are good candidates for the transmission of epigenetic information due to their influence on the changes in chromatin structure and are powerful modulators of almost all cellular functions involving DNA, such as gene expression, chromosome replication, repair, and recombination. In addition, some of the canonical histones are replaced by testis-specific variants. Both H2AX and its phosphorylated form, γ-H2AX, have been shown to associate with the sex body of meiotic spermatocytes which is involved in the inactivation of the sex chromosomes during spermatogenesis (reviewed in Dadoune 2009). H2A.Lap1 (lack of acidic patch) is expressed in late pachytene and round spermatids and binds to the transcription start site of active genes to facilitate transcription. Together with H2A. Lap1, H2A.Z creates a unique chromatin environment that is important for gene activation in late spermatid (Nekrasov *et al.* 2012). Two other H2A testicular variants — H2AL1 and H2AL2 — are involved in structures that protect DNA fragments in condensing spermatids (Govin *et al.* 2013). TH2B is expressed in spermatocytes and spermatids, and its presence inversely correlates with the presence of canonical H2B (Montellier *et al.* 2013). This suggests that TH2B gradually replaces H2B on chromatin, making it the first histone variant to be deposited at a genome-wide scale during spermiogenesis. Histone H3 is replaced by several histone variants which respective roles remain to be clarified (Reviewed in Filipescu *et al.* 2014). Finally, epigenetic inheritance is illustrated by centromere identity which is determined by the presence of the H3 variant CENP-A (Reviewed in Valente *et al.* 2012).

4.2.2.3 *Regulatory regions of genes*

Nucleo-protamine and nucleo-histone complexes are not distributed randomly in sperm head but are assigned to specific regions. For instance, protamines are scattered across the stationary core and histones are localized to the posterior annulus (Li *et al.* 2008). Furthermore, a specific arrangement of chromosomes in human sperm nuclei has been described which may have a biological significance in early fertilization events. It has indeed been hypothesized that at fertilization, a differential spatio-temporal exposure of sperm chromosomes and of selected chromosomal domains to the ooplasm components takes place. Proximity of sperm chromosomes to the nucleus periphery, i.e. to the point of the first contact of the sperm nucleus with the ooplasm, was suggested to be part of the epigenetic mechanisms driving differential transcription and replication during early embryogenesis (Mudrak *et al.* 2012). Importantly, the retained nucleosomes exhibit a non-random distribution within the sperm genome and were notably shown to be enriched at imprinted gene loci, microRNA and HOX gene clusters as well at promoters of genes important for early embryo development (Vavouri and Lehner 2011, Hammoud *et al.* 2011).

4.2.3 *Chromatin condensation defects*

Protamine 1 (PRM1) and Protamine 2 (PRM2) are normally expressed in a ratio of approximately 1:1 in healthy men. In a fertile normozoospermic population, the PRM1/PRM2 ratio actually ranges from 0.54 to 1.43 (Nanassy *et al.* 2011). Several studies have also established a relationship between abnormal protamine expression, resulting in an altered (increased or decreased) PRM1/PRM2 ratio and male infertility, as a consequence of chromatin condensation defects. This ratio was thus subsequently proposed as a biomarker of infertility since its modification is associated with impaired sperm quality. In addition to impaired sperm parameters (count, motility and morphology), patients with abnormal PRM1/PRM2 ratios also exhibit altered integrity of sperm DNA as a result of insufficient condensation of sperm nucleus and increased susceptibility to damage. It also seems that the ability of sperm to fertilize the egg is lessened in such patients with potential detrimental consequences on the development of deriving embryos (Francis *et al.* 2014). Premature protamine messenger RNA (mRNA) translation causes abnormal chromatin condensation which leads to male infertility (Reviewed in Dadoune, 2009). Furthermore, chromatin packaging may affect methylation process with a risk of impaired sperm genome contribution to early embryogenesis (Aoki *et al.* 2006).

4.2.4 *Transcripts in mature spermatozoa*

Given that mature spermatozoon is a highly differentiated cell with virtually no cytoplasm, it was assumed for decades that it does not contain RNA. The presence of mRNA in ejaculated mature spermatozoa was however evidenced for the first time through the detection of c-MYC, β-actin and protamine-2 transcripts (Kumar *et al.* 1993, Miller *et al.* 1994, Wykes *et al.* 1997). Several studies were conducted afterwards to establish the significance of these findings and it is now accepted that mature spermatozoa nuclear matrix contains a complex panel of mRNAs (Miller *et al.* 1994, Miller *et al.* 1999, Ostermeier *et al.* 2002, Reviewed in Krawetz 2005, Montjean *et al.* 2012). Furthermore, changes in abundance for such sperm RNAs were identified in asthenozoospermic (Jodar *et al.* 2012) or oligozoospermic (Montjean *et al.* 2012) patients.

In addition to containing mRNA, human and mouse spermatozoa were found to be enriched in small non-coding RNAs (sncRNAs) (Ostermeier *et al.* 2005, Liu *et al.* 2004, Amanai *et al.* 2006). These molecules are key regulators of gene expression at the post-transcriptional and translational level (Reviewed in Du and Zamore 2005), including during spermatogenesis (Gou *et al.* 2014). They can regulate gene expression by binding to complementary mRNA and subsequently inducing their degradation and are categorized into three major classes according to their biogenesis, size, mechanism of action and function:

- Small interfering RNA (siRNA): About 21 nucleotides (nt) long, double stranded. siRNAs can silence the expression of genes by binding to complementary mRNA and triggering their degradation (Reviewed in Ahlquist 2002).
- MicroRNA (miRNA): About 22 nt long, single stranded. miRNA are synthesized from a pre-miRNA double-stranded hairpin precursor, and bind to target mRNA, leading to their cleavage and/or their translation inhibition (Plasterk 2006).
- Piwi-interacting RNAs (piRNA) are about 27 nt long and are generated from long precursor RNAs that play a similar role to miRNA in the regulation of retrotransposon mouse piwi-family proteins. piRNAs are expressed specifically in germ cells and may play a role in repression of retrotransposon during germ cell development and spermatogenesis (Reviewed in He *et al.* 2009).

Several genetically modified mouse models proved out sncRNAs to be essential for primordial germ cell development as well as spermatogenesis (Hayashi *et al.* 2008, Maatouk *et al.* 2008, Reviewed in He *et al.* 2009, Korhonen *et al.* 2011,

Romero *et al.* 2011, Wu *et al.* 2012, Greenlee *et al.* 2012, Reviewed in Kotaja 2014, Zimmermann *et al.* 2014). Spermatogenesis is indeed governed by phase-specific gene expression that may be controlled by sncRNAs likely at post-transcriptional level (Reviewed in He *et al.* 2009, Reviewed in Kim *et al.* 2009). Furthermore, sncRNAs were also proposed to regulate DNA methylation since their absence in mouse models leads to impaired methylation profile in male germ cells (Kuramochi-Miyagawa *et al.* 2008, Watanabe *et al.* 2011).

A link between sncRNAs and male infertility has been described in the Chinese population where single nucleotides polymorphisms in piRNA pathway genes were shown to contribute to spermatogenic disruption (Gu *et al.* 2010). Furthermore, recent studies identified changes in expression profiles of sncRNAs in infertile men with impaired sperm parameters as compared to fertile normospermic controls (Liu *et al.* 2012, Abu-Halima *et al.* 2013, Abu-Halima *et al.* 2014). Further studies are required to provide deeper understanding of the functional role of sncRNAs in spermatogenesis, their contribution to male infertility and to early embryo development.

4.3 Environment, Sperm Epigenetics and Male Infertility

Several environmental factors have been shown to modify the sperm epigenome. Altered environment such as bacterial infection may modify histone methylation profile in the sperm (Dottermusch-Heidel *et al.* 2014). Another important factor is age: indeed, numerous genes in human sperm were shown to display DNA methylation changes that correlate with age and were enriched in regions known to contribute to diseases with increased incidence in offspring of older fathers (Jenkins *et al.* 2014). Given that even apparently innocuous habits such as exercise training have been shown to modify the sperm methylome (Denham *et al.* 2015), it is thus very likely that many other factors and exposures in our day-to-day life actually impact negatively the quality of the sperm epigenome and, consequently, male fertility.

Additionally, the foetal environment, which includes placental function, maternal metabolism and lifestyle factors (e.g. maternal smoking), is known to influence testicular development and function later in life (Reviewed in Juul *et al.* 2014). This notably involves alterations of gene expression and of the epigenetic programming of the germ line following exposure to toxic nutritional, chemical and physical factors around the period of gonadal sex determination. For instance, smoking, alcohol consumption or diets in humans have been suggested to influence offspring disease susceptibility, such as infertility, neuropsychiatric disorders or cancers.

Indeed, any alteration of the germline epigenome might result not only in potential sperm defects and direct fertility troubles, but also in the epigenetic transgenerational inheritance of various abnormal traits to the offspring. As described in animal models, epigenetic modifications can be maintained over several generations. For example, endocrine disruptors may induce transgenerational pathogenesis in rats, resulting in spermatogenic defects, male infertility, breast cancer, kidney disease, prostate disease and immune abnormalities (Anway *et al.* 2006). Potential transgenerational disorders are also induced by phthalate exposure, such as epigenetic inheritance of reproductive disease or sperm epimutations (Manikkam *et al.* 2013). These transgenerational diseases are transmitted to the majority of progeny over four generations and, although the transgenerational effect is transmitted principally through the male germ line, both males and females are affected. Since the paternal germ line transgenerational inheritance of various disease susceptibilities observed in human (Pembrey *et al.* 2006, Kaati *et al.* 2007) is likely to involve epigenetic mechanisms, it is also probable that such environmentally-induced transgenerational epigenetic alterations may be a component of the aetiology of human male infertility (Guerrero-Bosagna *et al.* 2014). Thus, the germline epigenetic landscape is not only important in ensuring normal sperm function as it also contributes to normal embryonic development and offspring health.

4.4 Consequences of Assisted Reproductive Techniques on Sperm Epigenome

During gametogenesis, the male genome undergoes waves of demethylation and remethylation where epigenetic defects induced by the environment may occur. As mentioned above, the exposure to various factors from fetal life to adulthood may thus impair the germ line epigenome, possibly leading to male infertility or even to transgenerational transmission of various disease susceptibilities. However, the recovery and handling of spermatozoa for IVF/ICSI take place after the reprogramming of the male-specific methylation and the establishment of the histone/protamine landscape. Even though it cannot be absolutely excluded that the manipulation of sperm for ART purposes may alter their nuclear integrity, it is also plausible, if not legitimate, to assume that most of the epigenetic alterations observed in sperm of infertile patients actually result from an impaired spermatogenesis. Indeed, infertile male patients have significant methylation alterations and these differ between oligozoospermic and abnormal protamine ratio patients, suggesting different risks of transmission of epigenetic alterations for such patients

(Hammoud *et al.* 2010). Furthermore, given that both imprinting errors and DNA sequence variants are more prevalent in patients with oligospermia (Kobayashi *et al.* 2007), it is very likely that the increased incidence of imprinting disorders observed in individuals born by ART (reviewed in Lazaraviciute *et al.* 2014) are due to the use of sperm with intrinsic epimutation. The alteration of the sperm nuclear integrity may therefore also explain most of ICSI failures as well as some poor in vitro fertilization successes.

4.5 Perspectives and Conclusion

The mature spermatozoon carries multiple epigenetic factors that constitute the sperm epigenome. This includes male-specific genomic imprinting related to DNA methylation, chromatin remodelling with protamines, histone modifications and a large population of mRNAs and sncRNAs. The mature sperm nucleus plays a role in the establishment of effective epigenetic information to be transmitted to the offspring. Whether paternal epigenetic factors could affect, directly or indirectly, the expression of genes during early embryogenesis requires further investigations.

Although a wide range of environment and genetic abnormalities have been associated with male infertility, the epigenetic basis of male infertility is still a matter of debate. Since the past decade, multiple studies have associated epimutations in mature sperm with male infertility. However, it is not clear where such errors occur during methylation pattern establishment in foetal or early post natal life or during the maintenance of methylation marks throughout spermatogenesis. Epigenetic factors and changes, such as changes in methylation patterns, histone/protamines or sncRNAs, are now to be explored in infertile individuals.

References

Abu-Halima, M., Hammadeh, M., Schmitt, J., Leidinger, P., Keller, A., Meese, E. and Backes, C. (2013). Altered microRNA expression profiles of human spermatozoa in patients with different spermatogenic impairments. *Fertility and Sterility* 99, 1249–1255.

Abu-Halima, M., Backes, C., Leidinger, P., Keller, A., Lubbad, A. M., Hammadeh, M. and Meese, E. (2014). MicroRNA expression profiles in human testicular tissues of infertile men with different histopathologic patterns. *Fertility and Sterility* 101, 78–86.

Ahlquist, P. (2002). RNA — dependent RNA polymerases, viruses, and RNA silencing. *Science* 296, 1270–1273.

Amanai, M., Brahmajosyula, M. and Perry, A. C. (2006). A restricted role for sperm — borne microRNAs in mammalian fertilization. *Biology of Reproduction* 75, 877–884.

Antequera, F. and Bird, A. (1993). CpG islands. In *DNA Methylation: Molecular Biology and Biological Significance*. Jost, J. P. and Saluz, H. P. eds. (Basel, Swizerland), 169–185.

Anway, M. D., Cupp, A. S., Uzumcu, M. and Skinner, M. K. (2005). Epigenetic transgenerational actions of endocrine disruptors and male fertility. *Science* 308, 1466–1469.

Anway, M. D. and Skinner, M. K. (2006). Epigenetic transgenerational actions of endocrine disruptors. *Endocrinology* 147, 43–49.

Anway, M. D. and Skinner, M. K. (2008). Epigenetic programming of the germ line: effects of endocrine disruptors on the development of transgenerational disease. *Reproductive Biomedicine online* 16, 23–25.

Aoki, V. W., Liu, L., Jones, K. P., Hatasaka, H. H., Gibson, M., Peterson, C. M. and Carrell, D. T. (2006). Sperm protamine 1/protamine 2 ratios are related to *in vitro* fertilization pregnancy rates and predictive of fertilization ability. *Fertility and Sterility* 86, 1408–1415.

Benkhalifa, M., Montjean, D., Belloc, S., Dalleac, A., Ducasse, M., Boyer, P., Merviel, P. and Copin, H. (2014). *Expert Review of Molecular Diagnostics* 14, 37–45.

Bird, A. P. and Wolffe, A. P. (1999). Methylation-induced repression – belts, braces, and chromatin. *Cell* 99, 451–454.

Björndahl, L. and Kvist, U. (2010). Human sperm chromatin stabilization: a proposed model including zinc bridges. *Molecular Human Reproduction* 16, 23–29.

Boissonnas, C. C., Abdalaoui, H. E., Haelewyn, V., Fauque, P., Dupont, J. M., Gut, I., Vaiman, D., Jouannet, P., Tost, J. and Jammes, H. (2010). Specific epigenetic alterations of IGF2 — H19 locus in spermatozoa from infertile men. *European Journal of Human Genetics* 18, 73–80.

Boivin, J., Bunting, L., Collins, J. A. and Nygren, K. G. (2007). International estimates of infertility prevalence and treatment-seeking: Potential need and demand for infertility medical care. *Human Reproduction* 22, 1506–1512.

Bonde, J. P. (2010). Male reproductive organs are at risk from environmental hazards. *Asian Journal of Andrology* 12, 152–156.

Bourc'his, D. and Bestor, T. H. (2004). Meiotic catastrophe and retrotransposon reactivation in male germ cells lacking Dnmt3L. *Nature* 431, 96–99.

Chang, H. S., Anway, M. D., Rekow, S. S. and Skinner, M. K. (2006). Transgenerational epigenetic imprinting of the male germline by endocrine disruptor exposure during gonadal sex determination. *Endocrinology* 147, 5524–5541.

Cooper, T. G., Noonan, E., von Eckardstein, S., Auger, J., Baker, H. W., Behre, H. M., Haugen, T. B., Kruger, T., Wang, C., Mbizvo, M. T. *et al.* (2010). World Health Organization reference values for human semen characteristics. *Human Reproduction Update* 16, 231–245.

Cox, G. F., Burger, J., Lip, V., Mau, U. A., Sperling, K., Wu, B. L. and Horsthemke, B. (2002). Intracytoplasmic sperm injection may increase the risk of imprinting defects. *American Journal of Human Genetics* 71, 162–164.

Dadoune, J. P. (2009). Spermatozoal RNAs: what about their functions? *Microscopy Research and Technique* 72, 536–551.

Davis, T. L., Trasler, J. M., Moss, S. B., Yang, G. J. and Bartolomei, M. S. (1999). Acquisition of the H19 methylation imprint occurs differentially on the parental alleles during spermatogenesis. *Genomics* 58(169), 18–28.

Davis, T. L., Yang, G. J., McCarrey, J. R. and Bartolomei, M. S. (2000). The H19 methylation imprint is erased and re-established differentially on the parental alleles during male germ cell development. *Human Molecular Genetics* 9, 2885–2894.

DeBaun, M. R., Niemitz, E. L. and Feinberg, A. P. (2003). Association of in vitro fertilization with Beckwith–Wiedemann syndrome and epigenetic alterations of LIT1 and H19. *American Journal of Human Genetics* 72, 156–160.

Denham, J., O'Brien, B. J., Harvey, J. T. and Charchar, F. J. (2015). Genome-wide sperm DNA methylation changes after 3 months of exercise training in humans. *Epigenomics* 13, 1–15.

Dhar, S., Thota, A. and Rao, M. R. (2012). Insights into role of bromodomain, testis-specific (Brdt) in acetylated histone H4-dependent chromatin remodeling in mammalian spermiogenesis. *Journal of Biological Chemistry* 287, 6387–405.

Doerksen, T. and Trasler, J. M. (1996). Developmental exposure of male germ cells to 5-azacytidine results in abnormal preimplantation development in rats. *Biology of Reproduction* 55, 1155–1162.

Doerksen, T., Benoit, G. and Trasler, J. M. (2000). Deoxyribonucleic acid hypomethylation of male germ cells by mitotic and meiotic exposure to 5-azacytidine is associated with altered testicular histology. *Endocrinology* 141, 3235–3244.

Dohle, G. R., Halley, D. J., Van Hemel, J. O., van den Ouwel, A. M., Pieters, M. H., Weber, R. F. and Govaerts, L. C. (2002). Genetic risk factors in infertile men with severe oligozoospermia and azoospermia. *Human Reproduction* 17, 13–16.

Dottermusch-Heidel, C., Klaus, E. S., Gonzalez, N. H., Bhushan, S., Meinhardt, A., Bergmann, M., Renkawitz-Pohl, R., Rathke, C. and Steger, K. (2014). H3K79 methylation directly precedes the histone-to-protamine transition in mammalian spermatids and is sensitive to bacterial infections. *Andrology* 2, 655–65.

Du, T. and Zamore, P. D. (2005). microPrimer: the biogenesis and function of microRNA. *Development* 132, 4645–4652.

Egger, G., Liang, G., Aparicio, A. and Jones, P. A. (2004). Epigenetics in human disease and prospects for epigenetic therapy. *Nature* 429, 457–463.

Filipescu, D., Müller, S. and Almouzni, G. (2014). Histone H3 variants and their chaperones during development and disease: Contributing to epigenetic control. *Annual Review of Cell and Developmental Biology* 30, 615–646.

Francis, S., Yelumalai, S., Jones, C. and Coward, K. (2014). Aberrant protamine content in sperm and consequential implications for infertility treatment. *Human Fertility (Camb)* 17, 80–89.

Fujimoto, S., Yoshida, N., Fukui, T., Amanai, M., Isobe, T., Itagaki, C., Izumi, T. and Perry, A. C. (2004). Mammalian phospholipase Czeta induces oocyte activation from the sperm perinuclear matrix. *Developmental Biology* 274, 370–383.

Gaucher, J., Boussouar, F., Montellier, E., Curtet, S., Buchou, T., Bertrand, S., Hery, P., Jounier, S., Depaux, A., Vitte, A. L. *et al.* (2012). Bromodomain-dependent stage-specific male genome programming by Brdt. *EMBO Journal* 31, 3809–3820.

Gekas, J., Thepot, F., Turleau, C., Siffroi, J. P., Dadoune, J. P., Briault, S., Rio, M., Bourouillou, G., Carré-Pigeon, F., Wasels, R. *et al.* (2001) Chromosomal factors of infertility in candidate couples for ICSI: an equal risk of constitutional aberrations in women and men. *Human Reproduction* 16, 82–90.

Gicquel, C., Gaston, V., Mandelbaum, J., Siffroi, J. P., Flahault, A. and LeBouc, Y. (2003). In vitro fertilization may increase the risk of Beckwith–Wiedemann syndrome related to the abnormal imprinting of the KCN1OT gene. *American Journal of Human Genetics* 72, 1338–1341.

Goll, M. G. and Bestor, T. H. (2005). Eukaryotic cytosine methyltransferases. *Annu Rev Biochem* 74, 481–514.

Gou, L. T., Dai, P., Yang, J. H., Xue, Y., Hu, Y. P., Zhou, Y., Kang, J. Y., Wang, X., Li, H., Hua, M. M. *et al.* (2014). Pachytene piRNAs instruct massive mRNA elimination during late spermiogenesis. *Cell Research* 24, 680–700.

Govin, J. and Khochbin, S. (2013). Histone variants and sensing of chromatin functional states. *Nucleus* 4, 438–442.

Greenlee, A. R., Shiao, M. S., Snyder, E., Buaas, F. W., Gu, T., Stearns, T. M., Sharma, M., Murchison, E. P., Puente, G. C. and Braun, R. E. (2012). Deregulated sex chromosome gene expression with male germ cell-specific loss of Dicer1. *PLoS One* 7, e46359.

Gu, A., Ji, G., Shi, X., Long, Y., Xia, Y., Song, L., Wang, S. and Wang, X. (2010). Genetic variants in Piwi-interacting RNA pathway genes confer susceptibility to spermatogenic failure in a Chinese population. *Human Reproduction* 25, 2955–2961.

Guerrero-Bosagna, C. M. and Skinner, M. K. (2009). Epigenetic transgenerational effects of endocrine disruptors on male reproduction. *Seminars Reproductive Medicine* 27, 403–408.

Guerrero-Bosagna, C., Covert, T. R., Haque, M. M., Settles, M., Nilsson, E. E., Anway, M. D. and Skinner, M. K. (2012). Epigenetic transgenerational inheritance of vinclozolin induced mouse adult onset disease and associated sperm epigenome biomarkers. *Reproductive Toxicology* 34, 694–707.

Guerrero-Bosagna, C. and Skinner, M. K. (2014). Environmentally induced epigenetic transgenerational inheritance of male infertility. *Current Opinion in Genetics & Development* 26, 79–88.

Hajkova, P., Erhardt, S., Lane, N., Haaf, T., El-Maarri, O., Reik, W., Walter, J. and Surani, M. A. (2002). Epigenetic reprogramming in mouse primordial germ cells. *Mechanisms of Development* 117, 15–23.

Hamada, A. J., Esteves, S. C. and Agarwal, A. (2013). A comprehensive review of genetics and genetic testing in azoospermia. *Clinics (Sao Paulo)* 68, 39–60.

Hammoud, S. S., Nix, D. A., Hammoud, A. O., Gibson, M., Cairns, B. R. and Carrell, D. T. (2011). Genome-wide analysis identifies changes in histone retention and epigenetic modifications at developmental and imprinted gene loci in the sperm of infertile men. *Human Reproduction* 26, 2558–2569.

Hammoud, S. S., Purwar, J., Pflueger, C., Cairns, B. R. and Carrell, D. T. (2010). Alterations in sperm DNA methylation patterns at imprinted loci in two classes of infertility. *Fertility and Sterility* 94, 1728–1733.

Hayashi, K., Chuva de Sousa Lopes, S. M., Kaneda, M., Tang, F., Hajkova, P., Lao, K., O'Carroll, D., Das, P. P., Tarakhovsky, A., Miska, E. A. *et al.* (2008). MicroRNA biogenesis is required for mouse primordial germ cell development and spermatogenesis. *PLoS One* 3, e1738.

He, Z., Kokkinaki, M., Pant, D., Gallicano, G. I. and Dym, M. (2009). Small RNA molecules in the regulation of spermatogenesis. *Reproduction* 137, 901–911.

Holley, R. W., Apgar, J., Everett, G. A., Madison, J. T., Marquisee, M., Merrill, S. H., Penswick, J. R. and Zamir, A. (1965). Structure of a ribonucleic acid. *Science* 147, 1462–1465.

Houshdaran, S., Cortessis, V. K., Siegmund, K., Yang, A., Laird, P. W. and Sokol, R. Z. (2007). Widespread epigenetic abnormalities suggest a broad DNA methylation erasure defect in abnormal human sperm. *PLoS One 2*, e1289.

Isles, A. R. and Holland, A. J. (2005). Imprinted genes and mother — offspring interactions. *Early Human Development* 81, 73–77.

Jenkins, T. G., Aston, K. I., Pflueger, C., Cairns, B. R. and Carrell, D. T. (2014). Age-associated sperm DNA methylation alterations: possible implications in offspring disease susceptibility. *PLoS Genetics* 10, e1004458.

Jodar, M., Kalko, S., Castillo, J., Ballescà, J. L. and Oliva, R. (2012). Differential RNAs in the sperm cells of asthenozoospermic patients. *Human Reproduction* 27, 1431–1438.

Jurewicz, J., Hanke, W., Radwan, M. and Bonde, J. P. (2009). Environmental factors and semen quality. *International Journal of Occupational Medicine and Environment Health* 22, 305–329.

Juul, A., Almstrup, K., Andersson, A. M., Jensen, T. K., Jørgensen, N., Main, K. M., Rajpert-De Meyts, E., Toppari, J. and Skakkebæk, N. E. (2014). Possible fetal determinants of male infertility. *Nature Reviews Endocrinology* 10, 553–562.

Kaati, G., Bygren, L. O., Pembrey, M. and Sjöström, M. (2007). Transgenerational response to nutrition, early life circumstances and longevity. *European Journal of Human Genetics* 15, 784–790.

Kato, Y., Kaneda, M., Hata, K., Kumaki, K., Hisano, M., Kohara, Y., Okano, M., Li, E., Nozaki, M. and Sasaki, H. (2007). Role of the Dnmt3 family in de novo methylation of imprinted and repetitive sequences during male germ cell development in the mouse. *Human Molecular Genetics* 16, 2272–2280.

Kelly, T. L., Li, E. and Trasler, J. M. (2003). 5-aza-2′-deoxycytidine induces alterations in murine spermatogenesis and pregnancy outcome. *Journal of Andrology* 24, 822–830.

Kim, V. N., Han, J. and Siomi, M. C. (2009). Biogenesis of small RNAs in animals. *Nature Reviews Molecular Cell Biology* 10, 126–139.

Klose, R. J. and Bird, A. P. (2006). Genomic DNA methylation: the mark and its mediators. *Trends in Biochemical Sciences* 31, 89–97.

Kobayashi, H., Sato, A., Otsu, E., Hiura, H., Tomatsu, C., Utsunomiya, T., Sasaki, H., Yaegashi, N. and Arima, T. (2007). Aberrant DNA methylation of imprinted loci in sperm from oligospermic patients. *Human Molecular Genetics* 16, 2542–2551.

Korhonen, H. M., Meikar, O., Yadav, R. P., Papaioannou, M. D., Romero, Y., Da Ros, M., Herrera, P. L., Toppari, J., Nef, S. and Kotaja, N. (2011). Dicer is required for haploid male germ cell differentiation in mice. *PLoS One* 6, e24821.

Kotaja, N. (2014). MicroRNAs and spermatogenesis. *Fertility and Sterility* 101, 1552–1562.

Krawetz, S. A. (2005). Paternal contribution: New insights and future challenges. *Nature Reviews Genetics* 6, 633–642.

Kuhtz, J., Schneider, E., El Hajj, N., Zimmermann, L., Fust, O., Linek, B., Seufert, R., Hahn, T., Schorsch, M. and Haaf, T. (2014). Epigenetic heterogeneity of developmentally important genes in human sperm: Implications for assisted reproduction outcome. *Epigenetics* 9, 1648–1658.

Kumar, G., Patel, D. and Naz, R. K. (1993). c-MYC mRNA is present in human sperm cells. *Cellular Molecular Biology Research* 39, 111–117.

Kuramochi-Miyagawa, S., Watanabe, T., Gotoh, K., Totoki, Y., Toyoda, A., Ikawa, M., Asada, N., Kojima, K., Yamaguchi, Y., Ijiri, T. W. *et al.* (2008). DNA methylation of retrotransposon genes is regulated by Piwi family members MILI and MIWI2 in murine fetal testes. *Genes & Development* 22, 908–917.

La Salle, S., Mertineit, C., Taketo, T., Moens, P. B., Bestor, T. H. *et al.* (2004). Windows for sex-specific methylation marked by DNA methyltransferase expression profiles in mouse germ cells. *Developmental Biology* 268, 403–415.

La Salle, S. and Trasler, J. M. (2006). Dynamic expression of DNMT3a and DNMT3b isoforms during male germ cell development in the mouse. *Developmental Biology* 296, 71–82.

La Salle, S., Oakes, C. C., Neaga, O. R., Bourc'his, D., Bestor, T. H. and Trasler, J. M. (2007). Loss of spermatogonia and wide-spread DNA methylation defects in newborn male mice deficient in DNMT3L. *BMC Developmental Biology* 18, 104–107.

Laurentino, S., Beygo, J., Nordhoff, V., Kliesch, S., Wistuba, J., Borgmann, J., Buiting, K., Horsthemke, B. and Gromoll, J. (2015). Epigenetic germline mosaicism in infertile men. *Human Molecular Genetics* 24, 1295–1304.

Lazaraviciute, G., Kauser, M., Bhattacharya, S., Haggarty, P. and Bhattacharya, S. (2014). A systematic review and meta-analysis of DNA methylation levels and imprinting disorders in children conceived by IVF/ICSI compared with children conceived spontaneously. *Human Reproduction Update* 20, 840–852.

Li, Y., Lalancette, C., Miller, D. and Krawetz, S. A. (2008). Characterization of nucleohistone and nucleoprotamine components in the mature human sperm nucleus. *Asian Journal of Andrology* 10, 535–541.

Liu, C. G., Calin, G. A., Meloon, B., Gamliel, N., Sevignani, C., Ferracin, M., Dumitru, C. D., Shimizu, M., Zupo, S., Dono, M. *et al.* (2004). An oligonucleotide microchip for genome — wide microRNA profiling in human and mouse tissues. *Proceedings of the National Academy of Sciences USA* 101, 9740–9744.

Liu, T., Cheng, W., Gao, Y., Wang, H. and Liu, Z. (2012). Microarray analysis of microRNA expression patterns in the semen of infertile men with semen abnormalities. *Molecular Medicine Reports* 6, 535–542.

Maatouk, D. M., Loveland, K. L., McManus, M. T., Moore, K. and Harfe, B. D. (2008). Dicer1 is required for differentiation of the mouse male germline. *Biology of Reproduction* 79, 696–703.

Maduro, M. R., Lo, K. C., Chuang, W. W. and Lamb, D. J. (2003). Genes and male infertility: What can go wrong? *Journal of Andrology* 24, 485–493.

Maher, E. R. and Reik, W. (2000). Beckwith–Wiedemann syndrome: Imprinting in clusters revisited. *Journal of Clinical Investigation* 105, 247–252.

Maher, E. R., Brueton, L. A., Bowdin, S. C., Luharia, A., Cooper, W., Cole, T. R., Macdonald, F., Sampson, J. R., Barratt, C. L., Reik, W. *et al.* (2003). Beckwith–Wiedemann syndrome and assisted reproduction technology (ART). *Journal of Medical Genetics* 40, 62–64.

Manikkam, M., Tracey, R., Guerrero-Bosagna, C. and Skinner, M. K. (2013). Plastics derived endocrine disruptors (BPA, DEHP and DBP) induce epigenetic transgenerational inheritance of obesity, reproductive disease and sperm epimutations. *PLoS One* 8, e55387.

Marques, C. J., Carvalho, F., Sousa, M. and Barros, A. (2004). Genomic imprinting in disruptive spermatogenesis. *Lancet* 363, 1700–1702.

Marques, C. J., Costa, P., Vaz, B., Carvalho, F., Fernandes, S., Barros, A. and Sousa, M. (2008). Abnormal methylation of imprinted genes in human sperm is associated with oligozoospermia. *Molecular Human Reproduction* 14, 67–74.

Matzuk, M. M. and Lamb, D. J. (2008). The biology of infertility: Research advances and clinical challenges. *Nature Medicine* 14, 1197–1213.

Miller, D., Tang, P. Z., Skinner, C. and Lilford, R. (1994). Differential RNA fingerprinting as a tool in the analysis of spermatozoal gene expression. *Human Reproduction* 9, 864–869.

Miller, D., Briggs, D., Snowden, H., Hamlington, J., Rollinson, S., Lilford, R. and Krawetz, S. A. (1999). A complex population of RNAs exists in human ejaculate spermatozoa: Implications for understanding molecular aspects of spermiogenesis. *Gene* 237, 385–392.

Montellier, E., Boussouar, F., Rousseaux, S., Zhang, K., Buchou, T., Fenaille, F., Shiota, H., Debernardi, A., Héry, P., Curtet, S. *et al.* (2013). Chromatin-to-nucleoprotamine transition is controlled by the histone H2B variant TH2B. *Genes & Development* 27, 1680–1692.

Montjean, D., De La Grange, P., Gentien, D., Rapinat, A., Belloc, S., Cohen-Bacrie, P., Menezo, Y. and Benkhalifa, M. (2012). Sperm transcriptome profiling in oligozoospermia. *Journal of Assisted Reproduction Genetics* 29, 3–10.

Montjean, D., Ravel, C., Benkhalifa, M., Cohen-Bacrie, P., Berthaut, I., Bashamboo, A. and McElreavey, K. (2013). Methylation changes in mature sperm deoxyribonucleic acid from oligozoospermic men: Assessment of genetic variants and assisted reproductive technology outcome. *Fertility and Sterility* 100, 1241–1247.

Montjean, D., Zini, A., Ravel, C., Belloc, S., Dalleac, A., Copin, H., Boyer, P., McElreavey, K. and Benkhalifa, M. (2015). Sperm global DNA methylation level: Association with semen parameters and genome integrity. *Andrology* 3, 235–240.

Morison, I. M., Ramsay, J. P. and Spencer, H. G. (2005). A census of mammalian imprinting. *Trends in Genetics* 21, 457–465.

Mudrak, O. S., Nazarov, I. B., Jones, E. L. and Zalensky, A. O. (2012). Positioning of chromosomes in human spermatozoa is determined by ordered centromere arrangement. *PLoS One* 7, e52944.

Nanassy, L., Liu, L., Griffin, J. and Carrell, D. T. (2011). The clinical utility of the protamine 1/protamine 2 ratio in sperm. *Protein & Peptide Letters* 18, 772–777.

Nekrasov, M., Amrichova, J., Parker, B. J., Soboleva, T. A., Jack, C., Williams, R., Huttley, G. A. and Tremethick, D. J. (2012). Histone H2A.Z inheritance during the cell cycle and its impact on promoter organization and dynamics. *Nature Structural Molecular & Biology* 19, 1076–1083.

Oakes, C. C., La Salle, S., Smiraglia, D. J., Robaire, B. and Trasler, J. M. (2007). Developmental acquisition of genome-wide DNA methylation occurs prior to meiosis in male germ cells. *Developmental Biology* 307, 368–379.

Oakes, C. C., Kelly, T. L., Robaire, B. and Trasler, J. M. (2007). Adverse effects of 5-aza-2′-deoxycytidine on spermatogenesis include reduced sperm function and selective inhibition of de novo DNA methylation. *Journal of Pharmacology and Experiment Therapeutics* 322, 1171–1180.

O'Doherty, A. M. and McGettigan, P. A. (2014). Epigenetic processes in the male germline. *Reproduction Fertility and Development* 27, 725–738.

Okano, M., Bell, D. W., Haber, D. A. and Li, E. (1999). DNA methyltransferases Dnmt3a and Dnmt3b are essential for de novo methylation and mammalian development. *Cell* 99, 247-257.

Orstavik, K. H., Eiklid, K., van der Hagen, C. B., Spetalen, S., Kierulf, K., Skjeldal, O. and Buiting, K. (2003). Another case of imprinting defect in a girl with Angelman syndrome who was conceived by intracytoplasmic semen injection. *American Journal of Human Genetics* 72, 218–219.

Ostermeier, G. C., Dix, D. J., Miller, D., Khatri, P. and Krawetz, S. A. (2002). Spermatozoal RNA profiles of normal fertile men. *Lancet* 360, 772–777.

Ostermeier, G. C., Goodrich, R. J., Moldenhauer, J. S., Diamond, M. P. and Krawetz, S. A. (2005). A suite of novel human spermatozoal RNAs. *Journal of Andrology* 26, 70–74.

Pembrey, M. E., Bygren, L. O., Kaati, G., Edvinsson, S., Northstone, K., Sjöström, M., Golding, J., ALSPAC Study Team. (2006). Sex-specific, male-line transgenerational responses in humans. *European Journal of Human Genetics* 14, 159–166.

Plasterk, R. H. (2006). Micro RNAs in animal development. *Cell* 124, 877–881.

Poongothai, J., Gopenath, T. S. and Manonayaki, S. (2009). Genetics of human male infertility. *Singapore Medical Journal* 50, 336–347.

Reik, W. and Murrell, A. (2000). Genomic imprinting. Silence across the border. *Nature* 405, 408–409.

Reik, W., Dean, W. and Walter, J. (2001). Epigenetic reprogramming in mammalian development. *Science* 293, 1089–1093.

Ribas-Maynou, J., Gawecka, J. E., Benet, J. and Ward, W. S. (2014). Double-stranded DNA breaks hidden in the neutral Comet assay suggest a role of the sperm nuclear matrix in DNA integrity maintenance. *Molecular Human Reproduction* 20, 330–340.

Romero, Y., Meikar, O., Papaioannou, M. D., Conne, B., Grey, C., Weier, M., Pralong, F., De Massy, B., Kaessmann, H., Vassalli, J. D. *et al.* (2011). Dicer1 depletion in male germ cells leads to infertility due to cumulative meiotic and spermiogenic defects. *PLoS One* 6, e25241.

Rousseaux, S., Gaucher, J., Thevenon, J., Caron, C., Vitte, A. L., Curtet, S., Derobertis, C., Faure, A. K., Levy, R., Aknin-Seifer, I. *et al.* (2009). Spermiogenesis: Histone acetylation triggers male genome reprogramming. *Gynecologie Obstetrique Fertilite* 37, 519–522.

Skaletsky, H., Kuroda-Kawaguchi, T., Minx, P. J., Cordum, H. S., Hillier, L., Brown, L. G., Repping, S., Pyntikova, T., Ali, J., Bieri, T., Chinwalla, A. *et al.* (2003). The male-specific region of the human Y chromosome is a mosaic of discrete sequence classes. *Nature* 423, 825–837.

Takashima, S., Takehashi, M., Lee, J., Chuma, S., Okano, M., Hata, K., Suetake, I., Nakatsuji, N., Miyoshi, H., Tajima, S. *et al.* (2009). Abnormal DNA methyltransferase expression in mouse germline stem cells results in spermatogenic defects. *Biology of Reproduction* 81, 155–164.

Thonneau, P., Marchand, S., Tallec, A., Ferial, M. L., Ducot, B., Lansac, J., Lopes, P., Tabaste, J. M. and Spira, A. (1991). Incidence and main causes of infertility in a resident population (1,850,000) of three French regions (1988–1989). *Human Reproduction* 6, 811–816.

Thonneau, P. and Spira, A. (1991). Prevalence of infertility: international data and problems of measurement. *European Journal of Obstetrics Gynecology and Reproductive Biology* 38, 43–52.

Tiepolo, L., Zuffardi, O. (1976). Localization of factors controlling spermatogenesis in the nonfluorescent portion of the human Y chromosome long arm. *Human Genetics* 34, 119–124.

Tropberger, P. and Schneider, R. (2013). Scratching the (lateral) surface of chromatin regulation by histone modifications. *Nature Structural Molecular Biology* 20, 657–661.

Urdinguio, R. G., Bayón, G. F., Dmitrijeva, M., Toraño, E. G., Bravo, C., Fraga, M. F., Bassas, L., Larriba, S. and Fernández, A. F. (2015). Aberrant DNA methylation patterns of spermatozoa in men with unexplained infertility. *Human Reproduction* 30, 1014–1028.

Valente, L. P., Silva, M. C. and Jansen, L. E. (2012). Temporal control of epigenetic centromere specification. *Chromosome Research* 20, 481–492.

Vavouri, T. and Lehner, B. (2011). Chromatin organization in sperm may be the major functional consequence of base composition variation in the human genome. *PLoS Genetics* 7, e1002036.

Watanabe, T., Tomizawa, S., Mitsuya, K., Totoki, Y., Yamamoto, Y., Kuramochi-Miyagawa, S., Iida, N., Hoki, Y., Murphy, P. J., Toyoda, A. *et al.* (2011). Role for piRNAs and noncoding RNA in de novo DNA methylation of the imprinted mouse Rasgrf1 locus. *Science* 332, 848–852.

Webster, K. E., O'Bryan, M. K., Fletcher, S., Crewther, P. E., Aapola, U., Craig, J., Harrison, D. K., Aung, H., Phutikanit, N., Lyle, R. *et al.* (2005). Meiotic and epigenetic defects in Dnmt3L knockout mouse spermatogenesis. *Proceedings of the National Academy of Sciences USA* 102, 4068–4073.

Wei, Y., Schatten, H. and Sun, Q. Y. (2015). Environmental epigenetic inheritance through gametes and implications for human reproduction. *Human Reproduction Update* 21, 194–208.

Wood, A. J. and Oakey, R. J. (2006). Genomic imprinting in mammals: Emerging themes and established theories. *PLoS Genetics* 2, e147.

World Health Organization. (1999). *WHO Laboratory Manual for the Examination of Human Semen and Sperm-cervical Mucus Interaction*, 4th edn. Cambridge: Cambridge University Press, 14–22.

Wu, Q., Song, R., Ortogero, N., Zheng, H., Evanoff, R., Small, C. L., Griswold, M. D., Namekawa, S. H., Royo, H., Turner, J. M. and Yan, W. (2012). The RNase III enzyme DROSHA is essential for microRNA production and spermatogenesis. *Journal of Biological Chemistry* 287, 25173–25190.

Wykes, S. M., Visscher, D. W. and Krawetz, S. A. (1997). Haploid transcripts persist in mature human spermatozoa. *Molecular Human Reproduction* 3, 15–19.

Zimmermann, C., Romero, Y., Warnefors, M., Bilican, A., Borel, C., Smith, L. B., Kotaja, N., Kaessmann, H. and Nef, S. (2014). Germ cell-specific targeting of DICER or DGCR8 reveals a novel role for endo siRNAs in the progression of mammalian spermatogenesis and male fertility. *PLoS One* 22, 9:e107023.

Epigenetic Regulation of X-Chromosome Inactivation

5

Bernhard Payer

Centre for Genomic Regulation (CRG),
The Barcelona Institute for Science and Technology and
Universitat Pompeu Fabra (UPF)
Dr. Aiguader 88, 08003 Barcelona, Spain

X-chromosome inactivation is a mammalian epigenetic phenomenon, which occurs during female embryogenesis and has the purpose to equalize expression of X-chromosomal genes between females and males. It is controlled by long non-coding RNAs, most prominently *Xist* and involves a plethora of epigenetic silencing mechanisms from histone modifications over histone variants to DNA-methylation and conformational changes of the inactive X-chromosome. Although X-inactivation is a classical field of epigenetics, advances of *in vitro* mouse and human model systems and epigenomic analysis methods increased our knowledge exponentially in the recent years. In this review chapter, I am providing an update on the current state of the X-inactivation field and the questions, which still remain to be answered.

Preface

This review chapter is dedicated to Mary Frances Lyon (1925–2014), who in 1961 founded the X-inactivation field with her seminal one page X-inactivation hypothesis published in Nature (Lyon 1961). She based her hypothesis on the mosaic coat color of female mice with X-linked gene mutations and on findings by Ohno and colleagues that one of the two X-chromosomes in female mammals is heteropyknotic (Ohno and Hauschka 1960), which has been initially identified cytologically in female cat neurons by Barr and Bertram (Barr and Bertram 1949). X-inactivation has since then become one of the best-characterized examples of epigenetic gene regulation and more than 50 years of research have revealed its multifaceted control mechanisms. Here, I give an update on the current state of this fascinating topic, which since Lyon's initial postulation has captivated many Scientists' minds and continues to remain on the forefront of epigenetics research due to its intriguing complexity.

Abbreviations

E	Embryonic day of development
ESCs	Embryonic Stem Cells
ICM	Inner Cell Mass
iPSCs	induced Pluripotent Stem Cells
PE	Primitive Endoderm
PGCs	Primordial Germ Cells
TAD	Topologically Associated Domain
TE	Trophectoderm
XCI	X-Chromosome Inactivation
XCR	X-Chromosome Reactivation
Xa	active X-Chromosome
Xi	inactive X-Chromosome
Xm	Maternally Inherited X-Chromosome
Xp	Paternally Inherited X-Chromosome
Xic	X-Inactivation Center
Xist	X-inactive Specific Transcript — Non-coding RNA and Master Regulator of X-Chromosome Inactivation

5.1 Introduction

Due to genetic sex determination, male and female mammals differ regarding their sex chromosome composition — females have two X-chromosomes and males have one X- and one Y-chromosome. This results in a potential imbalance of X-linked gene dosage between the sexes and female mammals compensate for this by transcriptional inactivation of one of their two X-chromosomes (Brockdorff and Turner 2015, Deng *et al.* 2014, Payer and Lee 2008, Schulz and Heard 2013). The way in which X-chromosome inactivation (XCI) is accomplished is best known in the mouse model system, where XCI has been extensively studied both *in vivo* during embryogenesis and *in vitro/ex vivo* during embryonic stem cell (ESC) differentiation. The discovery of the long non-coding RNA *Xist* (Borsani *et al.* 1991, Brockdorff *et al.* 1991, Brown *et al.* 1991), which sits on the top of the X-inactivation hierarchy in eutherian (placental) mammals including mice and humans, opened up the field to mechanistic studies regarding the regulation and epigenetic changes associated with XCI. Extensive transgenic and deletion studies of *Xist* and its regulatory elements at the *X-inactivation center* (*Xic*) defined the *cis*-requirements on the X-chromosome, which are important for its inactivation and helped with identification of the *trans*-acting factors, which bind and regulate these critical elements. Many turned out to be pluripotency factors, which provided a mechanistic link between XCI-regulation and pluripotency (Ohhata and Wutz 2012, Payer and Lee 2014). As XCI is a chromosome-wide event, epigenetic features like silencing-specific histone modifications can be readily visualized by microscopy. In combination with genomics-based high-resolution analysis methods, a great deal of information about the epigenetic regulation of XCI has been recently acquired. Further advances in identification of *Xist*-RNA bound proteins and information on the conformation and physical interactions within the X-chromosome provided a more complete picture on the mechanisms associated with XCI. Finally, new *in vitro* systems have been developed for studying XCI in human cells, which will be important, as not every aspect is conserved among species.

5.2 The *Xic* — the Regulatory Hub of X-Inactivation

The *Xic* is the region on the X-chromosome, which contains the *cis*-regulatory elements responsible for controlling *Xist* and X-inactivation (Figure 5.1). The region contains a number of long non-coding RNA genes, which have repressive or activating influence on *Xist*. Another feature of the *Xic* is accumulations of binding sites for trans-acting factors — in particular for pluripotency factors, linking XCI and X-chromosome reactivation (XCR) to the differentiated and pluripotent state, respectively. The most well-studied regulator of *Xist* is its repressor *Tsix*, a non-coding RNA gene which overlaps in mouse with *Xist* in antisense (Lee 2000, Lee and Lu 1999, Lee *et al.* 1999, Sado *et al.* 2001). *Tsix* is expressed in ESCs and becomes upregulated during differentiation on the active X-chromosome (Xa) whereas being downregulated on the inactive X-chromosome (Xi), allowing upregulation of *Xist* on the Xi. In female *Tsix*-heterozygous cells, X-inactivation is skewed towards the X-chromosome with the mutant *Tsix*-allele,

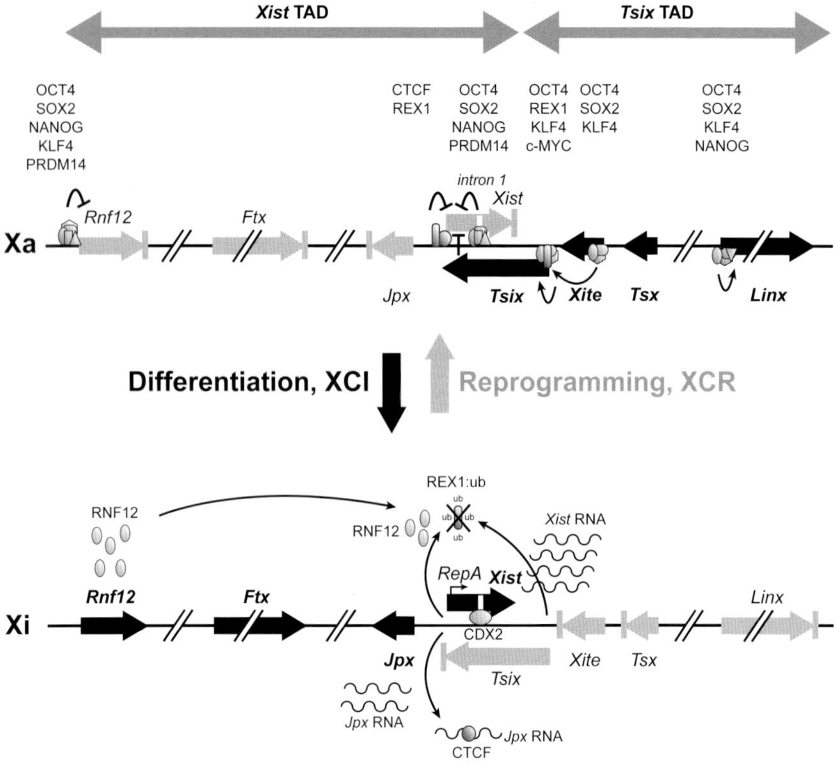

Figure 5.1. The *Xic* and its regulation during XCI and XCR. In undifferentiated pluripotent stem cells, several locations along the *Xic* are bound by pluripotency factors (listed above binding hubs), which activate transcription of genes (black arrows) within the *Tsix* TAD (topologically associated domain). Genes within the *Xist* TAD (grey arrows), including *Xist* itself, are repressed by those factors, keeping the X-chromosome in an active state (Xa).

During differentiation, pluripotency factors become downregulated and dissociate from the *Xic*. This leads to expression of the ubiquitin ligase RNF12, which ubiquitinates the pluripotency factor REX1 (REX1:ub), which induces its degradation and thereby releases its repressive action on *Xist* and its activating function on *Tsix*. Furthermore, the long non-coding RNA *Jpx* evicts CTCF from the *Xist* promoter, thereby removing another repressor of *Xist*. In addition, OCT4 is displaced by CDX2 from *Xist intron 1* and also the *Xist*-activating transcript Rep A becomes expressed. In summary, on the inactive X (Xi), *Tsix* and its activators become repressed, while *Xist* activators become upregulated, initiating *Xist* RNA expression and thereby inactivation of Xi.

Distances are approximately drawn to scale and diagonal parallel lines symbolize longer elements that have been shortened to fit the space.

which will upregulate *Xist* and become the Xi, showing the function for *Tsix* in choice of the Xa (Lee and Lu 1999). A recent study, however, observed that early during differentiation, *Tsix*-heterozygous cells sometimes express *Xist* from both alleles, after which they are removed by selection, ensuring that only cells with monoallelic *Xist* expression from the *Tsix*-mutant chromosome remain (Gayen *et al.* 2015). This suggests that *Tsix* is not responsible for the primary choice, which X-chromosome becomes active or inactive, but rather has the function to prevent *Xist* from being ectopically expressed on the Xa after the choice has been made stochastically (Monkhorst *et al.* 2008). Independent of the precise time point of *Tsix* action, it is clear that the antagonists *Xist* and *Tsix* decide which X-chromosome becomes the Xi and which the Xa.

Tsix itself is activated by its enhancers *DXPas34* and *Xite*, which also produce non-coding transcripts downstream and upstream of the *Tsix* promoter (Cohen *et al.* 2007, Ogawa and Lee 2003). Further 5′ of *Tsix* and *Xite* lie the noncoding RNA genes *Tsx* and *Linx*, which also have been implicated in *Tsix* regulation. While *Tsx* seems to be an activator of *Tsix* (Anguera *et al.* 2011), the proposed role of *Linx* as either a *Tsix*-activator (Nora *et al.* 2012) or a competitor for common regulatory elements with *Tsix* (Giorgetti *et al.* 2014) still needs to be tested by deletion studies. *Tsix* represses *Xist in cis* by being transcribed antisense across the *Xist* promoter and thereby establishing a repressive chromatin state including DNA methylation (Luikenhuis *et al.* 2001, Navarro *et al.* 2005, Ohhata *et al.* 2008, Sado *et al.* 2005, Sun *et al.* 2006). Furthermore, *Tsix* facilitates binding of the pluripotency factor PRDM14 to *Xist intron 1*, which might contribute to the repression of *Xist* in pluripotent stem cells (Payer *et al.* 2013). Another factor recruited by *Tsix* is the genome architecture protein CTCF (Kung *et al.* 2015). CTCF binds directly *Tsix* and *Xite* RNAs whose knockdown abrogates CTCF binding to multiple sites at the *Xic*, suggesting RNA-dependent recruitment. CTCF has been described as necessary for pairing of the two X-chromosomes during the onset of random XCI (Xu *et al.* 2007). X-chromosome pairing occurs at regions around *Tsix* and *Xite* and has been proposed to be involved in symmetry breaking leading to the choice; which X-chromosome becomes the Xi and which the Xa (Bacher *et al.* 2006, Lee 2011, Masui *et al.* 2011, Nicodemi and Prisco 2007, Xu *et al.* 2006). Another pairing region called *Xpr* has been described 5′of *Xist* (Augui *et al.* 2007), although its function has been under debate (Sun *et al.* 2010). A recent study, however, showed that XCI can also occur in the absence of pairing (Barakat *et al.* 2014). The single male X in male/female heterokaryons, which form after cell fusion and retain separate male and female nuclei, can also undergo XCI, without being able to pair with the female X-chromosomes. Furthermore, deletion of the entire region of proposed pairing elements on one X-chromosome in female cells still allows upregulation of *Xist* and XCI on the second X-chromosome without pairing. This does not rule out a symmetry breaking function for pairing during XCI, however, the exact functional role and if it is a regulator or a consequence of the XCI-process will need to be reexamined.

While the genes at the *Xic* downstream of *Xist* including *Tsix* and its activators predominantly have a repressive role on *Xist*, the opposite seems to be the case for genes localized upstream of *Xist* (Figure 5.1). Immediately 5′ of *Xist* lies the non-coding RNA gene *Jpx/Enox*, which becomes expressed during differentiation from both X-chromosomes and is required for upregulation of *Xist* during XCI (Tian *et al.* 2010). *Jpx* RNA conveys its function through repelling CTCF, which not only has a function in X–X-pairing, but also has a direct regulatory role on *Xist* by binding to the *Xist* promoter region (Navarro *et al.* 2006, Pugacheva *et al.* 2005, Sun *et al.* 2013). While CTCF blocks *Xist* upregulation on the Xa, it gets titrated away by *Jpx* RNA from the Xi, allowing the upregulation of *Xist* (Sun *et al.* 2013). The dosage equilibrium between the X-linked *Jpx* RNA and the autosomally encoded CTCF protein appear to be crucial to determine how many X-chromosomes are inactivated in relation to autosomal number, ensuring one X to be inactivated in XX females and none in XY males. *Jpx* acts both *in cis* and *in trans*, as transgenic *Jpx* expression can rescue the *Xist* upregulation defect in *Jpx* +/− cells (Tian *et al.* 2010) and can cause ectopic upregulation of *Xist* in wild-type male and female cells (Sun *et al.* 2013). Another recent study has challenged this view that *Jpx* can act *in trans*, by failing to rescue the X-inactivation defect of a large X-chromosome deletion, which includes *Jpx*, by transgenes (Barakat *et al.* 2014). If the discrepancy between the results is due to the size of the deletion, due to different genetic backgrounds, or due to other reasons, will need to be addressed in future studies. Further upstream of *Jpx* lies *Ftx*, another non-coding RNA gene, which has been put forward as potential *Xist*-activator (Chureau *et al.* 2011). While *Ftx*-deletion in male ESCs led to slight changes in gene expression at the *Xic* including *Xist, Jpx and Tsix* (Chureau *et al.* 2011), *Ftx* is dispensable for *Xist* upregulation during imprinted XCI in female mouse embryos (Soma *et al.* 2014). If *Ftx* has a role in random XCI still needs to be tested.

Located upstream of *Ftx* lies the E3 ubiquitin ligase gene *Rnf12/Rlim*, which has also been described as an *Xist*-activator (Barakat *et al.* 2011, Jonkers *et al.* 2009). RNF12 is upregulated during ESC differentiation and acts *in trans* to activate *Xist* by ubiquitinating and thereby degrading the pluripotency factor REX1, which binds as a repressor at the *Xist* promoter (Gontan *et al.* 2012). When overexpressed, RNF12 can induce ectopic X-inactivation in male and female ESCs and *Rnf12* mutant cells show *Xist*-upregulation defects in a dose-dependent manner (Barakat *et al.* 2011, Jonkers *et al.* 2009). RNF12 is essential for imprinted XCI in preimplantation embryos and in extraembryonic tissues and thereby for postimplantation survival of female embryos (Shin *et al.* 2010). However, random XCI progresses normally in *Rnf12*-null embryos and in ESCs derived from them (Shin *et al.* 2010, 2014). Whether the difference in phenotype for random XCI between different labs is due to different *Rnf12* knockout alleles, genetic backgrounds, or cell culture conditions, will need to be determined.

REX1 is not the only pluripotency factor, which binds along the *Xic*, providing a direct link between XCI regulation and the pluripotent stem cell state (Payer and Lee 2014). In fact there are multiple binding hubs along the *Xic*, where pluripotency factors accumulate in pluripotent stem cells (Figure 5.1). One of the most prominent regions, showing the strongest pluripotency factor occupancy, lies within *Xist intron 1*. *Xist intron 1* is bound by OCT4, SOX2, NANOG and PRDM14 and has been implicated in direct repression of *Xist* by pluripotency factors in ESCs (Donohoe *et al.* 2009, Ma *et al.* 2010, Navarro *et al.* 2008, Nesterova *et al.* 2011, Payer *et al.* 2013). In addition, *Xist intron 1* is bound by the trophectoderm regulator CDX2 in a mutually exclusive manner with OCT4, which could be involved in switching between imprinted XCI and XCR depending on cell lineage (Erwin *et al.* 2012). Depletion of the pluripotency factors OCT4, NANOG or PRDM14 results in *Xist* upregulation but also accelerates differentiation (Ma *et al.* 2010, Navarro *et al.* 2008), which could indirectly lead to *Xist* induction independent of pluripotency factors dissociation from *Xist intron 1*. Indeed, *Xist intron 1* deletion does neither perturb *Xist* repression in ESCs, nor does it greatly affect XCI *in vitro* or *in vivo* (Barakat *et al.* 2011, Minkovsky *et al.* 2013). Therefore redundant control mechanisms likely compensate to repress *Xist* in the absence of *intron 1*. Such compensatory mechanisms could either be other *Xist*-repressors or the absence of *Xist* activators in pluripotent cells. *Tsix* is one of the other *Xist*-repressors and is itself controlled by pluripotency factors. Like *Xist*, *Tsix* control elements in its 5′ region and its enhancers *DXPas34* and *Xite* are bound by OCT4, SOX2, REX1, KLF4 and c-MYC (Donohoe *et al.* 2009, Navarro *et al.* 2010). While pluripotency factor binding has a repressive effect on *Xist*, it has an activating role on *Tsix*, but it is unknown by which mechanism those factors regulate opposite transcriptional outcomes for *Xist* and *Tsix*. The reason why even combined deletion of both repressive *Xist intron 1* and *Tsix* is not sufficient to lead to full *Xist* upregulation (Minkovsky *et al.* 2013, Nesterova *et al.* 2011) can likely be explained with the low expression of *Xist* activators RNF12, *Ftx* and *Jpx* in pluripotent stem cells (Chureau *et al.* 2011, Jonkers *et al.* 2009,

Tian *et al.* 2010). While the mechanism for *Ftx* and *Jpx* regulation is unknown, the 5′ region of *Rnf12* is also bound by the pluripotency factors OCT4, SOX2, NANOG and PRDM14 (Navarro *et al.* 2011, Payer *et al.* 2013). Depletion of OCT4, NANOG or PRDM14 in ESCs leads to significant *Rnf12* upregulation, indicating that also *Rnf12*, like *Xist*, is directly repressed by pluripotency factors. PRDM14 represses *Rnf12* by recruitment of PRC2 (polycomb repressive complex 2) and the associated H3K27me3 mark to the *Rnf12* upstream region (Payer *et al.* 2013), which is consistent with PRDM14's action on its other repressed target genes (Chan *et al.* 2012, Yamaji *et al.* 2013). Like for *Xist intron 1*, deletional analysis of the pluripotency factor binding hubs along the *Xic* will be needed, to determine their functional significance. Additionally, the individual roles and functional interactions of the different pluripotency factors warrant further investigation.

In conclusion, the *Xic* consists of a large number of regulatory elements of predominantly non-coding RNA genes, whose purpose is to ensure the faithful and tight regulation of *Xist* and thereby XCI. The interplay with trans-acting autosomal factors, in particular pluripotency factors, connects the XCI and XCR process with the cellular differentiation state and allows for appropriate dosage compensation in the right developmental context. A recently discovered feedback of X-linked gene dosage on cell differentiation, in which female cells first need to undergo XCI in order to differentiate efficiently, is further evidence linking X-dosage regulation with pluripotency/differentiation (Schulz *et al.* 2014).

5.3 3D-Conformation of the X-Chromosome

The combination of newly emerging imaging-based and chromosome conformation capture (3C) technologies to study genomic interactions (Dekker *et al.* 2013, 2002, Pombo and Dillon 2015) allowed to gather information on the structure of the X-chromosome and its influence on gene regulation. Studies have focused both on local interactions impacting the genes at *Xic* but also on the conformation of the X-chromosome on a global scale, which influences *Xist*-RNA spreading (Figure 5.2(b)) and the kinetics of gene silencing during XCI. At the *Xic*, *Xist* and its regulators seem to fall into two discrete topologically associated domains (TAD) (Nora *et al.* 2012, Tsai *et al.* 2008). The first TAD encompasses *Xist* and its 5′-activators *Jpx*, *Ftx* and *Rnf12*, while the second TAD contains *Tsix* and its 5′ regulatory sequences including the *Tsix* activators *Xite* and *Tsx* and the putative *Tsix* regulator *Linx* (Figure 5.1). Within the *Xist* TAD, *Jpx*, *Ftx* and *Xist* show physical interactions (Nora *et al.* 2012, Tsai *et al.* 2008), potentially explaining the relative *cis*-preference, which *Jpx* and *Ftx* show in activating *Xist* expression (Barakat *et al.* 2014, Tian *et al.* 2010). On the other hand, *Tsix* and its enhancer *Xite* show interactions before and during XCI, which explains the activating role of *Xite* on *Tsix* expression (Ogawa and Lee 2003, Tsai *et al.* 2008). The TADs are separated by a boundary element which is bound by CTCF, and deletion of this boundary leads to ectopic interactions between the *Xist* and *Tsix* TADs and a failure of *Xist*-upregulation during differentiation (Nora *et al.* 2012, Spencer *et al.* 2011). Nevertheless, CTCF binds also to sites within TADs and deleting the boundary does not completely abrogate TAD separation, showing that additional factors play a role in keeping the *Xist* and *Tsix* chromatin hubs separated from each other. Although interactions within the TADs seem to fluctuate on a cell to cell level (Giorgetti *et al.* 2014), mapping TADs and their boundaries through chromatin interaction studies allowed to identify functional domains within the *Xic* and potential regulatory elements.

Regarding the general structure of the X-chromosome, chromosome conformation capture experiments have shown that the Xi is folded in a different fashion than the Xa or autosomes (Figure 5.2(c)) (Splinter *et al.* 2011). While genes on the Xa are engaged in multiple long-range interactions forming TADs, which are needed for efficient enhancer-promoter interactions and gene transcription, genes on the Xi do not form specific contacts with other regions but display a rather random organization. Genes escaping XCI, however, often colocalize with each other and are generally separated from the silent internal compartment of the Xi (Splinter *et al.* 2011). This is consistent with an earlier imaging-based study, which showed that X-linked genes get localized during XCI into the silencing territory occupied by *Xist* RNA, which is depleted of the transcriptional machinery, whereas escapee genes stay outside the silencing domain (Chaumeil *et al.* 2006). The specific structure of the Xi is dependent on *Xist* RNA and its role to repel cohesins from the Xi, as *Xist* RNA depletion leads to cohesin relocalization to the Xi and its refolding into a structure resembling the Xa (Figure 5.2(c)) (Minajigi *et al.* 2015, Splinter *et al.* 2011). In sum, the 3D-structure of the X-chromosome and *Xist* show functional intersections during XCI on multiple levels: Firstly, TADs are locally involved in regulation of the *Xist* and *Tsix* genes, which is important for the choice, which X-chromosome is to be inactivated: Secondly, the global conformation and gene density of the X-chromosome determines where *Xist* RNA can spread first; and finally, *Xist* RNA itself is important to form and maintain the silencing compartment of the Xi.

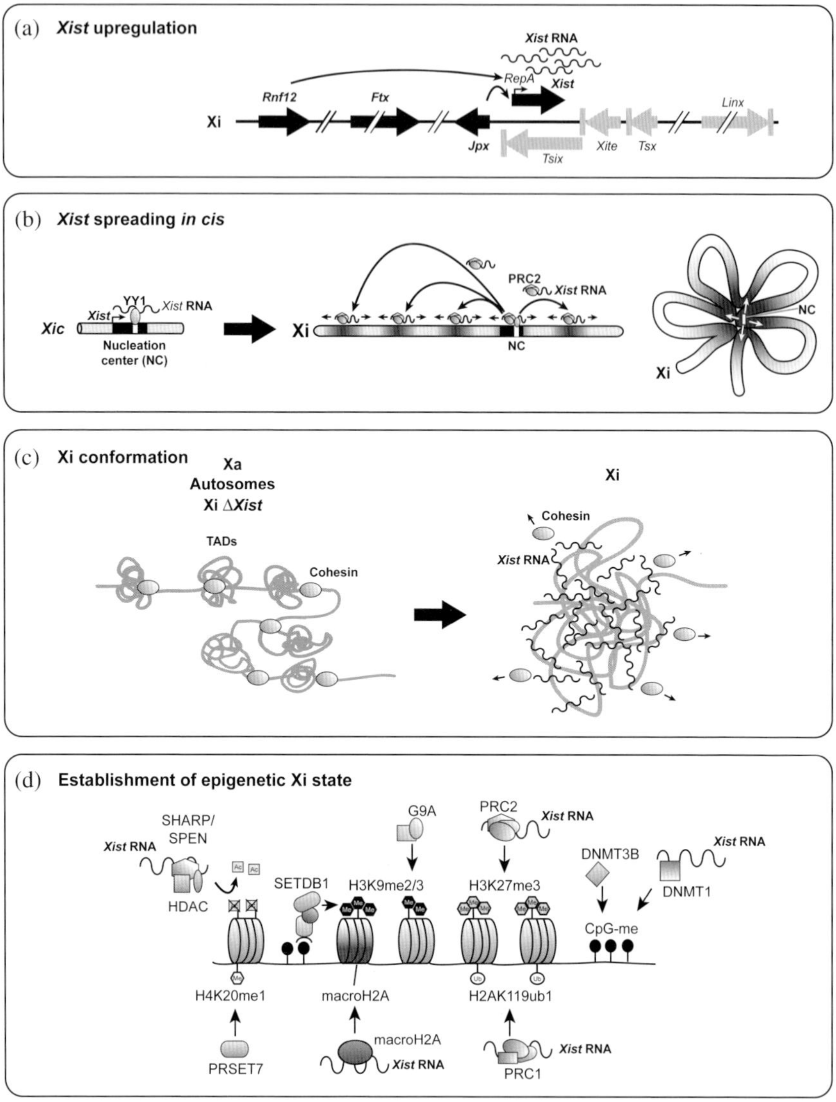
(a) Xist upregulation
Xist RNA
RepA
Xist
Rnf12
Ftx
Linx
Xi
Jpx
Xite
Tsx
Tsix
(b) Xist spreading in cis
PRC2
Xist RNA
YY1
Xist
Xist RNA
Xic
Nucleation center (NC)
Xi
NC
NC
Xi
(c) Xi conformation
Xa
Autosomes
Xi ΔXist
TADs
Cohesin
Xi
Cohesin
Xist RNA
(d) Establishment of epigenetic Xi state
SHARP/ SPEN
Xist RNA
HDAC
G9A
SETDB1
H3K9me2/3
PRC2
Xist RNA
H3K27me3
DNMT3B
Xist RNA
DNMT1
CpG-me
H4K20me1
PRSET7
macroH2A
macroH2A
Xist RNA
H2AK119ub1
Xist RNA
PRC1

Figure 5.2. Key events during XCI. (a) *Xist* upregulation is the initiator of XCI and is controlled by downregulation of *Xist* repressors like pluripotency factors and *Tsix* and upregulation of *Xist* activators like *Rnf12* and *Jpx*. For more details see Figure 5.1. (b) After *Xist* RNA becomes expressed, it is tethered *in cis* to the nucleation center (NC) within the *Xist* locus by YY1. This is followed by spreading of *Xist* and associated protein complexes like PRC2 to gene-rich regions across the Xi, from which it then spreads locally to the rest of the Xi. Spreading of *Xist* from the NC also takes advantage of the 3D-conformation of the X-chromosome and occurs initially to spatially closer regions on the future Xi (right). (c) The Xa has a similarly organized chromosome conformation like autosomes, which is divided into TADs, where spatial interactions occur (left). Cohesins play a role in organizing this structure. On the Xi, *Xist* RNA displaces cohesins from critical sites leading to a more disorganized chromosome structure without specific interactions (right). When *Xist* is deleted, the Xi reacquires an organized structure similar to the Xa. (d) As a consequence of *Xist* expression, the Xi acquires its distinctive chromatin state. *Xist* RNA thereby recruits some chromatin modifying complexes directly, like PRC2 (catalyzing H3K27me3), PRC1 (H2AK119ub1), SHARP/SPEN (part of HDAC — removes histone acetylations), DNMT1 (CpG DNA-methylation) and the histone variant macroH2A. Further histone modifiers and their associated marks are recruited not by *Xist* directly but by independent mechanisms (see text), like DNMT3B (CpG DNA-methylation), PRSET7 (H4K20me1) and H3K9-methylases G9A (H3K9me2) and SETDB1 (H3K9me3).

5.4 Coating of the Xi by *Xist* RNA

A long-standing question in the X-inactivation field has been the function of the long non-coding *Xist* RNA — how it is able to attach to and thereby "coat" the Xi and how this leads to recruitment of epigenetic silencing complexes and to X-linked gene silencing. Particularly challenging has it been to identify proteins, which directly interact with *Xist* RNA, but recently great progress has been made with the development of new biochemical methods.

A hallmark of *Xist* RNA is that it acts *in cis*, meaning that it only coats the X-chromosome, from which it is expressed but not the other X or autosomes (Clemson *et al.* 1996, Jonkers *et al.* 2008). Transgenic analysis revealed different functional domains of *Xist*, whereby the 5′ located Repeat A region is important for gene silencing while other repeat regions have roles in localization of *Xist* to the Xi (Wutz *et al.* 2002). Especially critical as an anchoring sequence is the Repeat C region on *Xist* RNA, as antisense locked nucleic acid (LNA) and peptide nucleic acid (PNA) antagomirs against Repeat C can displace *Xist* RNA from the Xi (Beletskii *et al.* 2001, Sarma *et al.* 2010). Repeat C is bound by the transcriptional regulator YY1, which recruits *Xist* RNA to its own chromosomal production site by binding to *Repeat F* on the Xi, a nucleation center within the *Xist* gene-locus (Figure 5.2(b)) (Jeon and Lee 2011). YY1 siRNA-knockdown or mutation of YY1 binding sites in the region abolishes *Xist* RNA-loading and chromosome coating, showing that YY1 acts as a bridging molecule by interacting both with *Xist* RNA and DNA at the same time. Interestingly, YY1 shows binding specificity for the Xi and does not bind to *Repeat F* on the Xa, ensuring that *Xist* RNA only attaches to the Xi, contributing to the *cis*-specificity of *Xist* RNA coating. How this allele specificity of YY1 binding is regulated will need to be investigated.

Another factor critical for anchoring *Xist* RNA to the Xi is the nuclear matrix protein hnRNP U/SAF-A (Hasegawa *et al.* 2010, Helbig and Fackelmayer 2003, Pullirsch *et al.* 2010), which directly interacts with *Xist* and which is important for *Xist*'s role in recruiting repressive chromatin marks and maintaining gene-silencing (Chu *et al.* 2015, Hasegawa *et al.* 2010, McHugh *et al.* 2015, Minajigi *et al.* 2015, Yamada *et al.* 2015). Interestingly, the gene-silencing domain of *Xist*, Repeat A, is not required for hnRNP U recruitment to the Xi, however, hnRNP U interacts with *Xist* RNA at *exon 1* and *7* and has a function in tethering *Xist* RNA to the Xi and the nuclear matrix and possibly for the structural organization of the Xi silencing compartment (Fackelmayer 2005, Hasegawa *et al.* 2010, Pullirsch *et al.* 2010, Smeets *et al.* 2014, Yamada *et al.* 2015). Besides with *Xist*, hNRNP U also interacts with another X-linked long noncoding RNA called *Firre*, which is required for chromosomal interactions and for localization of the Xi to the nucleolus during S-phase (Hacisuleyman *et al.* 2014, Yang *et al.* 2015). This perinucleolar localization of the Xi is necessary to replicate the Xi-specific heterochromatin configuration like the H3K27me3 mark, and either *Xist* or *Firre* depletion leads to dissociation of the Xi from the nucleolus and subsequently to a loss of H3K27me3 (Yang *et al.* 2015, Zhang *et al.* 2007). A similar role for Xi chromatin organization has been postulated for SATB1 and SATB2, which were shown to be involved in *Xist* localization and gene silencing during XCI in lymphoma cells and mouse embryonic stem cells (Agrelo *et al.* 2009). However, another knockout study could not detect abnormalities in female mouse development and XCI of *Satb1/Satb2* double-mutant mice (Nechanitzky *et al.* 2012). This does not exclude a role for these two proteins in the XCI process, but further studies will be needed to identify potential factors able to compensate for a loss of SATB1 and SATB2 (Wutz and Agrelo 2012).

After *Xist* RNA touches down on its nucleation site on the *Xist* DNA locus, it needs to spread across the X-chromosome in order to initiate chromosome-wide epigenetic changes and gene silencing (Figure 5.2(b)). While cytogenetic studies have shown that *Xist* RNA is not evenly distributed along the Xi (Clemson *et al.* 1996, Duthie *et al.* 1999), only the recent development of high resolution

RNA–DNA interaction detection methods provided insight into the detailed pattern how *Xist* RNA spreading occurs (Engreitz *et al.* 2013, Simon *et al.* 2013). In two similar studies, RNA Antisense Purification (RAP) (Engreitz *et al.* 2013) and Capture Hybridization Analysis of RNA Targets with deep sequencing (CHART-seq) (Simon *et al.* 2013) have been used to identify *Xist* RNA-interacting chromatin regions by pulling down *Xist* RNA with biotinylated antisense probes after crosslinking of cells. During initiation of XCI, *Xist* RNA spreads first to actively transcribed gene-rich regions (early domains) guided by the three-dimensional confirmation of the X-chromosome preferring regions in spatial proximity to the *Xist* locus, while later also filling in gene-poor, more distal regions (late domains) (Figure 5.2(b)). During maintenance of XCI, however, *Xist* spreads with equal kinetics to both early and late domains, which is likely needed to efficiently recoat the X-chromosome after mitosis, during which *Xist* RNA is less tightly associated with the Xi (Clemson *et al.* 1996, Jonkers *et al.* 2008).

5.5 The *Xist*-Interactome and Its Role in XCI

Similar methods as for isolating *Xist*-interacting chromatin have recently been deployed to identify *Xist* RNA-interacting proteins (Chu *et al.* 2015, McHugh *et al.* 2015, Minajigi *et al.* 2015). By capturing *Xist* RNA-protein complexes from either formaldehyde- or UV-crosslinked cells with complementary biotinylated oligos followed by mass spectrometry, comprehensive identification of RNA binding proteins by mass spectrometry (CHIRP-MS) (Chu *et al.* 2015), RNA antisense purification-mass spectrometry (RAP-MS) (McHugh *et al.* 2015) and identification of direct RNA interacting proteins (iDRiP) (Minajigi *et al.* 2015) were used to find 81, 10 or 80–250 *Xist*-interactors, respectively. In all studies, the previously known *Xist*-interactor hnRNP U/SAF-A (Hasegawa *et al.* 2010) was identified, confirming the validity of the methods. A further common found factor was the histone deacetylase (HDAC) complex subunit SHARP/SPEN, which interacts with the Repeat A domain of *Xist* and is required for the gene silencing function of *Xist* during XCI (Chu *et al.* 2015, McHugh *et al.* 2015, Minajigi *et al.* 2015, Moindrot *et al.* 2015, Monfort *et al.* 2015). The first and third study (Chu *et al.* 2015, Minajigi *et al.* 2015) also found other hnRNP proteins to be associated with *Xist*, like hnRNP M and hnRNP K, which has a function in gene silencing and is necessary for *Xist* to be able to attract polycomb-associated silencing marks to the Xi. Furthermore, these studies identified members of the polycomb repressive complex 1 (PRC1) (RNF2/RING1B, PCGF5, RYBP) to bind *Xist*, which is known to lead to the X-chromosome-wide deposition of the H2AK119ub1 histone mark during XCI (de Napoles *et al.* 2004, Fang *et al.* 2004, Plath *et al.* 2004, Schoeftner *et al.* 2006). The third study (Minajigi *et al.* 2015) additionally identified other repressive factors implicated in XCI such as RBBP4 + RBBP7 (PRC2), ATRX, macroH2A and SmcHD1, indicating that *Xist* orchestrates the recruitment of repressive complexes to the Xi. Another group of factors found by this study to interact with *Xist* were chromosome architecture proteins of the cohesin complex SMC1a, RAD21 and CTCF, which are repelled by *Xist* from the Xi, giving it its unique conformation devoid of TADs (Splinter *et al.* 2011) (Figure 5.2(c)). In conclusion, recent *Xist*-interactome studies (Chu *et al.* 2015, McHugh *et al.* 2015, Minajigi *et al.* 2015) have now identified *Xist*-binding factors with different functions for *Xist*-mediated gene silencing, chromatin modifications and chromosome conformation. Future studies will address further the mechanistic roles and functional interactions of the *Xist*-associated proteome at different stages of XCI.

5.6 Recruitment of Epigenetic Regulators to the Xi

A key function of *Xist* is the facilitation of the unique chromatin state of the Xi, both through direct and indirect recruitment mechanisms of epigenetic modifying complexes (Figure 5.2(d)). Surprisingly absent from the *Xist*-interactome studies mentioned above (Chu *et al.* 2015, McHugh *et al.* 2015, Minajigi *et al.* 2015) were the core members of the polycomb repressive complex 2 (PRC2) — EZH2, EED and SUZ12 — which lay down the X-chromosome-wide H3K27me3 mark, one of the earliest and most distinctive epigenetic features of the Xi (Erhardt *et al.* 2003, Plath *et al.* 2003, Silva *et al.* 2003). How coating of the Xi by *Xist* attracts PRC2 has been debated in the field and different, but not necessarily mutually exclusive models favoring either direct or indirect recruitment mechanisms have been put forward (Brockdorff 2013, Cifuentes-Rojas *et al.* 2014, Gendrel and Heard, 2014). There are multiple lines of evidence for a causal association between *Xist* RNA and PRC2. On one hand, *Xist* expression is both necessary and sufficient to recruit PRC2 to the Xi (Plath *et al.* 2003, Silva *et al.* 2003) and displacement of *Xist* RNA by LNA knock-off leads to simultaneous disappearance of PRC2 components from the Xi (Sarma *et al.* 2010). In addition, *Xist*- and PRC2-spreading patterns along the Xi occur with high spatiotemporal correlation (Engreitz *et al.* 2013, Pinter *et al.* 2012, Simon *et al.* 2013). Direct physical interaction between the *Xist* RNA Repeat A region and the PRC2 core subunits EZH2 and SUZ12 have been demonstrated both *in vivo*, *in vitro* and through computational prediction (Agostini *et al.* 2013, Cifuentes-Rojas *et al.* 2014, Davidovich *et al.* 2015, Maenner *et al.* 2010, Zhao *et al.* 2008). Repeat A is not only a part of *Xist* RNA but also forms an independent short transcript called *RepA*, which is required for the upregulation of *Xist* transcription during XCI (Hoki *et al.* 2009, Zhao *et al.* 2008). Repeat A forms two stem-loop structures, which are necessary for the interaction between *Xist* and the PRC2 complex (Agostini *et al.* 2013, Cifuentes-Rojas *et al.* 2014, Maenner *et al.* 2010) and this interaction is facilitated by the chromatin remodeling factor ATRX (Sarma *et al.* 2014). ATRX loads PRC2 onto *Xist* RNA by possibly modifying Repeat A structure and is necessary for establishment of the PRC2-dependent H3K27me3 mark on the Xi (Sarma *et al.* 2014). While PRC2 is being recruited by *Xist* RNA, it seems to be enzymatically inhibited by its interaction with *Xist* RNA and it needs the binding of the PRC2 cofactor JARID2 to release *Xist* RNA from PRC2 and to render the PRC2 complex active to lay down the H3K27me3 mark (Cifuentes-Rojas *et al.* 2014). In an alternative model, JARID2 acts as an intermediate recruitment factor for PRC2 to the Xi via the *Xist* Repeat B and F regions, which is supported by the observation that *Xist* RNA is also able to attract PRC2 in the absence of the Repeat A sequence (da Rocha *et al.* 2014, Plath *et al.* 2003), however less efficiently than in its presence (Jeon and Lee 2011, Kohlmaier *et al.* 2004). Another cofactor reported to be involved in PRC2 recruitment to both the Xi and autosomal target sites is the protein Polycomblike 2 (PCL2) (Casanova *et al.* 2011). Thus multiple, partially redundant mechanisms, dependent on separate *Xist* domains might be responsible for the recruitment of PRC2 to the Xi. Furthermore, a recent study using 3D-SIM super-resolution microscopy did not detect colocalization between *Xist* RNA and PRC2, suggesting that there is no physical interaction between them (Cerase *et al.* 2014). However, a second study employing STORM, another super-resolution microscopy method, explained this discrepancy through methodological reasons and still found statistically significant proximity between PRC2 and *Xist* signals in the Xi territory (Sunwoo *et al.* 2015). Thereby the authors favored a hit-and-run-model of recruitment of PRC2 by *Xist* to the Xi, in which relatively few (50–100) PRC2-*Xist* complexes at any given time are recruited to sites on the Xi, from which H3K27me3 methylation spreads outwards by PRC2, which becomes in the process dissociated from *Xist* RNA. Further investigations will be needed to clarify these issues and to settle the argument about the exact mechanisms of PRC2 recruitment during XCI.

The polycomb repressive 1 (PRC1) complex is responsible for monoubiquitylation of lysine 119 on histone H2A (H2AK119ub1), which is also a distinct silencing mark of the Xi (de Napoles *et al.* 2004, Fang *et al.* 2004, Plath *et al.* 2004). Like PRC2, PRC1 is recruited to the Xi in an *Xist*-dependent fashion early during XCI. Although PRC1 is attracted to many sites in the genome by recognizing H3K27me3 generated by PRC2, the X-chromosome-specific recruitment of PRC1 seems to occur independently of PRC2 (Schoeftner *et al.* 2006) and this pathway involves the non-canonical PRC1 subunit RYBP (Tavares *et al.* 2012). Like for PRC2, direct recruitment by *Xist* RNA has been proposed, as the PRC1 members RING1B, PCGF5 and RYBP have been identified as *Xist* RNA-interacting proteins (Chu *et al.* 2015, Minajigi *et al.* 2015). However also in this case, 3D-SIM super-resolution microscopy experiments could not detect a strong overlap of *Xist* and PRC1 staining (Cerase *et al.* 2014), suggesting a PRC1 recruitment mechanism without direct interaction with *Xist* RNA. However, like for PRC2, technical reasons could explain the observed distances (Sunwoo *et al.* 2015) and more studies will be needed to resolve the discrepancies between the results obtained from these new methods (between different super-resolution microscopy methods and biochemical RNA-interacting protein identification methods), before final conclusions can be drawn.

Functionally, neither PRC1 nor PRC2 is essential for initiation of gene silencing during XCI (Kalantry and Magnuson, 2006, Leeb and Wutz, 2007, Schoeftner *et al.* 2006), however overlapping roles cannot be excluded as they are recruited independently by *Xist* (Chu *et al.* 2015, Minajigi *et al.* 2015, Schoeftner *et al.* 2006). Nevertheless, PRC2 is required for faithful maintenance of X-linked gene silencing in the extraembryonic lineages (Kalantry *et al.* 2006, Wang *et al.* 2001), where fewer redundant silencing pathways exist than in the embryo, which could mask a potential repressive function of PRC2.

Another epigenetic mark prominently enriched on the Xi is H4K20me1, which is catalyzed by PRSET7 (Kohlmaier *et al.* 2004). The exact function for H4K20me1 during XCI is not yet clear, however it might be linked to the chromosome structure of the Xi, as male mutant mouse cells without H4K20me1 display an enlarged, decondensed Xa territory (Oda *et al.* 2009) and H4K20me1 influences chromosome compaction during dosage compensation in *C. elegans* (Lau and Csankovszki 2015). To which degree and in which aspect XCI in female mice might be perturbed in the absence of H4K20me1 remains to be investigated.

Further repressive epigenetic modifications associated with XCI are H3K9me2 and H3K9me3. H3K9me2 is present on the Xi after differentiation (Rougeulle *et al.* 2004) and is catalyzed by G9A, which is recruited by the chromodomain-containing repressor CDYL in a H3K27me3-dependent manner (Escamilla-Del-Arenal *et al.* 2013). CDYL-binding also relies on H3K9me2 itself and thereby provides a potential mechanism of propagation of the mark. The exact functional role for XCI, however, is not clear, as G9A knockout embryos appear to undergo and maintain XCI normally (Ohhata *et al.* 2004). H3K9me3 forms prominent domains especially on the human Xi (Chadwick and Willard, 2004), whereas in the mouse it is not as strongly enriched. Nevertheless, when SETDB1, which catalyzes the H3K9me3 mark, is knocked down in female mouse fibroblasts, it leads to upregulation of an

X-linked reporter gene, when also DNA-methyltransferases are inhibited at the same time (Minkovsky *et al.* 2014). Similar effects were seen when the SETDB1 complex partners ATF7IP or MBD1 are knocked down, which is interesting, as MBD1 binds specifically methylated DNA and ATF7IP is known to interact with the above mentioned CDYL, providing a potential link between DNA-methylation and H3K9 methylation on the Xi. Thus H3K9 methylation might be an additional silencing mark to stabilize maintenance of gene silencing on the Xi.

5.7 Epigenetic Gene Silencing and Its Maintenance

X-chromosome silencing is maintained by multiple layers of epigenetic regulation, safeguarding the silent state throughout the lifetime of the female (Figure 5.2(d)). Interference with individual epigenetic XCI regulators like *Xist*, DNA methylation or histone deacetylation does not result in full XCR due to synergism of the silencing pathways (Csankovszki *et al.* 2001, Minajigi *et al.* 2015). For example, after *Xist*-deletion in somatic cells, the subsequent epigenetic changes on the Xi, like loss of H3K27me3 and macroH2A and chromosomal refolding of the Xi are not sufficient to cause complete XCR (Brown and Willard 1994, Csankovszki *et al.* 1999, Minajigi *et al.* 2015, Splinter *et al.* 2011, Zhang *et al.* 2007). This is consistent with the fact that DNA-methylation remains on X-linked promoters keeping them transcriptionally inactive. Nevertheless, *Xist*-loss in somatic cells still results in stochastic reactivation of some X-linked genes (Zhang *et al.* 2007), which on the long term can lead to hematologic cancer in mice (Yildirim *et al.* 2013). Deletion of *Xist* however also affects X-chromosome replication timing and genetic stability, which might additionally contribute to the cancer-promoting effects of the *Xist*-deletion (Diaz Perez *et al.* 2006). Further evidence in favor of defective XCI-maintenance as being involved in cancer development comes from human cancers, where a loss of the Xi and gain of a second Xa either by chromosome segregation errors or by defective XCI maintenance has been commonly observed (Chaligné *et al.* 2015, Pageau *et al.* 2007). Although the mechanisms of Xi-loss and Xa-gain by chromosome segregation errors versus XCI-maintenance defects are different, the cancer-promoting end result of double dosage of X-linked genes is the same. The Xi furthermore tends to undergo hypermutations in human cancers due to its late replication in the cell cycle, providing another link between tumorigenesis and the inactivation state of the X-chromosome (Jäger *et al.* 2013).

DNA methylation of CpG islands at X-linked gene promoters is a late occurring event during XCI and is established by the *de novo* DNA methyltransferase DNMT3B (Gendrel *et al.* 2012). DNA methylation plays a role in repression of *Xist* on the Xa, but it does not seem to be essential for gene silencing during XCI

(Panning and Jaenisch 1996, Sado *et al.* 2004). However, the absence of DNA-methylation by deletion of the *de novo* DNA methyltransferase *Dnmt3b* (Gendrel *et al.* 2013) or the lack of its maintenance in mutants for the maintenance methyltransferase *Dnmt1* leads to sporadic reactivation of X-linked genes in embryonic and somatic cells (Csankovszki *et al.* 2001, Sado 2000). Only in the absence of DNA-methylation, *Xist*-deletion or depletion of other silencing factors leads to efficient reactivation of X-linked genes (Csankovszki *et al.* 2001, Minajigi *et al.* 2015), showing that DNA-methylation is an important factor to keep X-linked genes faithfully repressed during XCI-maintenance. In the placenta of *Dnmt1* knockout embryos on the other hand, imprinted X-inactivation is maintained normally, showing that DNA-methylation plays a role in maintaining embryonic but not as much in extraembryonic XCI (Sado 2000). Indeed, the lack of safeguarding of Xi-silencing by DNA-methylation might contribute to the relative instability of imprinted XCI in the placenta, where sporadic reactivation of X-linked genes frequently occurs (Corbel *et al.* 2013, Dubois *et al.* 2014, Hadjantonakis *et al.* 2001). In contrast to the Xi, the Xa is depleted of X-linked promoter-methylation but hypermethylated in gene bodies of X-linked genes (Hellman and Chess 2007), which either might have potential roles in suppressing repetitive elements or cryptic promoters or alternatively could be involved in the upregulation of genes on the Xa to compensate dosage with autosomes (Deng *et al.* 2014). In conclusion, DNA-methylation plays diverse important roles during XCI: ensuring the faithful maintenance of gene silencing on the Xi and repression of *Xist* and gene body methylation on the Xa.

A factor that has been studied for its role in attracting DNA-methylation during maintenance of XCI is the epigenetic regulator SmcHD1 (structural maintenance of chromosomes hinge domain containing 1), which is recruited to the Xi late during the XCI process (Blewitt *et al.* 2008, Gendrel *et al.* 2012). Female homozygous *SmcHD1*-mutant mouse embryos die due to a failure in XCI-maintenance both in the extraembryonic as well as embryonic compartments (Blewitt *et al.* 2008). SmcHD1 interacts with *Xist* RNA (Minajigi *et al.* 2015) and is involved in attracting DNA-methylation to CpG islands on the Xi by DNMT3b (Blewitt *et al.* 2008, Gendrel *et al.* 2012). Expression profiling has shown that around 10% of X-linked genes are de-repressed in *SmcHD1*-mutant embryos, however, they appear to be silenced independently of CpG-methylation as not all of them become fully reactivated in *Dnmt3b*-mutant embryos (Gendrel *et al.* 2013). Therefore, X-linked gene silencing by SmcHD1 occurs likely by a separate, CpG-methylation-independent mechanism, possibly related to SmcHD1's function in compaction of the human Xi (Nozawa *et al.* 2013). Further studies will be needed to identify the way in which SmcHD1 exerts its gene silencing function on the Xi.

Another late occurring event during XCI besides DNA-methylation is the incorporation of the histone variant macroH2A into the Xi (Costanzi and Pehrson 1998, Mermoud *et al.* 1999). MacroH2A accumulation on the Xi is *Xist*-dependent also during XCI-maintenance (Csankovszki *et al.* 1999) and might be therefore recruited by *Xist* RNA by direct interaction (Minajigi *et al.* 2015). MacroH2A is not essential for XCI as female knockout mice are viable (Changolkar *et al.* 2007) and its disappearance from the Xi after *Xist*-depletion does not lead to X-linked gene-reactivation (Csankovszki *et al.* 1999). However, macroH2A is one of the safeguards of XCI-maintenance, as macroH2A knockdown in combination with inhibition of DNA-methylation and histone deacetylases leads to an increase in reactivation of an X-linked GFP transgene (XGFP) (Hadjantonakis *et al.* 2001, Hernández-Muñoz *et al.* 2005). Furthermore, nuclear transfer from mouse epiblast stem cells (EpiSCs) to frog oocytes resulted in reactivation of the XGFP transgene, while XGFP was not reactivated when mouse somatic cell nuclei were transferred (Pasque *et al.* 2011). The difference regarding the Xi is its coating with macroH2A in somatic cells but not in EpiSCs, suggesting that macroH2A safeguards from XGFP-reactivation during nuclear reprogramming. MacroH2A is therefore likely one of the many epigenetic layers protecting the Xi from spontaneous reactivation during XCI-maintenance.

Overall, a number of factors have now been identified through different screening approaches, whose depletion by RNAi knockdown or pharmacological inhibition lead to defects in gene silencing during XCI in ESCs or to partial reactivation of X-linked genes in female somatic cells, indicating that they might play roles in gene silencing or maintenance of XCI (Bhatnagar *et al.* 2014, Chan *et al.* 2011, Minajigi *et al.* 2015, Moindrot *et al.* 2015, Monfort *et al.* 2015). While some of these factors have been also found to interact with *Xist* RNA and rely on continuous *Xist* expression to remain localized to the Xi or are involved in *Xist*-expression/localization, others seem to rather contribute to *Xist*-independent maintenance mechanisms of XCI. Reported gene silencing and maintenance factors include as diverse proteins as the maintenance DNA-methyltransferase DNMT1, topoisomerases TOP1 and TOP2a/b, aurora kinases AURKA + B, 3-phosphoinositide dependent protein kinase 1 (PDPK1), origin recognition complex 2 (ORC2), the HDAC interactors SHARP/SPEN and RBM15, the m6A RNA methyltransferase subunit WTAP and heterochromatin protein 1 alpha (HP1α) (Bhatnagar *et al.* 2014, Chan *et al.* 2011, Chu *et al.* 2015, McHugh *et al.* 2015, Minajigi *et al.* 2015, Moindrot *et al.* 2015, Monfort *et al.* 2015). To which degree and by which mechanisms these and other factors contribute to silencing establishment and maintenance, how specific they are for Xi-silencing and how the different silencing pathways are interconnected or independent still needs to be examined.

5.8 The X-Inactivation + X-Reactivation Cycle During Embryonic Development and During Germ Cell Differentiation

XCI and XCR are events closely linked to decision points during mouse development related to pluripotency, differentiation and the germ cell lineage (Figure 5.3). XCI and XCR occur thereby several times in an alternating fashion. XCI exists in two different forms — imprinted or random — and XCR is needed to switch

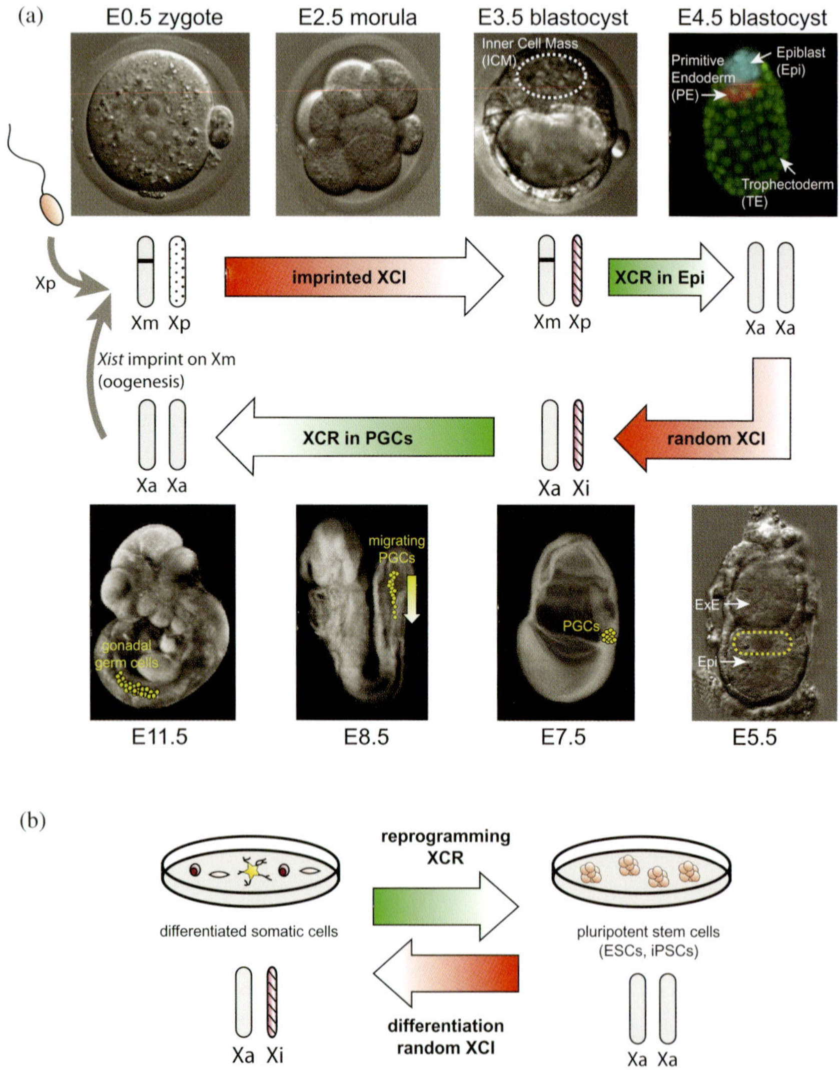

Figure 5.3. The X-inactivation (XCI) and X-reactivation (XCR) cycle during female mouse development and in *in vitro* model systems. (a) In the zygote both X-chromosomes are active, but there is a repressive *Xist* imprint on the Xm (black line) and an unknown paternal imprint on repeat elements of the Xp (dots). During preimplantation development, *Xist* becomes expressed on the Xp, which in turn becomes inactivated (pink). While imprinted XCI is maintained in the extraembryonic primitive endoderm (red) and trophectoderm (green) lineages, XCR in the epiblast (Cyan) of the E4.5 blastocyst leads to reactivation of the Xp. After implantation, the epiblast differentiates and all embryonic cells undergo random XCI including the proximal epiblast (Epi) from where PGCs delineate (yellow dotted area), adjacent to the extraembryonic ectoderm (ExE). During migration, PGCs gradually undergo XCR, which is completed after reaching the embryonic gonads. During oogenesis, finally the repressive *Xist* imprint is re-established, which will keep *Xist* repressed during imprinted XCI. (b) *In vitro*, XCI occurs during cell differentiation while XCR is linked to the pluripotent stem cell state in ESCs and iPSCs. Some images have been reproduced from the EMAP eMouse Atlas Project (http://www.emouseatlas.org) under a Creative Commons Attribution License (Richardson *et al.* 2014).

between these modes. In imprinted XCI, which occurs in the mouse in early embryos and the placenta, always the paternally inherited X-chromosome (Xp) is inactivated. In random XCI, which occurs in postimplantation embryos, either the Xp or the maternal X-chromosome (Xm) is inactivated and then clonally inherited during subsequent cell divisions into the adult. For X-linked mutations this random mode has the consequence that about half the cells of the female carry the mutation on the Xi and the wild-type copy on the Xa, meaning that they express the wild-type allele in those cells, making them phenotypically normal. Therefore females with random XCI have an advantage over males or females with imprinted XCI (like marsupials), where in all cells the same X is active and no compensation for a mutation on the Xa can occur. Thus X-linked diseases, such as hemophilia and color blindness, have frequently a milder phenotype in females than in males. As random XCI occurs also during ESC differentiation, most findings mentioned in the previous chapters are related to random XCI, and below I therefore focus mainly on imprinted XCI and the XCR process.

5.9 Imprinted XCI

The first instance of XCI happens during preimplantation development in mice, when specifically the paternally inherited Xp is inactivated in female embryos (Figure 5.3(a)) (Huynh and Lee 2003, Okamoto *et al.* 2004). This so-called imprinted XCI occurs in a stepwise fashion, in which first repeat sequences are silenced as early as the two-cell stage, followed by progressive accumulation of repressive chromatin marks and silencing of X-linked genes, which is dependent on *Xist* (Namekawa *et al.* 2010, Okamoto *et al.* 2004, Patrat *et al.* 2009). After embryo implantation, imprinted XCI is maintained in mice in extraembryonic tissues, which is critical for survival, as *Xist*- and *Eed*-mutant female embryos, which have defective imprinted XCI, die during early postimplantation stages (Marahrens *et al.* 1997, Wang *et al.* 2001). Despite its importance, imprinted XCI in the placenta is less stringently maintained than random XCI in the embryo, due to a less prominent involvement of DNA-methylation as an additional silencing safeguard (Sado 2000). Especially in trophoblast giant cells and trophoblast stem cells, X-linked genes can get sporadically reactivated (Corbel *et al.* 2013, Dubois *et al.* 2014, Hadjantonakis *et al.* 2001).

Early repeat silencing on the paternal Xp of preimplantation embryos does not require *Xist* and might be rooted in meiotic sex chromosome silencing (MSCI) during spermatogenesis (Namekawa *et al.* 2010). This early inactivation of non-genic sequences might contribute to formation of a silencing compartment, which enables efficient silencing of X-linked genes later on (Chaumeil *et al.* 2006, Namekawa *et al.* 2010). In one study, silencing of some X-linked genes during mouse preimplantation development in the absence of *Xist* by an unknown mechanism has been reported (Kalantry *et al.* 2009), however, stable silencing of X-linked genes generally requires *Xist*-RNA during imprinted XCI (Namekawa *et al.* 2010, Okamoto *et al.* 2005). While transmission of the Xp through the paternal germline predisposes repeats to be silenced, a maternal repressive imprint is established on the *Xist* gene on the Xm, to avoid its silencing during imprinted XCI. Nuclear transfer experiments have shown that this maternal imprint on *Xist* is established during oocyte maturation (Tada *et al.* 2000), as transfer of nuclei from other cells than fully grown oocytes into enucleated eggs led to upregulation of *Xist* from the transferred nuclei during preimplantation development (Oikawa *et al.* 2014). In fact, ectopic expression of *Xist* from the previously active X is one of the main causes for lethality in both male and female cloned embryos and can be rescued by using male or female donor cells with an *Xist*-mutation on the Xa (Inoue *et al.* 2010) or by *Xist* RNAi-knockdown in male embryos (Matoba *et al.* 2011). The molecular nature of the repressive maternal imprint on *Xist* has long been elusive. In contrast to autosomal imprints, it may be less dependent on DNA methylation, as even maternal double-mutant oocytes for the *de novo* DNA methyltransferases *Dnmt3a* and *Dnmt3b* produce embryos with normal *Xist* imprinting (Chiba *et al.* 2008). Nevertheless, female mouse embryos depleted for the DNA methyltransferase DNMT1o, which is normally maternally inherited and plays a role in maintaining DNA-methylation marks during preimplantation development, display defects in imprinted XCI in their placentae, indicating that some role of DNA-methylation also might exist for imprinted XCI (McGraw *et al.* 2013). This is consistent with differentially methylated domains being found in sperm and oocytes at the *Tsix* and *Xite* loci (Boumil *et al.* 2006), especially as *Tsix* is critical for suppressing *Xist* on the Xm during imprinted XCI in the extraembryonic lineages (Lee 2000, Sado *et al.* 2001). However, during preimplantation development, *Tsix* is not required to repress *Xist* on the Xm (Maclary *et al.* 2014, Payer *et al.* 2013) and a recent study uncovered that H3K9me3 is a mark enriched on the *Xist* promoter in early embryos, which is at least partially responsible for silencing of *Xist* on the Xm by making *Xist* non-responsive to its activator RNF12 (Fukuda *et al.* 2014). Further studies will be needed to fully understand the imprinting mechanisms of the X-chromosome and to which extent the paternal and maternal germlines contribute to its establishment.

5.10 XCR

In order to switch from imprinted XCI in preimplantation mouse embryos to random XCI, which occurs in the early postimplantation epiblast, the paternal Xi needs to be reactivated (Figure 5.3(a)) (Huynh and Lee 2003). This occurs in the inner cell mass of the mouse blastocyst, specifically in the epiblast lineage, which gives rise to the embryonic cell lineages, while imprinted XCI is maintained in the two extraembryonic lineages, primitive endoderm (PE) and trophectoderm (TE) (Mak *et al.* 2004, Okamoto *et al.* 2004). XCR takes place during blastocyst maturation between embryonic day (E)3.5 and E4.5 and is characterized by reversing the hallmarks of XCI: downregulation of *Xist*, removal of characteristic Xi chromatin marks like H3K27me3 and reactivation of X-linked genes (Mak *et al.* 2004, Okamoto *et al.* 2004). Surprisingly, gene-reactivation has been observed already before epigenetic changes during XCR became visible by microscopy (Williams *et al.* 2011). However, the specific chromatin configuration at promoters of reactivated genes is unknown, and it is possible that *Xist* and repressive chromatin marks initially disappear locally before chromosome-wide removal becomes apparent through imaging-based methods. It will be important to establish the exact sequence and kinetics of XCR steps in the blastocyst in order to determine the mechanism of reactivation of the imprinted Xp. Only few factors are known so far to play a role in XCR in blastocysts. One of them is the pluripotency factor and germline determinant PRDM14, which is expressed in the inner cell mass during XCR (Yamaji *et al.* 2008) and whose deletion causes a XCR defect as indicated by a failure to remove the chromosome-wide H3K27me3 mark from the Xp (Payer *et al.* 2013). The likely mechanisms is repression of *Xist* through PRDM14-binding to *Xist* intron 1 and to the promoter of the *Xist* activator *Rnf12*, to which it recruits the PRC2 complex and its associated H3K27me3 mark. Indeed, *Rnf12* gets downregulated in the ICM during XCR, which could explain the simultaneous *Xist* downregulation (Shin *et al.* 2014), which is further supported by the observation that *Rnf12*-mutant and *Rnf12*-knockdown embryos fail in *Xist* expression during preimplantation stages (Fukuda *et al.* 2014, Shin *et al.* 2010). PRDM14 is likely not the only pluripotency factor involved in XCR, as many bind at regulatory regions at the *Xic* and are expressed in the epiblast (Figure 5.1). NANOG, for example, which is also a repressor of *Xist* and *Rnf12* in ESCs (Navarro *et al.* 2008, 2011), switches from monoallelic to biallelelic expression in the epiblast (Miyanari and Torres-Padilla 2012) and is essential for XCR in the ICM (Silva *et al.* 2009). However it is not clear, to which degree this effect might be direct or indirect, as NANOG-null ICM cells do not form a correctly specified epiblast, which is the cell lineage, in which XCR occurs. Another gene involved in XCR in the blastocyst is *Tsix*. *Tsix* becomes biallelically expressed in the blastocyst ICM around the time of XCR (Debrand *et al.* 1999, Lee 2000) and induced expression of *Tsix* at earlier preimplantation stages is sufficient to downregulate *Xist* (Ohhata *et al.* 2011). Indeed, *Tsix*-mutant embryos display an early defect in XCR in blastocysts, which however is compensated later on, suggesting that *Tsix* plays a facilitating role to ensure correct timing of XCR but is not essential for the process (Maclary *et al.* 2014, Payer *et al.* 2013). All XCR-regulators in blastocysts studied so far have their function in *Xist* downregulation. Therefore other factors still need to be identified, which are responsible for the epigenetic changes and the gene reactivation process associated with XCR.

After XCR in the blastocyst, the epiblast undergoes differentiation, which induces random XCI during early postimplantation development (Figure 5.3(a)). This is maintained lifelong in all somatic cells and their descendants, but not in the germ cell lineage, where a second round of XCR takes place (Monk and McLaren 1981). XCR in the female germline is necessary to avoid inheritance of an inactive Xm, which would cause functional nullisomy for X-linked genes in the offspring, as the Xm is the only X-chromosome in male and the Xp is inactivated in female embryos. Mechanistically, XCR in germ cells might be different from XCR of the imprinted Xp in the blastocyst, which happens very rapidly within a day and without cell division in the epiblast (Mak *et al.* 2004, Okamoto *et al.* 2004, Payer *et al.* 2013, Williams *et al.* 2011). In Primordial germ cells (PGCs), which delineate from the early mouse proximal postimplantation epiblast around E6.5 (Ohinata *et al.* 2005), XCR is a much slower process, which takes several days and involves multiple cell divisions (Chuva de Sousa Lopes *et al.* 2008, Sugimoto and Abe 2007). PGCs migrate from their specification site through the hindgut towards the genital ridges (embryonic gonads) and while XCR initiates shortly after PGC specification, it is only completed in the gonads (Figure 5.3(a)). XCR in PGCs thereby begins with *Xist*-downregulation and loss of the H3K27me3 mark after which X-linked genes then get gradually reactivated. The efficient completion of XCR in PGCs seems to depend on a so far unknown signal from gonadal somatic cells, as coculture experiments have demonstrated (Chuva de Sousa Lopes *et al.* 2008). A possible reason for the difference in reactivation kinetics between the imprinted XCI in the blastocyst versus random XCI in PGCs might be that random XCI is more stringently maintained than imprinted XCI, as indicated by the leakiness of imprinted XCI in extraembryonic cells (Corbel *et al.* 2013, Dubois *et al.* 2014, Hadjantonakis *et al.* 2001). Also DNA-methylation marks on X-linked gene promoters need to be erased in PGCs during XCR, which might not be as important during XCR in the blastocyst. Indeed, the slow XCR-kinetics in PGCs mirrors the slow passive demethylation in migrating PGCs and the XCR completion and DNA-demethylation on the X-chromosome also coincides with the erasure of autosomal imprints when the PGCs enter the gonads (Seisenberger *et al.* 2012). The mechanisms of XCR *in vivo* and in particular in PGCs are still largely unknown and warrant further exploration.

A promising avenue to study XCR mechanisms are newly developed *in vitro* systems, which allow easy access to large number of cells for biochemical assays when compared to the low numbers in which XCR occurs in the embryo. Like the bulk of our knowledge about XCI has been derived from the ESC-differentiation system, in a similar fashion the dedifferentiation of somatic cells to pluripotent cells can be used to dissect the mechanisms of XCR (Figure 5.3(b)). The first such system described was the fusion of somatic cells with ESCs, which leads to the reactivation of the Xi in the somatic cell nucleus (Tada *et al.* 2001). These experiments have revealed that the factors required for XCR are not limited to female but also present in male ESCs, as their fusion with female somatic cells also leads to XCR. Another system to study XCR *in vitro* is the reprograming of somatic cells to induced pluripotent stem cells (iPSCs) through defined factors (Maherali *et al.* 2007). XCR during iPSC reprograming is a relatively late event and concurs approximately with the upregulation of endogenous pluripotency genes like *Nanog*. Careful study of the sequence of events has shown that downregulation of *Xist* and removal of DNA-methylation are critical for XCR (Pasque *et al.* 2014), which is consistent with our knowledge on maintenance mechanisms of XCI in somatic cells (Csankovszki *et al.* 2001). On the other hand, *Tsix* and *Xist intron 1* are dispensable for XCR to occur in iPSCs (Minkovsky *et al.* 2013, Pasque *et al.* 2014, Payer *et al.* 2013), which shows that compensatory pluripotency factor binding sites across the *Xic* and/or the downregulation of *Xist*-activators like *Rnf12* or *Jpx* might be sufficient to repress *Xist* during XCR in iPSCs. Indeed, the pluripotency factors NANOG and PRDM14, which repress *Rnf12* in pluripotent cells are required for efficient downregulation of *Xist* during XCR in iPSCs (Pasque *et al.* 2014, Payer *et al.* 2013). Still unclear is what mediates the epigenetic changes during XCR like removal of H3K27me3, macroH2A and DNA-methylation. To which degree this is due the lack of *Xist*-mediated recruitment after *Xist*-downregulation, or if active enzymatic processes are at play, will need to be determined. Active DNA-demethylation by TET1 and TET2 DNA-demethylases seems to be dispensable for X-linked gene reactivation and therefore other unknown DNA-demethylation mechanisms are involved (Pasque *et al.* 2014). Finally, also an *in vitro* system for studying XCR in the female germ cell lineage has been described, where ESCs are first differentiated towards an epiblast-like fate, whereafter PGC development is induced, mimicking the steps of PGC development *in vivo* (Hayashi *et al.* 2012). Like in embryos (Chuva de Sousa Lopes *et al.* 2008), efficient XCR only occurs after PGCs are aggregated with gonadal somatic cells, providing further evidence that complete XCR in germ cells requires a signal from the embryonic gonad. Now with several *in vitro* systems available, it will be possible to uncover the exact mechanisms of XCR.

5.11 The Human Perspective — XCI During Embryonic Development and in Human Pluripotent Stem Cells

While most of our knowledge on XCI and XCR in mammals stems from the mouse model system, it has become clear that not all findings can be extrapolated right away to humans. Although *XIST* remains the critical player in XCI also in humans (Brown *et al.* 1991), aspects of its expression pattern and transcriptional regulation are different from mice. For example, it is not clear, which role *TSIX* plays in regulation of the human *XIST* gene. In contrast to mice, where *Tsix* fully overlaps with the *Xist* transcriptional unit extending beyond the *Xist* promoter (Lee *et al.* 1999), human *TSIX* transcription stops within the *XIST* gene and does not reach the promoter region (Chow *et al.* 2003, Migeon *et al.* 2002). Furthermore, human *TSIX* seems to be coexpressed with *XIST* in several cell types and is missing a regulatory CpG island, which is critical for correct *Tsix* expression in mice (Chang and Brown 2010, Chow *et al.* 2003, Lee and Lu 1999, Migeon *et al.* 2002). Nevertheless, this does not exclude a different role or mechanism of action for human *TSIX* and the lack of functional studies and of data from the developmental time window, when XCI occurs in the human system makes it so far difficult to come to firm conclusions. Deletional analysis of human *XIC* elements, including human *TSIX*, will be needed to elucidate the regulatory mechanisms of human *XIST* and thereby human XCI.

Another aspect of species differences relates to the developmental control of XCI. In contrast to the mouse embryo, where a period of imprinted XCI is followed by XCR and then by random XCI (Figure 5.3(a)), the situation during human early development appears not to be as clear-cut. In one study, monoallelic *XIST* RNA enrichment in human preimplantation embryos has been detected in a similar fashion as in mouse embryos, beginning around the time of zygotic genome activation and maintained into the blastocyst stage (van den Berg *et al.* 2009). Also, the typical Xi-foci of H3K27me3 and macroH2A silencing marks were observed in blastocysts, while the *XIST* RNA signal overlapped with a silent X-linked gene allele and depleted areas of Cot1-signal indicating repression of X-linked repeats. These observations therefore suggested that human embryos display XCI in a similar fashion to mice, however it was not tested if this XCI was imprinted or random and if XCR occurred in the ICM of the blastocysts. A more recent study however, came to very different conclusions, that human *XIST* clouds were detected predominantly biallelically on both X-chromosomes in female and monoallelically in male preimplantation embryos (Okamoto *et al.* 2011). In addition, no accumulation of H3K27me3 or silencing of X-linked genes was seen, suggesting that *XIST* expression in human embryos is not imprinted and is not sufficient to induce XCI during early developmental stages. Support for this conclusion comes from a study on rhesus monkey preimplantation embryos, where methylation analysis and allele specific expression profiling by RT-PCR showed also biallelic expression of both *XIST* and X-linked genes (Tachibana *et al.* 2012). This suggested a lack of imprinted XCI, albeit no RNA-FISH was done to confirm this. It is not clear at the moment, if potential differences in embryo culture or RNA-FISH conditions/sensitivity are the reason for the contradicting results on human preimplantation XCI (Okamoto *et al.* 2011, van den Berg *et al.* 2009). As these studies show both evidence for and against a mouse-like XCI status in human early embryos, independent validation will be needed, to clarify, if imprinted XCI exists in early human embryos or not. Also regarding XCI in the human placenta, conflicting data exist with multiple studies being either in favor of imprinted or random XCI. While early studies focusing on skewing of few X-linked genes observed an imprinted XCI mode in the human placenta (Goto *et al.* 1997, Harrison, 1989), several more recent and comprehensive studies favored a random mechanism of XCI (Looijenga *et al.* 1999, Moreira de Mello *et al.* 2010, Peñaherrera *et al.* 2012). Overall, the combined evidence from studies on human and monkey embryos and human placentae speak against imprinted XCI in human development, which could be partially explained by the different structure of the *TSIX* gene in humans, which is a critical regulator of imprinted XCI in the mouse. Nevertheless, the results in human are not as clear-cut as in the mouse and thereby more work on human XCI models will be important in the future.

Until recently, human XCI research was mainly confined to *in vivo* studies, which are hampered due to sparse accessibility and was limited to descriptive, non-functional studies. In contrast to the mouse, where ESC differentiation and iPSC reprogramming provided powerful models both for XCI and XCR (Figure 5.3(b)), human pluripotent stem cells displayed complex XCI properties, limiting their application to recapitulate the XCI and XCR process *in vitro*. While in mouse female ESCs have two active X-chromosomes, of which one undergoes random XCI during differentiation, human pluripotent stem cells display a variety of X-inactivation states (reviewed in (Lessing *et al.* 2013)). Female human ESCs and iPSCs can be divided into three different classes (Silva *et al.* 2008): Class I: Pre-X-inactivation state similar to mouse ESCs, with two Xas — undergo XCI during differentiation; Class II: XCI-state with monoallelic *XIST* expression and gene silencing on the Xi — XCI is maintained during differentiation; Class III: Eroding XCI-state, in which previous *XIST* expression has been lost, but X-linked genes are partially still inactive on the Xi — *XIST* is not upregulated after differentiation. While Class I cells would be most useful to study human XCI and XCR *in vitro*, Class II and Class III cells are most frequently obtained in human pluripotent stem cell cultures (Lessing *et al.* 2013). Class III cells show progressive loss (erosion) of X-inactivation, likely due to their lack of *XIST* expression and associated silencing marks like H3K27me3 on the Xi (Mekhoubad *et al.* 2012). Gene-reactivation by erosion occurs specifically in H3K27me3 domains and is accompanied by upregulation of expression of *XACT*, a human-specific X-linked noncoding RNA, which is expressed on the Xa and which coats the Xa in human pluripotent stem cells (Vallot *et al.* 2013, 2015). It is an intriguing idea that *XACT* RNA might recruit activating epigenetic modifiers to the Xa in a similar fashion as *XIST* RNA recruits repressors to the Xi, however, if and how *XACT* is actively involved in the gene reactivation process on the Xa will need to be tested in *XACT*-KO cells. The epigenetic instability of Class III cells is associated with poor differentiation potential and cancer-like gene expression profiles, making them unsuitable for therapeutic approaches (Anguera *et al.* 2012). Therefore the X-inactivation status could be used, at least in female cells, as a diagnostic mark for human pluripotent stem cell quality. Further investigations into the mechanistic links between XCI and pluripotency will improve our assessment options, which cell lines should or should not be used for therapy and disease modeling. Class I cells, which represents the mouse ESC-like state with two Xa's, have been difficult to obtain and maintain in culture. Derivation of human ESCs under low physiological oxygen concentration improved the frequency of obtaining Class I cells (Lengner *et al.* 2010), but during prolonged culture or freeze-thaw cycles, also these cells would often lose their properties and become Class II or Class III cells

(Lessing *et al.* 2013). Nevertheless, recently developed culture conditions for naïve human pluripotent stem cell derivation seem to overcome this problem and result in predominantly Class I human ESCs and iPSCs with robust XCR (Barakat *et al.* 2015, Gafni *et al.* 2013, Takashima *et al.* 2014). These cells also show superior pluripotency qualities like contribution to mouse–human chimeric embryos, when compared to human pluripotent stem cells grown in conventional conditions (Gafni *et al.* 2013, Takashima *et al.* 2014). Conventionally grown cells represent a more primed pluripotent state, similar to mouse epiblast stem cells, which also show XCI (Nichols and Smith 2009). A different protocol to generate naïve cells did not result in XCR, showing that not all naïve cell derivation systems create cells of equal mouse-like XCR status (Theunissen *et al.* 2014). The naïve human pluripotent stem cells, which do undergo XCR, can be used to study human XCR mechanisms *in vitro*. As these cells undergo XCI when differentiated, they represent an ideal model system to study also human XCI. In addition, these cells have been successfully used to generate human primordial germ cell like cells (PGCLCs) *in vitro* (Irie *et al.* 2014), adding the capability to also study human XCR in the germ cell lineage, which recently has been demonstrated to occur in early human germ cells *in vivo* (Guo *et al.* 2015, Tang *et al.* 2015). Overall, naïve pluripotent stem cells have the capacity to provide the same toolset to study human XCI and XCR *in vitro*, which has been previously successfully applied in the mouse (Figure 5.3(b)), providing great promise to advance our knowledge on human XCI and XCR mechanisms in the near future. Furthermore, as the XCR-status correlates with the quality of human pluripotent stem cells (Class I high quality, Class III low quality), advancements in the understanding of the link between XCI/XCR and pluripotency will help to ensure better quality control of these cells, making them potentially more efficient and safer in clinical applications.

5.12 Concluding Remarks

The X-inactivation field has come a long way since Mary Lyon founded it more than 50 years ago. Key advances have been made by the discovery of *Xist* and its control elements at the X-inactivation center, thereby defining the *cis*-regulators of XCI and its trans-acting binding partners. For a long time it was unclear how *Xist* RNA mediates XCI, but the recent discovery of *Xist*-interacting factors has been a major leap forward. *Xist* RNA coats the X-chromosome, influences its 3D shape and attracts chromatin modifiers, which lock-in the silencing of X-linked genes. The whole process is tightly developmentally regulated and is reversed in pluripotent cells and the germ cell lineage by XCR. While the mouse model system has served well in dissecting the different XCI and XCR steps, very recently we also begin to slowly understand how it works in humans, although a lot more work needs to be done, as the level of knowledge is nowhere near to the level of mouse.

Now it will be important to fill in the gaps and to solve the remaining questions in the X-inactivation puzzle. These concern several areas of dispute, where different studies came to different conclusions. For example, the exact recruitment mechanisms and functional significance of the PRC2 complex to the Xi is such a topic. Several potentially important elements on the X-chromosome like *Ftx*, *Linx*, *XACT* and pluripotency-binding hubs still await deletional analysis in female ESCs to reveal their functional significance. The complexity of partially redundant epigenetic silencing marks still needs to be understood, to know, what specific role they play and how they are interconnected. Still relatively unresolved is the mechanism of XCR in the blastocysts, the germ line and during *in vitro* reprograming. Which are the critical players for the epigenetic remodeling of the X-chromosome and for gene reactivation, and are these mechanisms conserved or not in the different biological contexts? Also the role of the 3D-structure of the X-chromosome has only begun to be uncovered and needs more investigations regarding its functional significance for XCI and XCR. Finally, we still need to learn a lot more about the human system, which now became possible with the development of human *in vitro* XCI and XCR models. In summary, despite of the recent gain of knowledge on the XCI and XCR process, many new and old questions remain open to be answered.

Acknowledgments

I would like to thank the members of my lab, S.F Generoso, M. Guirola, and A. Tarruell for critical discussions. I acknowledge support of the Spanish Ministry of Economy and Competitiveness, 'Centro de Excelencia Severo Ochoa 2013–2017', SEV-2012-0208 and for a Plan Estatal Grant to B.P. (BFU2014-55275-P).

References

Agostini, F., Cirillo, D., Bolognesi, B. and Tartaglia, G. G. (2013). X-inactivation: quantitative predictions of protein interactions in the Xist network. *Nucleic Acids Research* 41, e31.

Agrelo, R., Souabni, A., Novatchkova, M., Haslinger, C., Leeb, M., Komnenovic, V., Kishimoto, H., Gresh, L., Kohwi-Shigematsu, T., Kenner, L. *et al.* (2009). SATB1 defines the developmental context for gene silencing by Xist in lymphoma and embryonic cells. *Developmental Cell* 16, 507–516.

Anguera, M. C., Ma, W., Clift, D., Namekawa, S. H., Kelleher, R. J. and Lee, J. T. (2011). Tsx produces a long noncoding RNA and has general functions in the germline, stem cells, and brain. *PLoS Genetics* 7, e1002248.

Anguera, M. C., Sadreyev, R., Zhang, Z., Szanto, A., Payer, B., Sheridan, S. D., Kwok, S., Haggarty, S. J., Sur, M., Alvarez, J. *et al.* (2012). Molecular signatures of human induced pluripotent stem cells highlight sex differences and cancer genes. *Cell Stem Cell* 11, 75–90.

Augui, S., Filion, G. J., Huart, S., Nora, E. P., Guggiari, M., Maresca, M., Stewart, A. F. and Heard, E. (2007). Sensing X chromosome pairs before X inactivation via a novel X-pairing region of the Xic. *Science* 318, 1632–1636.

Bacher, C. P., Guggiari, M., Brors, B., Augui, S., Clerc, P., Avner, P., Eils, R. and Heard, E. (2006). Transient colocalization of X-inactivation centres accompanies the initiation of X inactivation. *Nature Cell Biology* 8, 293–299.

Barakat, T. S., Ghazvini, M., de Hoon, B., Li, T., Eussen, B., Douben, H., van der Linden, R., van der Stap, N., Boter, M., Laven, J. S. *et al.* (2015). Stable X chromosome reactivation in female human induced pluripotent stem cells. *Stem Cell Reports* 4, 199–208.

Barakat, T. S., Gunhanlar, N., Pardo, C. G., Achame, E. M., Ghazvini, M., Boers, R., Kenter, A., Rentmeester, E., Grootegoed, J. A. and Gribnau, J. (2011). RNF12 activates Xist and is essential for X chromosome inactivation. *PLoS Genetics* 7, e1002001.

Barakat, T. S., Loos, F., van Staveren, S., Myronova, E., Ghazvini, M., Grootegoed, J. A. and Gribnau, J. (2014). The trans-activator RNF12 and cis-acting elements effectuate X chromosome inactivation independent of X-pairing. *Molecular Cell* 53, 965–978.

Barr, M. L. and Bertram, E. G. (1949). A morphological distinction between neurones of the male and female, and the behaviour of the nucleolar satellite during accelerated nucleoprotein synthesis. *Nature* 163, 676.

Beletskii, A., Hong, Y.-K., Pehrson, J., Egholm, M. and Strauss, W. M. (2001). PNA interference mapping demonstrates functional domains in the noncoding RNA Xist. *Proceedings of the National Academy of Sciences USA* 98, 9215–9220.

Bhatnagar, S., Zhu, X., Ou, J., Lin, L., Chamberlain, L., Zhu, L. J., Wajapeyee, N. and Green, M. R. (2014). Genetic and pharmacological reactivation of the mammalian inactive X chromosome. *Proceedings of the National Academy of Sciences USA* 111, 12591–12598.

Blewitt, M. E., Gendrel, A.-V., Pang, Z., Sparrow, D. B., Whitelaw, N., Craig, J. M., Apedaile, A., Hilton, D. J., Dunwoodie, S. L., Brockdorff, N. *et al.* (2008). SmcHD1, containing a structural-maintenance-of-chromosomes hinge domain, has a critical role in X inactivation. *Nature Genetics* 40, 663–669.

Borsani, G., Tonlorenzi, R., Simmler, M., Dandolo, L., Arnaud, D., Capra, V., Grompe, M., Pizzuti, A., Muzny, D., Lawrence, C. *et al.* (1991). Characterization of a murine gene expressed from the inactive X chromosome. *Nature* 351, 325–329.

Boumil, R. M., Ogawa, Y., Sun, B. K., Huynh, K. D. and Lee, J. T. (2006). Differential methylation of Xite and CTCF sites in Tsix mirrors the pattern of X-inactivation choice in mice. *Molecular and Cellular Biology* 26, 2109–2117.

Brockdorff, N. (2013). Noncoding RNA and polycomb recruitment. *RNA* 19, 429–442.

Brockdorff, N. and Turner, B. M. (2015). Dosage compensation in mammals. *Cold Spring Harbor Perspectives in Biology* 7, a019406.

Brockdorff, N., Ashworth, A., Kay, G. F., Cooper, P., Smith, S., McCabe, V. M., Norris, D. P., Penny, G. D., Patel, D. and Rastan, S. (1991). Conservation of position and exclusive expression of mouse Xist from the inactive X chromosome. *Nature* 351, 329–331.

Brown, C. J. and Willard, H. F. (1994). The human X-inactivation centre is not required for maintenance of X-chromosome inactivation. *Nature* 368, 154–156.

Brown, C. J., Ballabio, A., Rupert, J. L., Lafreniere, R. G., Grompe, M., Tonlorenzi, R. and Willard, H. F. (1991). A gene from the region of the human X inactivation centre is expressed exclusively from the inactive X chromosome. *Nature* 349, 38–44.

Casanova, M., Preissner, T., Cerase, A., Poot, R., Yamada, D., Li, X., Appanah, R., Bezstarosti, K., Demmers, J., Koseki, H. *et al.* (2011). Polycomblike 2 facilitates the recruitment of PRC2 Polycomb group complexes to the inactive X chromosome and to target loci in embryonic stem cells. *Development* 138, 1471–1482.

Cerase, A., Smeets, D., Tang, Y. A., Gdula, M., Kraus, F., Spivakov, M., Moindrot, B., Leleu, M., Tattermusch, A., Demmerle, J. *et al.* (2014). Spatial separation of Xist RNA and polycomb proteins revealed by superresolution microscopy. *Proceedings of the National Academy of Sciences USA* 111, 2235–2240.

Chadwick, B. P. and Willard, H. F. (2004). Multiple spatially distinct types of facultative heterochromatin on the human inactive X chromosome. *Proceedings of the National Academy of Sciences USA* 101, 17450–17455.

Chaligné, R., Popova, T., Mendoza-Parra, M.-A., Saleem, M.-A. M., Gentien, D., Ban, K., Piolot, T., Leroy, O., Mariani, O., Gronemeyer, H. *et al.* (2015). The inactive X chromosome is epigenetically unstable and transcriptionally labile in breast cancer. *Genome Research* 25, 488–503.

Chan, K. M., Zhang, H., Malureanu, L. A., van Deursen, J. and Zhang, Z. (2011). Diverse factors are involved in maintaining X chromosome inactivation. *Proceedings of the National Academy of Sciences USA* 108, 16699–16704.

Chan, Y.-S., Göke, J., Lu, X., Venkatesan, N., Feng, B., Su, I.-H. and Ng, H.-H. (2012). A PRC2 Dependent repressive role of PRDM14 in human embryonic stem cells and iPSC reprogramming. *Stem Cells* 31, 682–692.

Chang, S. C. and Brown, C. J. (2010). Identification of regulatory elements flanking human XIST reveals species differences. BMC *Molecular Biology* 11, 20.

Changolkar, L. N., Costanzi, C., Leu, N. A., Chen, D., McLaughlin, K. J. and Pehrson, J. R. (2007). Developmental changes in histone macroH2A1-mediated gene regulation. *Molecular Cell Biology* 27, 2758–2764.

Chaumeil, J., Le Baccon, P., Wutz, A. and Heard, E. (2006). A novel role for Xist RNA in the formation of a repressive nuclear compartment into which genes are recruited when silenced. *Genes & Development* 20, 2223–2237.

Chiba, H., Hirasawa, R., Kaneda, M., Amakawa, Y., Li, E., Sado, T. and Sasaki, H. (2008). De novo DNA methylation independent establishment of maternal imprint on X chromosome in mouse oocytes. *Genesis* 46, 768–774.

Chow, J. C., Hall, L. L., Clemson, C. M., Lawrence, J. B. and Brown, C. J. (2003). Characterization of expression at the human XIST locus in somatic, embryonal carcinoma, and transgenic cell lines. *Genomics* 82, 309–322.

Chu, C., Zhang, Q. C., da Rocha, S. T., Flynn, R. A., Bharadwaj, M., Calabrese, J. M., Magnuson, T. R., Heard, E. and Chang, H. Y. (2015). Systematic discovery of Xist RNA binding proteins. *Cell* 161, 404–416.

Chureau, C., Chantalat, S., Romito, A., Galvani, A., Duret, L., Avner, P. and Rougeulle, C. (2011). Ftx is a non-coding RNA which affects Xist expression and chromatin structure within the X-inactivation center region. *Human Molecular Genetics* 20, 705–718.

Chuva de Sousa Lopes, S. M., Hayashi, K., Shovlin, T. C., Mifsud, W., Surani, M. A. and McLaren, A. (2008). X chromosome activity in mouse XX primordial germ cells. *PLoS Genetics* 4, e30.

Cifuentes-Rojas, C., Hernandez, A. J., Sarma, K. and Lee, J. T. (2014). Regulatory Interactions between RNA and Polycomb Repressive Complex 2. *Molecular Cell* 55, 171–185.

Clemson, C. M., McNeil, J. A., Willard, H. F. and Lawrence, J. B. (1996). XIST RNA paints the inactive X chromosome at interphase: evidence for a novel RNA involved in nuclear/chromosome structure. *Journal of Cell Biology* 132, 259–275.

Cohen, D. E., Davidow, L. S., Erwin, J. A., Xu, N., Warshawsky, D. and Lee, J. T. (2007). The DXPas34 repeat regulates random and imprinted X inactivation. *Developmental Cell* 12, 57–71.

Corbel, C., Diabangouaya, P., Gendrel, A.-V., Chow, J. C. and Heard, E. (2013). Unusual chromatin status and organization of the inactive X chromosome in murine trophoblast giant cells. *Development* 140, 861–872.

Costanzi, C. and Pehrson, J. R. (1998). Histone macroH2A1 is concentrated in the inactive X chromosome of female mammals. *Nature News* 393, 599–601.

Csankovszki, G., Nagy, A. and Jaenisch, R. (2001). Synergism of Xist RNA, DNA methylation, and histone hypoacetylation in maintaining X chromosome inactivation. *The Journal of Cell Biology* 153, 773–784.

Csankovszki, G., Panning, B., Bates, B., Pehrson, J. R. and Jaenisch, R. (1999). Conditional deletion of Xist disrupts histone macroH2A localization but not maintenance of X inactivation. *Nature Genetics* 22, 323–324.

da Rocha, S. T., Boeva, V., Escamilla-Del-Arenal, M., Ancelin, K., Granier, C., Matias, N. R., Sanulli, S., Chow, J., Schulz, E., Picard, C. *et al.* (2014). Jarid2 Is Implicated in the Initial Xist-Induced Targeting of PRC2 to the Inactive X Chromosome. *Molecular Cell* 53, 301–316.

Davidovich, C., Wang, X., Cifuentes-Rojas, C., Goodrich, K. J., Gooding, A. R., Lee, J. T. and Cech, T. R. (2015). Toward a consensus on the binding specificity and promiscuity of PRC2 for RNA. *Molecular Cell* 57, 552–558.

de Napoles, M., Mermoud, J. E., Wakao, R., Tang, Y. A., Endoh, M., Appanah, R., Nesterova, T. B., Silva, J. C. R., Otte, A. P., Vidal, M. *et al.* (2004). Polycomb group proteins Ring1A/B link ubiquitylation of histone H2A to heritable gene silencing and X inactivation. *Developmental Cell* 7, 663–676.

Debrand, E., Chureau, C., Arnaud, D., Avner, P. and Heard, E. (1999). Functional analysis of the DXPas34 locus, a 3′ regulator of Xist expression. *Molecular Cell Biology* 19, 8513–8525.

Dekker, J., Marti-Renom, M. A. and Mirny, L. A. (2013). Exploring the three-dimensional organization of genomes: interpreting chromatin interaction data. *Nature Review Genetics* 14, 390–403.

Dekker, J., Rippe, K., Dekker, M. and Kleckner, N. (2002). Capturing chromosome conformation. *Science* 295, 1306–1311.

Deng, X., Berletch, J. B., Nguyen, D. K. and Disteche, C. M. (2014). X chromosome regulation: diverse patterns in development, tissues and disease. *Nature Review Genetics* 15, 367–378.

Diaz Perez, S. V., Ferguson, D. O., Wang, C., Csankovszki, G., Wang, C., Tsai, S.-C., Dutta, D., Perez, V., Kim, S., Eller, C. D. *et al.* (2006). A deletion at the mouse Xist gene exposes trans-effects that alter the heterochromatin of the inactive X chromosome and the replication time and DNA stability of both X chromosomes. *Genetics* 174, 1115–1133.

Donohoe, M. E., Silva, S. S., Pinter, S. F., Xu, N. and Lee, J. T. (2009). The pluripotency factor Oct4 interacts with Ctcf and also controls X-chromosome pairing and counting. *Nature* 460, 128–132.

Dubois, A., Deuve, J. L., Navarro, P., Merzouk, S., Pichard, S., Commere, P.-H., Louise, A., Arnaud, D., Avner, P. and Morey, C. (2014). Spontaneous reactivation of clusters of x-linked genes is associated with the plasticity of x-inactivation in mouse trophoblast stem cells. *Stem Cells* 32, 377–390.

Duthie, S. M., Nesterova, T. B., Formstone, E. J., Keohane, A. M., Turner, B. M., Zakian, S. M. and Brockdorff, N. (1999). Xist RNA exhibits a banded localization on the inactive X chromosome and is excluded from autosomal material in cis. *Human Molecular Genetics* 8, 195–204.

Engreitz, J. M., Pandya-Jones, A., McDonel, P., Shishkin, A., Sirokman, K., Surka, C., Kadri, S., Xing, J., Goren, A., Lander, E. S. *et al.* (2013). The Xist lncRNA exploits three-dimensional genome architecture to spread across the X chromosome. *Science* 341, 1237973.

Erhardt, S., Su, I.-H., Schneider, R., Barton, S. C., Bannister, A. J., Perez-Burgos, L., Jenuwein, T., Kouzarides, T., Tarakhovsky, A. and Surani, M. A. (2003). Consequences of the depletion of zygotic and embryonic enhancer of zeste 2 during preimplantation mouse development. *Development* 130, 4235–4248.

Erwin, J. A., Del Rosario, B., Payer, B. and Lee, J. T. (2012). An *Ex Vivo* model for imprinting: mutually exclusive binding of Cdx2 and Oct4 as a switch for imprinted and random X-inactivation. *Genetics* 192, 857–868.

Escamilla-Del-Arenal, M., da Rocha, S. T., Spruijt, C. G., Masui, O., Renaud, O., Smits, A. H., Margueron, R., Vermeulen, M. and Heard, E. (2013). Cdyl: a new partner of the inactive X chromosome and potential reader of H3K27me3 and H3K9me2. *Molecular and Cellular Biology* 33, 5005–5020.

Fackelmayer, F. O. (2005). A stable proteinaceous structure in the territory of inactive X chromosomes. *Journal of Biological Chemistry* 280, 1720–1723.

Fang, J., Chen, T., Chadwick, B. P., Li, E. and Zhang, Y. (2004). Ring1b-mediated H2A ubiquitination associates with inactive X chromosomes and is involved in initiation of X inactivation. *Journal of Biological Chemistry* 279, 52812–52815.

Fukuda, A., Tomikawa, J., Miura, T., Hata, K., Nakabayashi, K., Eggan, K., Akutsu, H. and Umezawa, A. (2014). The role of maternal-specific H3K9me3 modification in establishing imprinted X-chromosome inactivation and embryogenesis in mice. *Nature Communications* 5, 5464.

Gafni, O., Weinberger, L., Mansour, A. A., Manor, Y. S., Chomsky, E., Ben-Yosef, D., Kalma, Y., Viukov, S., Maza, I., Zviran, A. *et al.* (2013). Derivation of novel human ground state naive pluripotent stem cells. *Nature* 504, 282–286.

Gayen, S., Maclary, E., Buttigieg, E., Hinten, M. and Kalantry, S. (2015). A primary role for the Tsix lncRNA in maintaining random X-chromosome inactivation. *Cell Rep* 11, 1251–1265.

Gendrel, A.-V. and Heard, E. (2014). Noncoding RNAs and epigenetic mechanisms during X-chromosome inactivation. *Annual Review Cell Developmental Biology* 30, 561–580.

Gendrel, A.-V., Apedaile, A., Coker, H., Termanis, A., Zvetkova, I., Godwin, J., Tang, Y. A., Huntley, D., Montana, G., Taylor, S. *et al.* (2012). Smchd1-dependent and -independent pathways determine developmental dynamics of CpG island methylation on the inactive X chromosome. *Developmental Cell* 23, 265–279.

Gendrel, A.-V., Tang, Y. A., Suzuki, M., Godwin, J., Nesterova, T. B., Greally, J. M., Heard, E. and Brockdorff, N. (2013). Epigenetic functions of smchd1 repress gene clusters on the inactive x chromosome and on autosomes. *Molecular and Cellular Biology* 33, 3150–3165.

Giorgetti, L., Galupa, R., Nora, E. P., Piolot, T., Lam, F., Dekker, J., Tiana, G. and Heard, E. (2014). Predictive polymer modeling reveals coupled fluctuations in chromosome conformation and transcription. *Cell* 157, 950–963.

Gontan, C., Achame, E. M., Demmers, J., Barakat, T. S., Rentmeester, E., van Ijcken, W., Grootegoed, J. A. and Gribnau, J. (2012). RNF12 initiates X-chromosome inactivation by targeting REX1 for degradation. *Nature* 485, 386–390.

Goto, T., Wright, E. and Monk, M. (1997). Paternal X-chromosome inactivation in human trophoblastic cells. *Molecular Human Reproduction* 3, 77–80.

Guo, F., Yan, L., Guo, H., Li, L., Hu, B., Zhao, Y., Yong, J., Hu, Y., Wang, X., Wei, Y. *et al.* (2015). The transcriptome and DNA methylome landscapes of human primordial germ cells. *Cell* 161, 1437–1452.

Hacisuleyman, E., Goff, L. A., Trapnell, C., Williams, A., Henao-Mejia, J., Sun, L., McClanahan, P., Hendrickson, D. G., Sauvageau, M., Kelley, D. R., *et al.* (2014). Topological organization of multichromosomal regions by the long intergenic noncoding RNA Firre. *Nature Structural & Molecular Biology* 21, 198–206.

Hadjantonakis, A. K., Cox, L. L., Tam, P. P. and Nagy, A. (2001). An X-linked GFP transgene reveals unexpected paternal X-chromosome activity in trophoblastic giant cells of the mouse placenta. *Genesis* 29, 133–140.

Harrison, K. (1989). X-chromosome inactivation in the human cytotrophoblast. *Cytogenetic Cell Genetics* 52, 37–41.

Hasegawa, Y., Brockdorff, N., Kawano, S., Tsutui, K., Tsutui, K. and Nakagawa, S. (2010). The matrix protein hnRNP U is required for chromosomal localization of Xist RNA. *Developmental Cell* 19, 469–476.

Hayashi, K., Ogushi, S., Kurimoto, K., Shimamoto, S., Ohta, H. and Saitou, M. (2012). Offspring from oocytes derived from in vitro primordial germ cell-like cells in mice. *Science* 338, 971–975.

Helbig, R. and Fackelmayer, F. O. (2003). Scaffold attachment factor A (SAF-A) is concentrated in inactive X chromosome territories through its RGG domain. *Chromosoma* 112, 173–182.

Hellman, A. and Chess, A. (2007). Gene body-specific methylation on the active X chromosome. *Science* 315, 1141–1143.

Hernández-Muñoz, I., Lund, A. H., van der Stoop, P., Boutsma, E., Muijrers, I., Verhoeven, E., Nusinow, D. A., Panning, B., Marahrens, Y. and Van Lohuizen, M. (2005). Stable X chromosome inactivation involves the PRC1 Polycomb complex and requires histone MACROH2A1 and the CULLIN3/SPOP ubiquitin E3 ligase. *Proceedings of National Academy of Sciences USA* 102, 7635–7640.

Hoki, Y., Kimura, N., Kanbayashi, M., Amakawa, Y., Ohhata, T., Sasaki, H. and Sado, T. (2009). A proximal conserved repeat in the Xist gene is essential as a genomic element for X-inactivation in mouse. *Development* 136, 139–146.

Huynh, K. D. and Lee, J. T. (2003). Inheritance of a pre-inactivated paternal X chromosome in early mouse embryos. Nature 426, 857–862.

Inoue, K., Kohda, T., Sugimoto, M., Sado, T., Ogonuki, N., Matoba, S., Shiura, H., Ikeda, R., Mochida, K., Fujii, T. *et al.* (2010). Impeding Xist expression from the active X chromosome improves mouse somatic cell nuclear transfer. *Science* 330, 496–499.

Irie, N., Weinberger, L., Tang, W. W.C., Kobayashi, T., Viukov, S., Manor, Y. S., Dietmann, S., Hanna, J. H. and Surani, M. A. (2014). SOX17 is a critical specifier of human primordial germ cell fate. *Cell* 160, 253–268.

Jäger, N., Schlesner, M., Jones, D. T.W., Raffel, S., Mallm, J.-P., Junge, K. M., Weichenhan, D., Bauer, T., Ishaque, N., Kool, M. *et al.* (2013). Hypermutation of the inactive X chromosome is a frequent event in cancer. *Cell* 155, 567–581.

Jeon, Y. and Lee, J. T. (2011). YY1 tethers xist RNA to the inactive X nucleation center. *Cell* 146, 119–133.

Jonkers, I., Barakat, T. S., Achame, E. M., Monkhorst, K., Kenter, A., Rentmeester, E., Grosveld, F., Grootegoed, J. A. and Gribnau, J. (2009). RNF12 is an X-encoded dose-dependent activator of X chromosome inactivation. *Cell* 139, 999–1011.

Jonkers, I., Monkhorst, K., Rentmeester, E., Grootegoed, J. A., Grosveld, F. and Gribnau, J. (2008). Xist RNA is confined to the nuclear territory of the silenced X chromosome throughout the cell cycle. *Molecular and Cellular Biology* 28, 5583–5594.

Kalantry, S. and Magnuson, T. R. (2006). The Polycomb group protein EED is dispensable for the initiation of random X-chromosome inactivation. *PLoS Genetics* 2, e66.

Kalantry, S., Purushothaman, S., Bowen, R. B., Starmer, J. and Magnuson, T. R. (2009). Evidence of Xist RNA-independent initiation of mouse imprinted X-chromosome inactivation. *Nature* 460, 647–651.

Kalantry, S., Mills, K. C., Yee, D., Otte, A. P., Panning, B. and Magnuson, T. R. (2006). The Polycomb group protein Eed protects the inactive X-chromosome from differentiation-induced reactivation. *Nature Publishing Group* 8, 195–202.

Kohlmaier, A., Savarese, F., Lachner, M., Martens, J., Jenuwein, T. and Wutz, A. (2004). A chromosomal memory triggered by Xist regulates histone methylation in X inactivation. *Plos Biology* 2, E171.

Kung, J. T., Kesner, B., An, J. Y., Ahn, J. Y., Cifuentes-Rojas, C., Colognori, D., Jeon, Y., Szanto, A., del Rosario, B. C., Pinter, S. F. *et al.* (2015). Locus-specific targeting to the X chromosome revealed by the RNA interactome of CTCF. *Molecular Cell* 57, 361–375.

Lau, A. C. and Csankovszki, G. (2015). Balancing up and downregulation of the C. elegans X chromosomes. *Current Opinion in Genetics & Development* 31, 50–56.

Lee, J. T. (2000). Disruption of imprinted X inactivation by parent-of-origin effects at Tsix. *Cell* 103, 17–27.

Lee, J. T. (2011). Gracefully ageing at 50, X-chromosome inactivation becomes a paradigm for RNA and chromatin control. *Nature Reviews Molecular Cell Biology* 12, 815–826.

Lee, J. T. and Lu, N. (1999). Targeted mutagenesis of Tsix leads to nonrandom X inactivation. *Cell* 99, 47–57.

Lee, J. T., Davidow, L. and Warshawsky, D. (1999). Tsix, a gene antisense to Xist at the X-inactivation centre. *Nature Genetics* 21, 400–404.

Leeb, M. and Wutz, A. (2007). Ring1B is crucial for the regulation of developmental control genes and PRC1 proteins but not X inactivation in embryonic cells. *Journal of Cell Biology* 178, 219–229.

Lengner, C. J., Gimelbrant, A. A., Erwin, J. A., Cheng, A. W., Guenther, M. G., Welstead, G. G., Alagappan, R., Frampton, G. M., Xu, P., Muffat, J. *et al.* (2010). Derivation of pre-X inactivation human embryonic stem cells under physiological oxygen concentrations. *Cell* 141, 872–883.

Lessing, D., Anguera, M. C. and Lee, J. T. (2013). X chromosome inactivation and epigenetic responses to cellular reprogramming. *Annual Review Genomics Human Genetics* 14, 85–110.

Looijenga, L. H., Gillis, A. J., Verkerk, A. J., van Putten, W. L. and Oosterhuis, J. W. (1999). Heterogeneous X inactivation in trophoblastic cells of human full-term female placentas. *American Journal of Human Genetics* 64, 1445–1452.

Luikenhuis, S., Wutz, A. and Jaenisch, R. (2001). Antisense transcription through the Xist locus mediates Tsix function in embryonic stem cells. *Molecular Cell Biology* 21, 8512–8520.

Lyon, M. F. (1961). Gene action in the X-chromosome of the mouse (Mus musculus L.). *Nature* 190, 372–373.

Ma, Z., Swigut, T., Valouev, A., Rada-Iglesias, A. and Wysocka, J. (2010). Sequence-specific regulator Prdm14 safeguards mouse ESCs from entering extraembryonic endoderm fates. *Nature Structure and Molecular Biology* 18, 120–127.

Maclary, E., Buttigieg, E., Hinten, M., Gayen, S., Harris, C., Sarkar, M. K., Purushothaman, S. and Kalantry, S. (2014). Differentiation-dependent requirement of Tsix long non-coding RNA in imprinted X-chromosome inactivation. *Nature Communications* 5, 4209.

Maenner, S., Blaud, M., Fouillen, L., Savoye, A., Marchand, V., Dubois, A., Sanglier-Cianférani, S., Van Dorsselaer, A., Clerc, P., Avner, P. *et al.* (2010). 2-D structure of the A region of Xist RNA and its implication for PRC2 association. *Plos Biology* 8, e1000276.

Maherali, N., Sridharan, R., Xie, W., Utikal, J., Eminli, S., Arnold, K., Stadtfeld, M., Yachechko, R., Tchieu, J., Jaenisch, R. *et al.* (2007). Directly reprogrammed fibroblasts show global epigenetic remodeling and widespread tissue contribution. *Cell Stem Cell* 1, 55–70.

Mak, W., Nesterova, T. B., de Napoles, M., Appanah, R., Yamanaka, S., Otte, A. P. and Brockdorff, N. (2004). Reactivation of the paternal X chromosome in early mouse embryos. *Science* 303, 666–669.

Marahrens, Y., Panning, B., Dausman, J., Strauss, W. and Jaenisch, R. (1997). Xist-deficient mice are defective in dosage compensation but not spermatogenesis. *Genes and Development* 11, 156–166.

Masui, O., Bonnet, I., Le Baccon, P., Brito, I., Pollex, T., Murphy, N., Hupé, P., Barillot, E., Belmont, A. S. and Heard, E. (2011). Live-cell chromosome dynamics and outcome of X chromosome pairing events during ES cell differentiation. *Cell* 145, 447–458.

Matoba, S., Inoue, K., Kohda, T., Sugimoto, M., Mizutani, E., Ogonuki, N., Nakamura, T., Abe, K., Nakano, T., Ishino, F. *et al.* (2011). RNAi-mediated knockdown of Xist can rescue the impaired postimplantation development of cloned mouse embryos. *Proceedings of National Academy of Sciences USA* 108, 20621–20626.

McGraw, S., Oakes, C. C., Martel, J., Cirio, M. C., de Zeeuw, P., Mak, W., Plass, C., Bartolomei, M. S., Chaillet, J. R. and Trasler, J. M. (2013). Loss of DNMT1o disrupts imprinted X chromosome inactivation and accentuates placental defects in females. *PLoS Genetics* 9, e1003873.

McHugh, C. A., Chen, C.-K., Chow, A., Surka, C. F., Tran, C., Mcdonel, P., Pandya-Jones, A., Blanco, M., Burghard, C., Moradian, A. *et al.* (2015). The Xist lncRNA interacts directly with SHARP to silence transcription through HDAC3. *Nature* 521, 232–236.

Mekhoubad, S., Bock, C., de Boer, A. S., Kiskinis, E., Meissner, A. and Eggan, K. (2012). Erosion of dosage compensation impacts human iPSC disease modeling. *Cell Stem Cell* 10, 595–609.

Mermoud, J. E., Costanzi, C., Pehrson, J. R. and Brockdorff, N. (1999). Histone macroH2A1.2 relocates to the inactive X chromosome after initiation and propagation of X-inactivation. *Journal of Cell Biology* 147, 1399–1408.

Migeon, B. R., Lee, C. H., Chowdhury, A. K. and Carpenter, H. (2002). Species differences in TSIX/Tsix reveal the roles of these genes in X-chromosome inactivation. *American Journal of Human Genetics* 71, 286–293.

Minajigi, A., Froberg, J. E., Wei, C., Sunwoo, H., Kesner, B., Colognori, D., Lessing, D., Payer, B., Boukhali, M., Haas, W. *et al.* (2015). A comprehensive Xist interactome reveals cohesin repulsion and an RNA-directed chromosome conformation. *Science* 349, aab2276.

Minkovsky, A., Barakat, T. S., Sellami, N., Chin, M. H., Gunhanlar, N., Gribnau, J. and Plath, K. (2013). The pluripotency factor-bound intron 1 of xist is dispensable for X chromosome inactivation and reactivation *in vitro* and *in vivo*. *Cell Representation* 3, 905–918.

Minkovsky, A., Sahakyan, A., Rankin-Gee, E., Bonora, G., Patel, S. and Plath, K. (2014). The Mbd1-Atf7ip-Setdb1 pathway contributes to the maintenance of X chromosome inactivation. *Epigenetics & Chromatin* 7, 12.

Miyanari, Y. and Torres-Padilla, M.-E. (2012). Control of ground-state pluripotency by allelic regulation of Nanog. *Nature* 483, 470–473.

Moindrot, B., Cerase, A., Coker, H., Masui, O., Grijzenhout, A., Pintacuda, G., Schermelleh, L., Nesterova, T. B. and Brockdorff, N. (2015). A pooled shRNA screen identifies Rbm15, Spen, and Wtap as factors required for Xist RNA-Mediated silencing. *Cell Representation* 12, 562–572.

Monfort, A., Di Minin, G., Postlmayr, A., Freimann, R., Arieti, F., Thore, S. and Wutz, A. (2015). Identification of Spen as a crucial factor for Xist function through forward genetic screening in haploid embryonic stem cells. *Cell Representation* 12, 554–561.

Monk, M. and McLaren, A. (1981). X-chromosome activity in foetal germ cells of the mouse. *Journal of Embryology and Experimental Morphology* 63, 75–84.

Monkhorst, K., Jonkers, I., Rentmeester, E., Grosveld, F. and Gribnau, J. (2008). X inactivation counting and choice is a stochastic process: evidence for involvement of an X-linked activator. *Cell* 132, 410–421.

Moreira de Mello, J. C., de Araújo, E. S.S., Stabellini, R., Fraga, A. M., de Souza, J. E. S., Sumita, D. R., Camargo, A. A. and Pereira, L. V. (2010). Random X inactivation and extensive mosaicism in human placenta revealed by analysis of allele-specific gene expression along the X chromosome. *PLoS ONE* 5, e10947.

Namekawa, S. H., Payer, B., Huynh, K. D., Jaenisch, R. and Lee, J. T. (2010). Two-step imprinted X inactivation: repeat versus genic silencing in the mouse. *Molecular and Cellular Biology* 30, 3187–3205.

Navarro, P., Chambers, I., Karwacki-Neisius, V., Chureau, C., Morey, C., Rougeulle, C. and Avner, P. (2008). Molecular coupling of Xist regulation and pluripotency. *Science* 321, 1693–1695.

Navarro, P., Moffat, M., Mullin, N. P. and Chambers, I. (2011). The X-inactivation trans-activator Rnf12 is negatively regulated by pluripotency factors in embryonic stem cells. *Human Genetics* 130, 255–264.

Navarro, P., Oldfield, A., Legoupi, J., Festuccia, N., Dubois, A., Attia, M., Schoorlemmer, J., Rougeulle, C., Chambers, I. and Avner, P. (2010). Molecular coupling of Tsix regulation and pluripotency. *Nature* 468, 457–460.

Navarro, P., Page, D. R., Avner, P. and Rougeulle, C. (2006). Tsix-mediated epigenetic switch of a CTCF-flanked region of the Xist promoter determines the Xist transcription program. *Genes Development* 20, 2787–2792.

Navarro, P., Pichard, S., Ciaudo, C., Avner, P. and Rougeulle, C. (2005). Tsix transcription across the Xist gene alters chromatin conformation without affecting Xist transcription: implications for X-chromosome inactivation. *Genes Development* 19, 1474–1484.

Nechanitzky, R., Dávila, A., Savarese, F., Fietze, S. and Grosschedl, R. (2012). Satb1 and satb2 are dispensable for x chromosome inactivation in mice. *Developmental Cell* 23, 866–871.

Nesterova, T. B., Senner, C. E., Schneider, J., Alcayna-Stevens, T., Tattermusch, A., Hemberger, M. and Brockdorff, N. (2011). Pluripotency factor binding and Tsix expression act synergistically to repress Xist in undifferentiated ES cells. *Epigenetics & Chromatin* 4, 17.

Nichols, J. and Smith, A. (2009). Naive and primed pluripotent states. *Cell Stem Cell* 4, 487–492.

Nicodemi, M. and Prisco, A. (2007). Symmetry-breaking model for X-chromosome inactivation. *Physical Review Letters* 98, 108104.

Nora, E. P., Lajoie, B. R., Schulz, E. G., Giorgetti, L., Okamoto, I., Servant, N., Piolot, T., van Berkum, N. L., Meisig, J., Sedat, J. *et al.* (2012). Spatial partitioning of the regulatory landscape of the X-inactivation centre. *Nature* 485, 381–385.

Nozawa, R.-S., Nagao, K., Igami, K.-T., Shibata, S., Shirai, N., Nozaki, N., Sado, T., Kimura, H. and Obuse, C. (2013). Human inactive X chromosome is compacted through a PRC2-independent SMCHD1-HBiX1 pathway. *Nature Structural and Molecular Biology* 20, 566–573.

Oda, H., Okamoto, I., Murphy, N., Chu, J., Price, S. M., Shen, M. M., Torres-Padilla, M.-E., Heard, E. and Reinberg, D. (2009). Monomethylation of histone H4-lysine 20 is involved in chromosome structure and stability and is essential for mouse development. *Molecular and Cellular Biology* 29, 2278–2295.

Ogawa, Y. and Lee, J. T. (2003). Xite, X-inactivation intergenic transcription elements that regulate the probability of choice. *Molecular Cell* 11, 731–743.

Ohhata, T. and Wutz, A. (2012). Reactivation of the inactive X chromosome in development and reprogramming. Cell. Mol. *Life Science* 70, 2443–2461.

Ohhata, T., Hoki, Y., Sasaki, H. and Sado, T. (2008). Crucial role of antisense transcription across the Xist promoter in Tsix-mediated Xist chromatin modification. *Development* 135, 227–235.

Ohhata, T., Senner, C. E., Hemberger, M. and Wutz, A. (2011). Lineage-specific function of the noncoding Tsix RNA for Xist repression and Xi reactivation in mice. *Genes Development* 25, 1702–1715.

Ohhata, T., Tachibana, M., Tada, M., Tada, T., Sasaki, H., Shinkai, Y. and Sado, T. (2004). X-inactivation is stably maintained in mouse embryos deficient for histone methyl transferase G9a. *Genesis* 40, 151–156.

Ohinata, Y., Payer, B., O'carroll, D., Ancelin, K., Ono, Y., Sano, M., Barton, S. C., Obukhanych, T., Nussenzweig, M., Tarakhovsky, A. *et al.* (2005). Blimp1 is a critical determinant of the germ cell lineage in mice. *Nature* 436, 207–213.

OHNO, S. and HAUSCHKA, T. S. (1960). Allocycly of the X-chromosome in tumors and normal tissues. *Cancer Research* 20, 541–545.

Oikawa, M., Inoue, K., Shiura, H., Matoba, S., Kamimura, S., Hirose, M., Mekada, K., Yoshiki, A., Tanaka, S., Abe, K., *et al.* (2014). Understanding the X chromosome inactivation cycle in mice: a comprehensive view provided by nuclear transfer. *Epigenetics : Official Journal of the DNA Methylation Society* 9, 204–211.

Okamoto, I., Arnaud, D., Le Baccon, P., Otte, A. P., Disteche, C. M., Avner, P. and Heard, E. (2005). Evidence for de novo imprinted X-chromosome inactivation independent of meiotic inactivation in mice. *Nature* 438, 369–373.

Okamoto, I., Otte, A. P., Allis, C. D., Reinberg, D. and Heard, E. (2004). Epigenetic dynamics of imprinted X inactivation during early mouse development. *Science* 303, 644–649.

Okamoto, I., Patrat, C., Thépot, D., Peynot, N., Fauque, P., Daniel, N., Diabangouaya, P., Wolf, J.-P., Renard, J.-P., Duranthon, V. *et al.* (2011). Eutherian mammals use diverse strategies to initiate X-chromosome inactivation during development. *Nature* 472, 370–374.

Pageau, G. J., Hall, L. L., Ganesan, S., Livingston, D. M. and Lawrence, J. B. (2007). The disappearing Barr body in breast and ovarian cancers. Nat. Rev. *Cancer* 7, 628–633.

Panning, B. and Jaenisch, R. (1996). DNA hypomethylation can activate Xist expression and silence X-linked genes. *Genes Development* 10, 1991–2002.

Pasque, V., Gillich, A., Garrett, N. and Gurdon, J. B. (2011). Histone variant macroH2A confers resistance to nuclear reprogramming. *Embo Journal* 30, 2373–2387.

Pasque, V., Tchieu, J., Karnik, R., Uyeda, M., Sadhu Dimashkie, A., Case, D., Papp, B., Bonora, G., Patel, S., Ho, R. *et al.* (2014). X chromosome reactivation dynamics reveal stages of reprogramming to pluripotency. *Cell* 159, 1681–1697.

Patrat, C., Okamoto, I., Diabangouaya, P., Vialon, V., Le Baccon, P., Chow, J. C. and Heard, E. (2009). Dynamic changes in paternal X-chromosome activity during imprinted X-chromosome inactivation in mice. *Proceedings of the National Academy of Sciences USA* 106, 5198–5203.

Payer, B. and Lee, J. T. (2008). X chromosome dosage compensation: how mammals keep the balance. *Annual Review Genetics* 42, 733–772.

Payer, B. and Lee, J. T. (2014). Coupling of X-Chromosome reactivation with the pluripotent stem cell state. *RNA Biology* 11, 798–807.

Payer, B., Rosenberg, M., Yamaji, M., Yabuta, Y., Koyanagi-Aoi, M., Hayashi, K., Yamanaka, S., Saitou, M. and Lee, J. T. (2013). Tsix RNA and the germline factor, PRDM14, link X reactivation and stem cell reprogramming. *Molecular Cell* 52, 805–818.

Peñaherrera, M. S., Jiang, R., Avila, L., Yuen, R. K.C., Brown, C. J. and Robinson, W. P. (2012). Patterns of placental development evaluated by X chromosome inactivation profiling provide a basis to evaluate the origin of epigenetic variation. *Human Reproduction* 27, 1745–1753.

Pinter, S. F., Sadreyev, R. I., Yildirim, E., Jeon, Y., Ohsumi, T. K., Borowsky, M. and Lee, J. T. (2012). Spreading of X chromosome inactivation via a hierarchy of defined Polycomb stations. *Genome Research* 22, 1864–1876.

Plath, K., Fang, J., Mlynarczyk-Evans, S., Cao, R., Worringer, K., Wang, H., la Cruz, de, C., Otte, A., Panning, B. and Zhang, Y. (2003). Role of histone H3 lysine 27 methylation in X inactivation. *Science* 300, 131–135.

Plath, K., Talbot, D., Hamer, K., Otte, A., Yang, T., Jaenisch, R. and Panning, B. (2004). Developmentally regulated alterations in Polycomb repressive complex 1 proteins on the inactive X chromosome. *Journal of Cell Biology* 167, 1025–1035.

Pombo, A. and Dillon, N. (2015). Three-dimensional genome architecture: players and mechanisms. *Nature Reviews Molecular Cell Biology* 16, 245–257.

Pugacheva, E. M., Tiwari, V. K., Abdullaev, Z., Vostrov, A. A., Flanagan, P. T., Quitschke, W. W., Loukinov, D. I., Ohlsson, R. and Lobanenkov, V. V. (2005). Familial cases of point mutations in the XIST promoter reveal a correlation between CTCF binding and pre-emptive choices of X chromosome inactivation. *Human Molecular Genetics* 14, 953–965.

Pullirsch, D., Härtel, R., Kishimoto, H., Leeb, M., Steiner, G. and Wutz, A. (2010). The Trithorax group protein Ash2l and Saf-A are recruited to the inactive X chromosome at the onset of stable X inactivation. *Development* 137, 935–943.

Richardson, L., Venkataraman, S., Stevenson, P., Yang, Y., Moss, J., Graham, L., Burton, N., Hill, B., Rao, J., Baldock, R. A. *et al.* (2014). EMAGE mouse embryo spatial gene expression database: 2014 update. *Nucleic Acids Research* 42, D835–D844.

Rougeulle, C., Chaumeil, J., Sarma, K., Allis, C. D., Reinberg, D., Avner, P. and Heard, E. (2004). Differential histone H3 Lys-9 and Lys-27 methylation profiles on the X chromosome. *Molecular Cell Biology* 24, 5475–5484.

Sado, T. (2000). X inactivation in the mouse embryo deficient for Dnmt1: distinct effect of hypomethylation on imprinted and random X inactivation. *Developmental Biology* 225, 294–303.

Sado, T., Hoki, Y. and Sasaki, H. (2005). Tsix silences Xist through modification of chromatin structure. *Developmental Cell* 9, 159–165.

Sado, T., Okano, M., Li, E. and Sasaki, H. (2004). De novo DNA methylation is dispensable for the initiation and propagation of X chromosome inactivation. *Development* 131, 975–982.

Sado, T., Wang, Z., Sasaki, H. and Li, E. (2001). Regulation of imprinted X-chromosome inactivation in mice by Tsix. *Development* 128, 1275–1286.

Sarma, K., Cifuentes-Rojas, C., Ergun, A., Del Rosario, A., Jeon, Y., White, F., Sadreyev, R. and Lee, J. T. (2014). ATRX directs binding of PRC2 to Xist RNA and polycomb targets. *Cell* 159, 869–883.

Sarma, K., Levasseur, P., Aristarkhov, A. and Lee, J. T. (2010). Locked nucleic acids (LNAs) reveal sequence requirements and kinetics of Xist RNA localization to the X chromosome. *Proceedings of the National Academy of Sciences USA* 107, 22196–22201.

Schoeftner, S., Sengupta, A. K., Kubicek, S., Mechtler, K., Spahn, L., Koseki, H., Jenuwein, T. and Wutz, A. (2006). Recruitment of PRC1 function at the initiation of X inactivation independent of PRC2 and silencing. *Embo Journal* 25, 3110–3122.

Schulz, E. G. and Heard, E. (2013). Role and control of X chromosome dosage in mammalian development. *Current Opinion in Genetics & Development* 23, 109–115.

Schulz, E. G., Meisig, J., Nakamura, T., Okamoto, I., Sieber, A., Picard, C., Borensztein, M., Saitou, M., Blüthgen, N. and Heard, E. (2014). The two active X chromosomes in female ESCs block exit from the pluripotent state by modulating the ESC signaling network. *Cell Stem Cell* 14, 203–216.

Seisenberger, S., Andrews, S., Krueger, F., Arand, J., Walter, J., Santos, F., Popp, C., Thienpont, B., Dean, W. and Reik, W. (2012). The dynamics of genome-wide DNA methylation reprogramming in mouse primordial germ cells. *Molecular Cell* 48, 849–862.

Shin, J., Bossenz, M., Chung, Y., Ma, H., Byron, M., Taniguchi-Ishigaki, N., Zhu, X., Jiao, B., Hall, L. L., Green, M. R. *et al.* (2010). Maternal Rnf12/RLIM is required for imprinted X-chromosome inactivation in mice. *Nature* 467, 977–981.

Shin, J., Wallingford, M. C., Gallant, J., Marcho, C., Jiao, B., Byron, M., Bossenz, M., Lawrence, J. B., Jones, S. N., Mager, J. *et al.* (2014). RLIM is dispensable for X-chromosome inactivation in the mouse embryonic epiblast. *Nature* 511, 86–89.

Silva, J. C.R., Mak, W., Zvetkova, I., Appanah, R., Nesterova, T. B., Webster, Z., Peters, A., Jenuwein, T., Otte, A. P. and Brockdorff, N. (2003). Establishment of histone h3 methylation on the inactive X chromosome requires transient recruitment of Eed-Enx1 polycomb group complexes. *Developmental Cell* 4, 481–495.

Silva, J. C.R., Nichols, J., Theunissen, T. W., Guo, G., van Oosten, A. L., Barrandon, O., Wray, J., Yamanaka, S., Chambers, I. and Smith, A. (2009). Nanog is the gateway to the pluripotent ground state. *Cell* 138, 722–737.

Silva, S. S., Rowntree, R. K., Mekhoubad, S. and Lee, J. T. (2008). X-chromosome inactivation and epigenetic fluidity in human embryonic stem cells. *Proceedings of the National Academy of Sciences USA* 105, 4820–4825.

Simon, M. D., Pinter, S. F., Fang, R., Sarma, K., Rutenberg-Schoenberg, M., Bowman, S. K., Kesner, B. A., Maier, V. K., Kingston, R. E. and Lee, J. T. (2013). High-resolution Xist binding maps reveal two-step spreading during X-chromosome inactivation. *Nature* 504, 465–469.

Smeets, D., Markaki, Y., Schmid, V. J., Kraus, F., Tattermusch, A., Cerase, A., Sterr, M., Fiedler, S., Demmerle, J., Popken, J. *et al.* (2014). Three-dimensional super-resolution microscopy of the inactive X chromosome territory reveals a collapse of its active nuclear compartment harboring distinct Xist RNA foci. *Epigenetics & Chromatin* 7, 8.

Soma, M., Fujihara, Y., Okabe, M., Ishino, F. and Kobayashi, S. (2014). Ftx is dispensable for imprinted X-chromosome inactivation in preimplantation mouse embryos. *Scientific Reports* 4, 5181.

Spencer, R. J., del Rosario, B. C., Pinter, S. F., Lessing, D., Sadreyev, R. I. and Lee, J. T. (2011). A boundary element between Tsix and Xist binds the chromatin insulator Ctcf and contributes to initiation of X-chromosome inactivation. *Genetics* 189, 441–454.

Splinter, E., de Wit, E., Nora, E. P., Klous, P., van de Werken, H. J.G., Zhu, Y., Kaaij, L. J.T., van Ijcken, W., Gribnau, J., Heard, E. *et al.* (2011). The inactive X chromosome adopts a unique three-dimensional conformation that is dependent on Xist RNA. *Genes Development* 25, 1371–1383.

Sugimoto, M. and Abe, K. (2007). X chromosome reactivation initiates in nascent primordial germ cells in mice. *PLoS Genetics* 3, e116.

Sun, B. K., Deaton, A. M. and Lee, J. T. (2006). A transient heterochromatic state in Xist preempts X inactivation choice without RNA stabilization. *Molecular Cell* 21, 617–628.

Sun, S., del Rosario, B. C., Szanto, A., Ogawa, Y., Jeon, Y. and Lee, J. T. (2013). Jpx RNA Activates Xist by Evicting CTCF. *Cell* 153, 1537–1551.

Sun, S., Fukue, Y., Nolen, L., Sadreyev, R. and Lee, J. T. (2010). Characterization of Xpr (Xpct) reveals instability but no effects on X-chromosome pairing or Xist expression. *Transcription* 1, 46–56.

Sunwoo, H., Wu, J. Y. and Lee, J. T. (2015). The Xist RNA-PRC2 complex at 20-nm resolution reveals a low Xist stoichiometry and suggests a hit-and-run mechanism in mouse cells. *Proceedings of the National Academy of Sciences USA* 112, E4216-E4225.

Tachibana, M., Ma, H., Sparman, M. L., Lee, H.-S., Ramsey, C. M., Woodward, J. S., Sritanaudomchai, H., Masterson, K. R., Wolff, E. E., Jia, Y. *et al.* (2012). X-chromosome inactivation in monkey embryos and pluripotent stem cells. *Developmental Biology* 371, 146–155.

Tada, M., Takahama, Y., Abe, K., Nakatsuji, N. and Tada, T. (2001). Nuclear reprogramming of somatic cells by *in vitro* hybridization with ES cells. *Current Biology* 11, 1553–1558.

Tada, T., Obata, Y., Tada, M., Goto, Y., Nakatsuji, N., Tan, S., Kono, T. and Takagi, N. (2000). Imprint switching for non-random X-chromosome inactivation during mouse oocyte growth. *Development* 127, 3101–3105.

Takashima, Y., Guo, G., Loos, R., Nichols, J., Ficz, G., Krueger, F., Oxley, D., Santos, F., Clarke, J., Mansfield, W. *et al.* (2014). Resetting transcription factor control circuitry toward ground-state pluripotency in human. *Cell* 158, 1254–1269.

Tang, W. W.C., Dietmann, S., Irie, N., Leitch, H. G., Floros, V. I., Bradshaw, C. R., Hackett, J. A., Chinnery, P. F. and Surani, M. A. (2015). A unique gene regulatory network resets the human germline epigenome for development. *Cell* 161, 1453–1467.

Tavares, L., Dimitrova, E., Oxley, D., Webster, J., Poot, R., Demmers, J., Bezstarosti, K., Taylor, S., Ura, H., Koide, H. *et al.* (2012). RYBP-PRC1 complexes mediate H2A ubiquitylation at polycomb target sites independently of PRC2 and H3K27me3. *Cell* 148, 664–678.

Theunissen, T. W., Powell, B. E., Wang, H., Mitalipova, M., Faddah, D. A., Reddy, J., Fan, Z. P., Maetzel, D., Ganz, K., Shi, L. *et al.* (2014). Systematic identification of culture conditions for induction and maintenance of naive human pluripotency. *Cell Stem Cell* 15, 471–487.

Tian, D., Sun, S. and Lee, J. T. (2010). The long noncoding RNA, Jpx, is a molecular switch for X chromosome inactivation. *Cell* 143, 390–403.

Tsai, C.-L., Rowntree, R. K., Cohen, D. E. and Lee, J. T. (2008). Higher order chromatin structure at the X-inactivation center via looping DNA. *Developmental Biology* 319, 416–425.

Vallot, C., Huret, C., Lesecque, Y., Resch, A., Oudrhiri, N., Bennaceur-Griscelli, A., Duret, L. and Rougeulle, C. (2013). XACT, a long noncoding transcript coating the active X chromosome in human pluripotent cells. *Nature Genetics* 45, 239–241.

Vallot, C., Ouimette, J.-F., Makhlouf, M., Féraud, O., Pontis, J., Côme, J., Martinat, C., Bennaceur-Griscelli, A., Lalande, M. and Rougeulle, C. (2015). Erosion of X chromosome inactivation in human pluripotent cells initiates with XACT coating and depends on a specific heterochromatin landscape. *Cell Stem Cell* 16, 533–546.

van den Berg, I. M., Laven, J. S.E., Stevens, M., Jonkers, I., Galjaard, R.-J., Gribnau, J. and van Doorninck, J. H. (2009). X chromosome inactivation is initiated in human preimplantation embryos. *American Journal of Human Genetic* 84, 771–779.

Wang, J., Mager, J., Chen, Y., Schneider, E., Cross, J. C., Nagy, A. and Magnuson, T. R. (2001). Imprinted X inactivation maintained by a mouse Polycomb group gene. *Nature Genetics* 28, 371–375.

Williams, L. H., Kalantry, S., Starmer, J. and Magnuson, T. R. (2011). Transcription precedes loss of Xist coating and depletion of H3K27me3 during X-chromosome reprogramming in the mouse inner cell mass. *Development* 138, 2049–2057.

Wutz, A. and Agrelo, R. (2012). Response: the diversity of proteins linking xist to gene silencing. *Developmental Cell* 23, 680.

Wutz, A., Rasmussen, T. P. and Jaenisch, R. (2002). Chromosomal silencing and localization are mediated by different domains of Xist RNA. *Nature Genetics* 30, 167–174.

Xu, N., Donohoe, M. E., Silva, S. S. and Lee, J. T. (2007). Evidence that homologous X-chromosome pairing requires transcription and Ctcf protein. *Nature Genetics* 39, 1390–1396.

Xu, N., Tsai, C.-L. and Lee, J. T. (2006). Transient homologous chromosome pairing marks the onset of X inactivation. *Science* 311, 1149–1152.

Yamada, N., Hasegawa, Y., Yue, M., Hamada, T., Nakagawa, S. and Ogawa, Y. (2015). Xist exon 7 contributes to the stable localization of Xist RNA on the inactive X-chromosome. *PLoS Genetics* 11, e1005430.

Yamaji, M., Seki, Y., Kurimoto, K., Yabuta, Y., Yuasa, M., Shigeta, M., Yamanaka, K., Ohinata, Y. and Saitou, M. (2008). Critical function of Prdm14 for the establishment of the germ cell lineage in mice. *Nature Genetics* 40, 1016–1022.

Yamaji, M., Ueda, J., Hayashi, K., Ohta, H., Yabuta, Y., Kurimoto, K., Nakato, R., Yamada, Y., Shirahige, K. and Saitou, M. (2013). PRDM14 ensures naive pluripotency through dual regulation of signaling and epigenetic pathways in mouse embryonic stem cells. *Cell Stem Cell* 12, 368–382.

Yang, F., Deng, X., Ma, W., Berletch, J. B., Rabaia, N., Wei, G., Moore, J. M., Filippova, G. N., Xu, J., Liu, Y. *et al.* (2015). The lncRNA Firre anchors the inactive X chromosome to the nucleolus by binding CTCF and maintains H3K27me3 methylation. *Bipartite Structure of the Inactive Mouse X Chromosome*. 16, 52.

Yildirim, E., Kirby, J. E., Brown, D. E., Mercier, F. E., Sadreyev, R. I., Scadden, D. T. and Lee, J. T. (2013). Xist RNA is a potent suppressor of hematologic cancer in mice. *Cell* 152, 727–742.

Zhang, L.-F., Huynh, K. D. and Lee, J. T. (2007). Perinucleolar targeting of the inactive X during S phase: evidence for a role in the maintenance of silencing. *Cell* 129, 693–706.

Zhao, J. C., Sun, B. K., Erwin, J. A., Song, J.-J. and Lee, J. T. (2008). Polycomb proteins targeted by a short repeat RNA to the mouse X chromosome. *Science* 322, 750–756.

6 Placentation and Genomic Imprinting

Courtney W. Hanna and Wendy P. Robinson*[†,‡]

**Epigenetics Programme, Babraham Institute, Cambridge, UK;*
[†]Department of Medical Genetics, University of British Columbia, Vancouver, Canada;
[‡]BC Children's Hospital Research, Vancouver, Canada

Genomic imprinting is an epigenetic phenomenon that permits a gene to be expressed from a single allele based on the parent-of-origin. While only a small subset of genes is imprinted in the human genome, the correct regulation of these loci is essential for healthy pregnancy. Imprinted loci not only contribute to the development of the fetus, but are required for placental function. This chapter will explore our current understanding of genomic imprinting, its effects on placentation and its contribution to complications of human pregnancy.

6.1 Introduction

The maternal and paternal copies of the genome are both required for mammalian development; this is primarily due to the existence of imprinted genes, those that are mono-allelically expressed based on parent-of-origin. Imprinted genes are involved in growth and development, and imprinting disorders characteristically are associated with fetal growth and placental abnormalities (Frost and Moore 2010). The development of nuclear transfer techniques in mouse zygotes, more than 30 years ago, allowed researchers to first show that the two parental contributions to the genome were not equivalent, as the uniparental embryos died early in development (McGrath and Solter 1984, Surani *et al.* 1984). Furthermore, this imbalance of parental genomes resulted in dramatic placental phenotypes, such that embryos with two maternal genomes (parthenogenetic) largely lacked placental tissue, while embryos with two paternal genomes (androgenetic) did not develop embryonic tissues and had an overgrowth of the placenta (McGrath and Solter 1984, Surani *et al.* 1984). This work set the base for the continuing investigation of how imprinted genes are important for normal fetal and placental development. In this chapter, we will discuss (i) the unique characteristics of the placenta, (ii) epigenetics of imprinting, (iii) how imprinting is hypothesised to have evolved, (iv) the role of imprinting specifically in the placenta and (v) how aberrant imprinting may influence the health of pregnancy.

6.2 Features of the Human Placenta

The placenta is a transient fetal organ that plays many roles in establishing and maintaining a pregnancy. Trophoblast, in the form of trophectoderm, is the first cell type to differentiate during embryonic development, composing the outer layer of the blastocyst (Rossant and Cross 2001). These cells are involved in adhering to and invading into the maternal endometrium allowing the embryo to implant, and producing a cascade of maternal effects that prevent immune rejection of the embryo and promote pregnancy-associated physiological changes (Rossant and Cross 2001). The trophoblast also plays a role in regulating oxygen/blood flow, first by plugging maternal vessels to maintain a hypoxic environment needed for trophoblast proliferation, and ultimately remodelling maternal vessels to allow blood to bathe the placental villi (Enders *et al.* 1996). This enables gas and nutrients to be extracted from the maternal blood and transported across the trophoblast layer and inner mesenchymal core of the chorionic villi to the fetal vasculature (Carter 2012). As pregnancy progresses, nutrient delivery to the fetus is tightly regulated by the placenta through a variety of metabolic processes (Diaz *et al.* 2014). The placenta also serves as a key endocrine organ with production of such hormones as human chorionic gonadotropin, placental lactinogen, estrogen, progesterone, and leptin (Carter 2012). Throughout gestation, placental macrophages and other immune cells are abundant, offering protection from infectious agents and possibly playing a role in angiogenesis (Seval *et al.* 2007, Tang *et al.* 2011).

While the function of the placenta is similar across mammalian species, its structure is remarkably diverse (Carter and Enders 2004, Enders and Carter 2012). The human placenta is discoid and consists of 50–60 villous trees that are connected at the chorionic plate; the outer trophoblast layer of the chorionic villi is in contact with the maternal blood to enable nutrient and oxygen exchange (haemochorial). Most carnivores have an endothelial placenta, in which the chorionic villi are juxtaposed to maternal endothelium but not in direct contact with maternal blood, and it is often in the form of a continuous band of tissue that surrounds the fetus. In many ruminant species, the fetal vessels are further separated from the maternal blood supply by a uterine epithelial layer (referred to as epithieliochorial). In some ruminants (horses and pigs), the placenta is thin and diffuse and covers almost the entire allantochorionic membrane, while in others (cattle, sheep, goats), there are multiple distinct sites of attachment called cotyledons.

The mouse also has a discoid, haemochorial placenta and is often used as a model for understanding early gene expression in placental development; however, it is important to appreciate that there are some major differences in both structure, cell types present, and gene expression (Dilworth and Sibley 2013, Rossant and Cross 2001). For example, in the mouse, there are three trophoblast layers that separate maternal and fetal circulation, whereas in humans there is one layer (Dilworth and Sibley 2013). Some functionally important genes are uniquely expressed in one or the other species, which may reflect divergent placental processes or that other genes may have evolved to replace the function of another (Cox *et al.* 2009). Nonetheless, most regulatory processes are conserved across species and over 80% of genes known to cause placental phenotypes in mouse are also expressed in human (Cox *et al.* 2009).

6.3 Genomic Imprinting

Imprinted genes are those that are expressed from a single allele of a chromosome pair, based on the parent of origin. This mono-allelic expression is regulated by a differentially methylated region (DMR) either at the gene promoter or at a nearby imprinting control region (ICR). This pattern of differential DNA methylation can be established in the gametes (germline DMR) or in somatic tissues during development (secondary DMR). Imprinted DMRs, including ICRs, are characteristically GC rich regions and are distributed throughout the genome (Kobayashi *et al.* 2006, Schulz *et al.* 2010). They can occur in isolation regulating a single gene, or more commonly in clusters. Clusters of imprinted genes are regulated by an ICR that is established in the germline and is commonly associated with the expression of one or more non-coding RNAs that are postulated to regulate the mono-allelic DNA methylation and/or expression of genes in *cis* (on the same chromosome) (Wan and Bartolomei 2008).

Imprinting has been extensively studied in mouse, with efforts to systematically identify imprinted loci and understand the mechanisms of imprinting during development. In mouse, 23 germline ICRs have been described, 20 maternally and 3 paternally-methylated regions (Arnaud 2010), and 18 of these are also conserved in humans (Table 6.1) (Court *et al.* 2014). The comprehensive study of imprinting

Table 6.1. Germline imprinted ICRs that are conserved between mouse (Arnaud 2010) and human (Court *et al.* 2014) with associated human phenotypes, co-ordinately regulated genes (http://igc.otago.ac.nz/) and mouse phenotypes (Varmuza and Miri 2015).

Human locus	Parental DNAm	Human DMR (hg19)	Human phenotype	Associated imprinted genes	Mouse DMR	Mouse phenotype
ZDBF2	Paternal	chr2:207114583-207136544		*GPR1-AS*	*Zdbf2*	No KO model
NAP1L5	Maternal	chr4:89618184-89619237			*Nap1L5*	No KO model
PLAGL1	Maternal	chr6:144328078-144329888	Transient neonatal diabetes mellitus	HYMAI, PHACTR2	*Plagl1*	KO has placental growth restriction
IGF2R	Maternal*	chr6:160426558-160427561		*SLC22A2*	*Igf2r*	KO shows placentomegaly
GRB10	Maternal	chr7:50848726-50851312			*Grb10*	KO has placental and fetal overgrowth
PEG10	Maternal	chr7:94285537-94287960		*CALCR*, *TFPI2*, *SGCE*	*Peg10*	*Peg10* KO shows placental defects *Calcr* KO is embryonic lethal *Sgce* KO has myoclonus dystonia
MEST	Maternal	chr7:130130122-130134388		*CPA4*, *MESTIT1*, *KLF14*	*Mest*	KO has growth restricted placenta and embryo
KCNK9	Maternal	chr8:141108147-141111081			*Kcnk9*	KO has mild neural impairment
INPP5F	Maternal	chr10:121578046-121578727			*Inpp5f*	Heart phenotype
H19/IGF2	Paternal	chr11:2018812-2024740	Silver-Russell/Beckwith-Wiedemann syndrome	*IGF2AS*, *INS*	*H19/Igf2*	*H19* KO has placentomegaly and increased birth weight Placental-specific *Igf2* KO shows fetal growth restriction

KCNQ1OT1	Maternal	chr11:2719948-2722259	Beckwith Wiedemann syndrome	*KCNQ1*, *KCNQ1DN*, *CDKN1C*, *SLC22A18*, *PHLDA2*	*Kcnqlot1*	*Kcnq1ot1* ICR KO shows fetal and placental growth restriction *Cdkn1c* and *Phlda2* KO placentomegaly, *Kcnq1* KO has deafness
DLK1/ MEG3	Paternal	chr14:101275427-101278058	Uniparental Disomy 14 syndrome	*RTL1*, *MEG8*	*Dlk1/Meg3*	*Dlk1* KO is growth restricted with placental defects *Meg3* and *Rtl1* paternal KO has IUGR and maternal KO has placentomegaly
SNRPN	Maternal	chr15:25200004-25201976	Prader-Willi/Angelman syndrome	*MKRN3*, *MAGEL2*, *NDN*, *SNURF*, *UBE3A*	*Snrpn*	*Snrpn* KO has no phenotype KO of *Snrpn* ICR causes IUGR *Ube3a* KO has neural impairment *Magel2* KO is perinatally lethal
PEG3	Maternal	chr19:57348493-57353271		*ZIM2*, *MIMT1*	*Peg3*	KO shows placental growth restriction
PSIMCT-1	Maternal	chr20:30134663-30135933			*Mcts2*	No KO model
NNAT	Maternal	chr20:36148604-36150528		*BLCAP*	*Nnat*	No phenotype
GNAS	Maternal	chr20:57425649-57428033	Pseudohypoparathyroidism (Albright hereditary osteodystrophy)	*NESP55*, *GNASXL*, *NESPAS*	*Gnas*	*Gnas* and *Nesp55* KO have IUGR and die postnatally

*Polymorphic and placental-specific

in humans has been challenging due to the lack of genetic tools that are available in mouse models and the technical and ethical limitations of studying early human development and gametes. Despite these challenges, evidence thus far shows that the synteny of imprinted gene clusters, parental imprinting patterns and phenotypes associated with imprinted regions are largely conserved between mouse and human (Table 6.1) (Onyango *et al.* 2000).

6.3.1 *Imprinting in development*

Imprinted DMRs are erased and, largely, re-established in the gametes, allowing for the inheritance of the parent-of-origin specific pattern in each subsequent generation. There is widespread erasure of DNA methylation following the migration and division of primordial germ cells in the early embryo (Seisenberger *et al.* 2012). This is followed by an asynchronous establishment of germline DMRs (gDMRs) that is dependent on the sex of the individual (Figure 6.1). In males this DNA

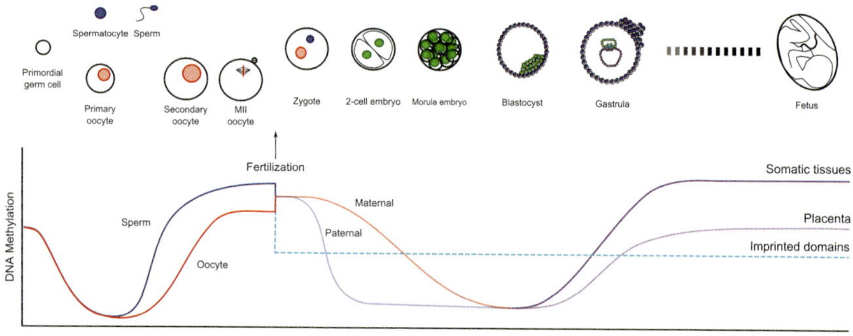

Figure 6.1. DNA methylation is reprogrammed in germ cells and the early embryo. DNA methylation is erased upon the specification of primordial germ cells in both males and females. During the latter stages of fetal development, sperm reacquires substantial methylation throughout the genome. Oocytes become remethylated in adulthood during oocyte maturation and have overall less methylation than sperm. Upon fertilization, there is active demethylation of the paternal pronuclei before the first cell division. Subsequently, there is primarily passive demethylation of the maternal pronuclei through the early cell divisions. At the blastocyst stage most of the DNA methylation has been removed with the exception of some repetitive elements and imprinted domains, which maintain methylation throughout this embryonic reprograming. During gastrulation, the somatic lineages acquire tissue-specific methylation patterns necessary for differentiation and while the placenta accumulates some methylation, but it largely remains intermediately methylated throughout gestation.

methylation occurs before birth starting in prospermatogonia prior to meiosis, while in females this occurs primarily during adulthood in oocytes that have been meiotically-arrested since fetal development and undergo maturation concomitant with the menstrual cycle (Lee *et al.* 2014). The established pattern of DNA methylation is dramatically different between the two germ cells; with oocytes showing enrichment of DNA methylation in gene bodies, while sperm are methylated extensively at both inter-genic regions and gene bodies (Smallwood *et al.* 2011). Consistently, the characteristics of maternal and paternal gDMRs differ; while maternal DMRs are associated with CpG islands at gene promoters, paternal DMRs are inter-genic and have been associated with GC rich repeats (Kobayashi *et al.* 2006, Schulz *et al.* 2010). Upon fertilisation, maternal and paternal imprints are protected throughout the widespread reprogramming of DNA methylation that shortly follows and during subsequent embryonic and extra-embryonic differentiation.

The enzymes responsible for the establishment of imprints in the male and female germ cells are the *de novo* DNA methyltransferases (DNMTs) 3A, 3L and 3B (Hata *et al.* 2002, Kaneda *et al.* 2004, Kato *et al.* 2007). These proteins form hetero-tetramers that target specific regions of the genome for DNA methylation (Jia *et al.* 2007). The differential targeting of these enzymes between sperm and oocytes is still not well understood, but it is hypothesized to be dependent on gene expression in the oocyte and is perhaps a widespread occurrence in sperm (Smallwood *et al.* 2011). The maintenance DNMT1 is required to faithfully copy DNA methylation patterns to the newly synthesized strand of DNA after cell division, as it preferentially recognizes hemi-methylated DNA (Lyko *et al.* 1999, Yoder *et al.* 1997). DNMT1 along with several cofactors, such as ZFP57, TRIM28 and PGC7, have been shown to be essential for the maintenance of imprints during early embryonic development (Li *et al.* 2008, Messerschmidt *et al.* 2012, Nakamura *et al.* 2007). This maintenance allows the imprinted DMRs, and correspondingly mono-allelic expression, to be conferred to all relevant tissue types in the developing fetus and into adulthood. Imprinted genes are particularly important in fetal growth and neural development, in addition to placentation (Plasschaert and Bartolomei 2014). After birth, imprinted genes may be particularly important in continued brain development and function (Keverne *et al.* 1996).

6.3.2 *Evolution of imprinting*

Imprinting is present in all eutherians (placental mammals) studied to date, but not in egg-laying mammals (monotremes) (Killian *et al.* 2000) and only to a limited degree in marsupials (Renfree *et al.* 2008). Hence imprinting may have evolved with placentation and viviparity. The exception to this is endospermic plants, although they may have evolved imprinting in a related manner (Moore and Haig 1991). While there are several hypotheses to explain how imprinting evolved in the literature, the most prominent is the parental conflict hypothesis or kinship theory (Haig and Westoby 1989).

The parental conflict hypothesis suggests that imprinting arose due to the conflict between the maternal and paternal genomes transmitted to the offspring over maternal resource allocation (Haig and Westoby 1989, Moore and Haig 1991). In other words, the paternal genome promotes growth of the fetus, thereby demanding more resources from the mother through the maternal-fetal interface; while the maternal genome limits growth to allow survival of the mother and preservation of resources for subsequent reproduction. As the placenta mediates nutrient exchange between the fetus and mother, many imprinted genes may affect placental invasiveness and functioning under this hypothesis. Furthermore, it has been suggested that an influence of imprinted genes over metabolism and behavior postnatally could be an extension of the intrauterine conflict into adulthood (Lefebvre *et al.* 1998, Plagge *et al.* 2004).

Alternatively, the contrasting placental trophoblast phenotypes observed in the parthenogenetic and androgenetic mouse embryos (McGrath and Solter 1984, Surani *et al.* 1984) supported the development of the hypothesis that imprinting may be a mechanism to prevent parthenogenetic activation in eutherians (Hall 1990, Solter 1988). Parthenogenetic embryos can arise from activation of an unfertilized oocyte (Linder *et al.* 1975, Stevens and Varnum 1974) and can have serious complications if they develop into malignant ovarian teratomas. The silencing of the maternal copy of genes required for extra-embryonic development and implantation may provide a defence against spontaneous parthenogenetic pregnancies (Varmuza and Mann 1994).

Several additional hypotheses can be found in the literature centred on resource allocation between mother and fetus, including the juxtaposed maternal-fetal co-adaptation (Wolf and Hager 2006) and maternal-fetal coordination (Curley *et al.* 2004, Keverne and Curley 2008) hypotheses. An alternative mechanistic explanation is that imprinting arose from genomic defence against viral repetitive elements (Barlow 1993). There are many other less well-developed hypotheses that have been reviewed elsewhere (Spencer and Clark 2014). At present, none of the proposed hypotheses can explain all facets of imprinting, and it is possible that imprinting arose due to multiple contributing factors; therefore, as the study of imprinting continues this scientific discussion will likely be revisited.

6.3.3 *Placental-specific imprinting*

There are several genes that have been shown to only be imprinted in the placenta in mouse (Frost and Moore 2010) or human (Court *et al.* 2014); however, the majority of these placental-specific imprinted genes appear not to be conserved in methylation status. Furthermore, in mouse they have been associated with the periphery of larger imprinted clusters (Lewis *et al.* 2004); while in humans they appear to be isolated and dispersed throughout the genome (Court *et al.* 2014). Similarly, in humans there appears to be a diversity of placental-specific features that are not common between all DMRs.

For example, the largest microRNA cluster in humans, the chromosome 19 microRNA cluster (C19MC), is only expressed in placenta and is imprinted, being regulated by a maternally methylated placental-specific ICR (Noguer-Dance *et al.* 2010). These microRNAs are selectively packaged into exosomes and exported into maternal circulation where they appear to protect against viruses, amongst other functions (Ouyang *et al.* 2014). Other genes, such as DNA methyltransferase *DNMT1*, are ubiquitously expressed and have widespread functionality but are specifically imprinted in the placenta (Das *et al.* 2013). The function of placental-specific imprinting of *DNMT1* is still unclear. In addition, placental-specific transcripts have been identified for several genes, including *CDKN1C* and *IGF2* (Monk *et al.* 2006, Yuen *et al.* 2011). Polymorphic or variable imprinting has also been reported in placenta, such as for *IGF2R* (Lambertini *et al.* 2008); the nature of this polyorphic patterning is uncertain, but these loci may represent recently evolved imprinted loci where methylation patterns are modulated by genetic variants or genomic features (Hanna and Kelsey 2014).

The investigation of placental-specific imprinted loci in mouse suggests that DNA methylation may not always be required to maintain mono-allelic expression in the placenta. The placental-specific imprinted genes on the periphery of the *Kcnq1ot1* imprinted cluster, including *Osbpl5*, *Tssc4* and *Cd81*, maintain mono-allelic expression in the placenta in the absence of DNA methylation during embryonic development (Lewis *et al.* 2004). This is in contrast to the 'classically-behaving' imprinted genes in the cluster, whereby loss of DNA methylation results in bi-allelic expression in both the embryo and placenta (Lewis *et al.* 2004). These data suggest that placental-specific imprinting may also be mediated by another epigenetic mechanism, such as histone modifications or non-coding RNAs. Placental-specific imprints have germline DMRs in humans (Okae *et al.* 2014), but while the necessity of DNA methylation across development has not been directly explored, imprinting is largely conserved across gestation in the placenta (Bourque *et al.* 2011).

It might be expected that placental-specific imprinted genes should have an important role in placentation and/or embryonic development; however, among the several human genes, few have been reported to be directly involved in metabolism or growth, nor placental function (Table 6.2). This raises the question as to whether this placental-specific imprinting has a functional role or whether these DMRs may be the result of an 'innocent bystander' effect (Wang *et al.* 2011).

Table 6.2. Human placental-specific imprinted genes (Court *et al.* 2014) and their putative reported function (http://www.omim.org/).

Placental-specific imprinted gene	Gene function
GPR1-AS	Non-coding RNA
MCCC1	Biotin-dependent mitochondrial enzyme that catabolizes leucine
PDE4D	Mediates response to extracellular signals and regulate cyclic nucleotides, important for growth and female fertility in mice
LIN28B	Zinc finger protein that promotes cell growth
AIM1	Similar in structure to beta-crystalline genes and may interact with the cytoskeleton
AGBL3	Catalyzes deglutamylation of polyglutamated proteins
ZFAT/ZFAT-AS	Zinc finger protein, important for spongiotrophoblast development and hematopoiesis in mice/non-coding RNA
GLIS3	Zinc finger protein that regulates transcription, expressed in neural tissues during early mouse development and may be important for kidney and pancreatic function
DCAF10	Unknown
FAM196A/DOCK1	Unknown/membrane protein involved in recognition and engulfment of dead cells
ZC3H12C	Zinc finger protein that regulates macrophage activation
RGMA	Signaling peptide important in early embryonic development and neural cell function
FAM20A	Secreted glycoprotein that may be important for mouse growth and dental enamel formation
ZNF396	Zinc finger protein that may act as a transcriptional repressor
C19MC	MiRNA cluster involved in cell proliferation and placental immune response (Ouyang *et al.* 2014)
DNMT1	Maintenance DNA methyltransferase

6.4 Functional Role of Imprinted Genes in the Human Placental Development

6.4.1 *Placental abnormalities in imprinting disorders*

Evidence from mouse models suggests that the majority of imprinted gene clusters have at least one gene that is important in placentation and fetal growth (Table 6.1).

There are seven human imprinting disorders, the majority of which are associated with a growth phenotype. While the human literature is not as comprehensive on placental abnormalities, these disorders can be accompanied by defects in placentation.

Prader-Willi and Angelman syndrome are caused by reciprocal deletions on chromosome 15q11–13, uniparental disomies (UPD), genetic mutations or epimutations (erroneous loss or gain of methylation at an ICR) (Girardot *et al.* 2013). Prader–Willi syndrome results from the erroneous silencing of the cluster of paternally expressed genes in this region, including *MKRN3*, *MAGEL2*, *NDN*, *C15orf2*, *SNURF-SNRPN* and snoRNA genes; while suppression of the maternally expressed genes *UBE3A* and *ATP10A* leads to Angelman Syndrome (Cassidy *et al.* 2012, Horsthemke and Wagstaff 2008). All of these genes appear to be regulated from a common maternally methylated ICR located proximal to the *SNRPN* gene. Prader–Willi newborns are usually within the normal growth range at birth, though there is an increase in the reported rate of small for gestational age and preterm births (Lionti *et al.* 2015). The latter may be caused by early induction of labor due to the decreased fetal movement characteristic of fetuses with Prader-Willi syndrome. An increase in polyhydramnios, excessive fluid surrounding the fetus, has also been documented (Dudley and Muscatelli 2007), which is also associated with preterm birth. Increased birth weight and altered postnatal eating behaviour has been associated with Angelman syndrome in a few cases, but appears to be dependent on type of underlying genetic alteration (Mertz *et al.* 2014). There are no consistent reports of placental abnormalities associated with the 15q11–13 imprinting defects.

Beckwith–Weidemann (BWS) and Silver–Russell (SRS) syndromes are another example of reciprocal disorders caused by abnormalities at one of two ICRs located on chromosome 11p15. The paternally methylated ICR1 is located near *H19* and regulates the maternal expression of *H19*, a non-coding transcript, and the paternal expression of *IGF2*. The maternally methylated ICR2 is located within the *KCNQ1* gene at the promoter of *KCNQ1OT1*, another non-coding transcript, and co-ordinately regulates the imprinting of several nearby genes (Table 6.1). Similarly to Prader-Willi and Angelman, these disorders can result from deletions, UPDs, genetic mutations or epimutations that disrupt the imprinting patterns at ICR1 in SRS and at either the ICR1 or ICR2 clusters for BWS. BWS is characterised by overgrowth, while SRS by growth restriction both prenatally and postnatally (Girardot *et al.* 2013). Similarly, BWS is associated with placentomegaly (overgrowth of the placenta) (Lage 1991), while epimutations causing SRS have been associated with small placentas with hypoplastic chorionic villi (Yamazawa *et al.* 2008). In SRS, maternal UPD7 accounts for about 10% of cases (Monk *et al.* 2002, Penaherrera *et al.* 2010) and while several candidate imprinted genes on chromosome 7 have been implicated in placental and fetal growth (e.g. *MEST*, *GRB10*), they have not been directly implicated as causing SRS (Kobayashi *et al.* 2001, Monk *et al.* 2009).

Transient neonatal diabetes mellitus (TNDM) results from the bi-allelic expression of the paternally expressed imprinted gene *PLAGL1* on chromosome 6 and is associated with intrauterine growth restriction (IUGR) (Kamiya *et al.* 2000). A case study of paternal UPD6 found that the placenta was small for gestational age and showed abnormal vasculature (Cajaiba *et al.* 2011); however, placental abnormalities have not been widely reported.

Pseudohypoparathyroidism (PHP) results from resistance to the parathyroid hormone and is caused by abnormal imprinting at the GNAS locus (Rao *et al.* 1991). PHP has been associated with IUGR (Martin *et al.* 2012), but not placental abnormalities to date.

UPD14 syndromes are associated with abnormal imprinting at the *DLK1/MEG3* locus located near a cluster of genes on 14q32 (Kagami *et al.* 2008). Paternal expression of *DLK1* and *RTL1* and maternal expression of *MEG3*, *RTL1AS* and *MEG8*, as well as imprinting of several snoRNA and miRNA genes, are regulated by the inter-genic paternal ICR upstream of *DLK1* and at *MEG3* (Court *et al.* 2014, Kagami *et al.* 2008). Maternal UPD14 is characterized by growth restriction, while paternal UPD14 is characterized by dysmorphism and placentomegaly (Kagami *et al.* 2008, 2012).

6.4.2 *Genomic imbalance of parental contribution and placental pathologies*

Genomic imbalances observed in human pregnancies have been essential for understanding the importance of imprinting in placental and fetal development. In the obstetrical clinic, the phenotypic manifestations of imprinting defects are most commonly seen in the context of whole genomic imbalance as occurs in triploid or molar pregnancies. Triploidy, three haploid chromosome complements, occurs in ~1–2% of clinically recognized pregnancies, but most are miscarried in the first trimester and rarely survive to term (Lakovschek *et al.* 2011, Shepard *et al.* 1989). The phenotypic differences between triploids of digynic or diandric origin provided some of the first evidence for imprinting in humans (McFadden and Kalousek 1991). While digynic triploids tend to have small placentas, a subset of diandric triploids shows features of a partial hydatidiform mole (PHM) (Redline *et al.* 1998). PHM are defined by the presence of a developing fetus and variable degrees of cystic chorionic villi with regional trophoblast hyperplasia (McFadden and Pantzar

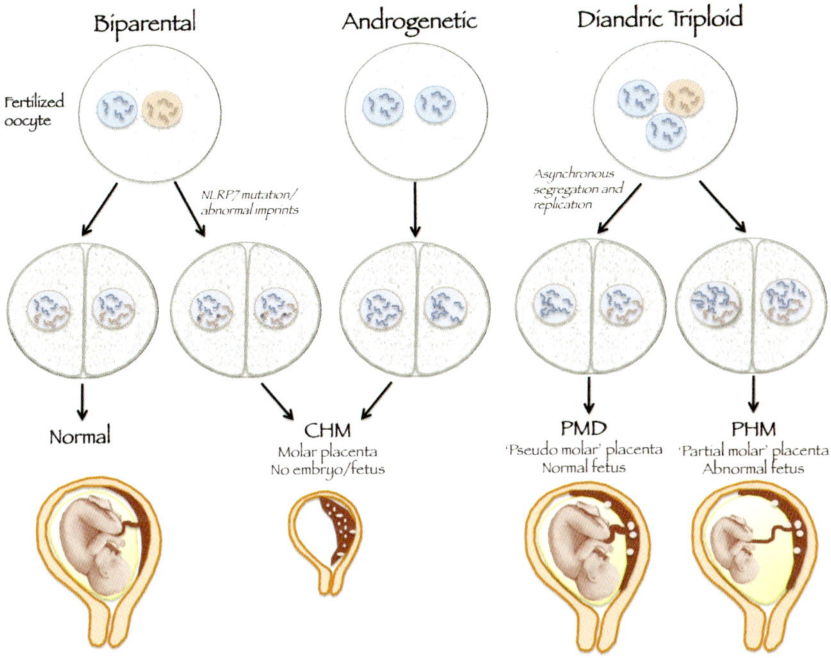

Figure 6.2. Molar-like changes in the placenta can result from several related mechanisms involving loss of maternal imprints at multiple loci. Complete hydatidiform mole (CHM) normally results from androgenetic conceptions (two paternal and no maternal genome contribution). However, CHM can occur in biparental conceptuses when the mother carries a mutation in *NLRP7 or KHDC3L* associated with aberrant setting or maintenance of maternal imprints. Partial hydatidiform mole (PHM) and placental mesenchymal dysplasia (PMD) show overlapping placental changes but are associated with different underlying mechanisms: PHM is due to androgenetic triploidy while PMD is usually due to mixed androgenetic and biparental diploid cell populations. Fetal outcomes are also different, with PMD often found in association with a normal healthy baby, though these pregnancies are at increased risk for features of BWS and/or stillbirth.

1996); whereas complete hydatidiform moles (CHM), caused by an androgenetic pregnancy (only paternal genomic contribution), is characterized by no fetal development and more widespread placental changes (Figure 6.2) (Wake *et al.* 1978).

A related pathology is that of placental mesenchymal dysplasia (PMD), which is normally attributable to the presence of mixed androgenetic and biparental diploid cell populations (Robinson *et al.* 2007a). Affected pregnancies are usually associated with developmentally normal fetuses but have large and bulky placentas with enlarged blood vessels (Paradinas *et al.* 2001). Some fetuses exhibit features of BWS, although PMD appears to be a relatively rare finding among BWS cases as a whole (Robinson *et al.* 2007b). In the rare cases of BWS reported with abnormal placental pathology that is either diagnosed as PMD or similar to this, the underlying mechanism may be due to androgenetic cells confined to the placenta (causing fetal overgrowth as a secondary consequence of placental overgrowth) or due to mosaic paternal duplications or UPD of chromosome 11p15.5. Considering other mutations leading to BWS (*CDKN1C*, or loss of imprinting at either ICR1 or ICR2) have not been reported with PMD, it may only occur when there is an imbalance between the two involved imprinted regions (Robinson *et al.* 2007b). There is evidence for cross-talk between the ICR1 and ICR2 regions, whereby their expression is co-dependent (Jacob *et al.* 2013).

While PMD, CHM and PHM have an imbalance of the entire genome (including all imprinted genes), it is conceivable that much of the phenotype is primarily due to a small subset of imprinted genes. Some CHMs have been shown to be biparental, but exhibit loss of imprinting (LOI) at multiple maternally imprinted loci (Judson *et al.* 2002). Some women experiencing recurrent biparental CHMs have been found to carry mutations in both copies of *NLRP7* (Abdalla *et al.* 2012, Wang *et al.* 2009) or *KHDC3L* (Reddy *et al.* 2013). Expression of both *NLRP7* and *KHDC3L* is similar with very high expression in mature oocytes, but undetectable by the late blastocyst stage (Parry *et al.* 2011, Zhang *et al.* 2008). These genes may affect either the setting of maternal imprints during late oocyte maturation or their maintenance in the first few cell divisions. As the loci affected by LOI are not identical in each case the specific imprinted loci accounting for the placental phenotype have not been fully delineated (El-Maarri *et al.* 2003, Hayward *et al.* 2009, Nguyen and Slim 2014).

The placental phenotypes observed in triploid, PMD and CHM conceptions are similar to the findings observed in androgenetic and parthenogenetic mouse embryos, which reinforces that imprinted gene expression, particularly at conserved loci, are essential in placentation.

6.4.3 *Contribution of abnormal imprinting to non-syndromic placental pathology*

While many imprinted genes are expressed in the placenta, and knockouts of many imprinted genes in mouse affects placental growth and function (Table 6.1) (Varmuza and Miri 2015), direct data in humans for non-syndromic LOI/mutations at specific imprinted loci and placental pathology is sparse. Investigations searching for such changes in humans have focused largely on altered fetal growth and LOI at only a few candidate genes.

Altered DNA methylation at imprinted DMRs for *PLAGL1* (Azzi *et al.* 2014, Diplas *et al.* 2009, Liu *et al.* 2012), *PEG10* (Lim *et al.* 2012), and *H19/IGF2* (Bourque *et al.* 2010, Bouwland-Both *et al.* 2013, Cordeiro *et al.* 2014, Guo *et al.* 2008, Koukoura *et al.* 2011, Liu *et al.* 2013, Murphy *et al.* 2014, St-Pierre *et al.* 2012) have been associated with fetal growth restriction and/or small for gestational age. However, these changes have not been consistently seen and the magnitudes of effects are often small. Most genome-wide studies on placentas associated with fetal growth restriction have failed to show significant changes at imprinted genes by expression (Sitras *et al.* 2009, Turan *et al.* 2012) or methylation analysis (Banister *et al.* 2011). There has been one report of increased *PHLDA2* and decreased *MEST*, *MEG3*, *GNAS* and *PLAGL1* expression (McMinn *et al.* 2006) and another of DNA methylation changes at the imprinted genes *GRB10* and *OSBPL5* amongst the significant hits (Turan *et al.* 2012).

Importantly, the causes of reduced fetal growth are heterogeneous and the presence of placental insufficiency (as opposed to fetal or maternal origins of reduced growth) is often not evaluated. Small for gestational age (SGA) (<10th percentile) as often used as a surrogate for IUGR, which is a pathological cause of reduced growth that represents only a minority of SGA births. Furthermore, many small babies are associated with pregnancies that deliver pre-term, which in turn can be due to a number of other complications of pregnancy, including but not limited to preeclampsia, chorioamnionitis and polyhydramnios. Altered expression of imprinted genes in the placenta could be a cause of IUGR, a consequence of IUGR or a compensatory effect of other defects leading to IUGR. Further studies with strict inclusion criteria are needed to clarify the relationship between altered placental imprinting and IUGR.

Maternal preeclampsia (hypertension in pregnancy in conjunction with signs of placental insufficiency) is frequently associated with IUGR. An early factor predisposing to preeclampsia is reduced placental invasion into the maternal endometrium (Meekins *et al.* 1994). A role for imprinting is suggested by the high rate of preeclampsia in women pregnant with CHMs or triploid pregnancies with placentomegaly or molar changes (and thus likely androgenetic) (Rahimpanah and Smoleniec 2000, Rijhsinghani *et al.* 1997). A mutation in *CDKN1C* was implicated in increased risk for preeclampsia in a case study (Romanelli *et al.* 2009), and a similar phenotype is seen in mice carrying this mutation (Kanayama *et al.* 2002). Together, this suggests that abnormal imprinting could play a role in preeclampsia.

Mutations in multiple genes have been implicated in the risk for preeclampsia in isolated pedigrees (van Dijk and Oudejans 2013); however, most of the identified gene mutations are not in imprinted genes. The exception is an inherited mutation in *STOX1*, a paternally imprinted gene that is associated with risk of preeclampsia in female carriers (van Dijk *et al.* 2010). The high-risk allele in *STOX1* negatively affects expression of *CTNNA3* and extravillous trophoblast proliferation.

DNA methylation profiling has repeatedly shown that placentas associated with early onset preeclampsia exhibit widespread changes in DNA methylation (Anton *et al.* 2014, Blair *et al.* 2013, Chu *et al.* 2014, Jia *et al.* 2012, Yuen *et al.* 2010). While some imprinted loci were among these changes (e.g. altered methylation at *MEST* in early onset preeclampsia) (Blair *et al.* 2013), imprinted loci do not appear to be preferentially involved. Many of these DNA methylation changes may be due to changes in cell composition, which may obscure more subtle changes at imprinted (or other) loci or those occurring in only a subset of individuals.

Since many imprinted genes may be critical in placental development, LOI may be more likely to lead to miscarriage than placental pathology in an ongoing pregnancy. More extreme values of DNA methylation at several imprinted loci were reported to be increased in muscle samples from miscarriages and stillbirths as compared to elective terminations (Pliushch *et al.* 2010). A similar observation has been made for placental samples from miscarriages (Hanna *et al.* 2013) and an increase in average methylation at ICR1. It is impossible to say, however, whether these observations are causative of loss or instead a consequence of poor placental function that contributes to miscarriage.

While there has been extensive study into the contribution of abnormal imprinting to the health of human pregnancy, there is not overwhelming evidence suggesting that one or multiple imprinted genes involved in placentation contribute to IUGR, preeclampsia or miscarriage. This is likely because defects in imprinting may only contribute to a subset of pregnancy complications. Therefore, subtle overall differences in expression or DNA methylation at imprinted loci associated with disease may in fact be a consequence of altered functionality of the maternal-fetal interface in these states.

6.5 Conclusions

Imprinted genes are essential for eutherian fetal and placental development and this has been firmly established from genetic studies in mouse and imprinting disorders and genomic imbalances in humans. The co-ordinate evolution of imprinted genes and placentation in mammals supports this link. As many imprinted genes do not appear to have a direct role in the maintenance of pregnancy, it also posits a function of imprinting beyond intrauterine fetal development, which still remains to be fully elucidated. While loss of imprinting can result in congenital disease in humans; paradoxically, the more common related complications of pregnancy do not appear to be generally attributable to abnormal imprinting in either the embryo or placenta. An interesting area of future study will be understating the function of placental-specific imprinted genes and their contribution to healthy pregnancy.

Acknowledgments

We thank Dr. Maria Peñaherrera and Samantha Wilson for comments on the manuscript and Dr. Irina Manokhina for assisting with Figure 6.2. WPR is supported by an Investigator award from the Child & Family Research Institute.

References

Abdalla, E. M., Hayward, B. E., Shamseddin, A. and Nawar, M. M. (2012). Recurrent hydatidiform mole: detection of two novel mutations in the NLRP7 gene in two Egyptian families. *European Journal of Obstetics Gynecology Reproductive Biology* 164, 211–215.

Anton, L., Brown, A. G., Bartolomei, M. S. and Elovitz, M. A. (2014). Differential methylation of genes associated with cell adhesion in preeclamptic placentas. *PLoS One* 9, e100148.

Arnaud P. (2010). Genomic imprinting in germ cells: imprints are under control. *Reproduction* 140, 411–423.

Azzi, S., Sas, T. C., Koudou, Y., Le Bouc, Y., Souberbielle, J. C., Dargent-Molina, P., Netchine, I. and Charles, M. A. (2014). Degree of methylation of ZAC1 (PLAGL1) is associated with prenatal and post-natal growth in healthy infants of the EDEN mother child cohort. *Epigenetics* 9, 338–345.

Banister, C. E., Koestler, D. C., Maccani, M. A., Padbury, J. F., Houseman, E. A. and Marsit, C. J. (2011). Infant growth restriction is associated with distinct patterns of DNA methylation in human placentas. *Epigenetics* 6, 920–927.

Barlow, D. P. (1993). Methylation and imprinting: from host defense to gene regulation? *Science* 260, 309–310.

Blair, J. D., Yuen, R. K., Lim, B. K., McFadden, D. E., von Dadelszen, P. and Robinson, W. P. (2013). Widespread DNA hypomethylation at gene enhancer regions in placentas associated with early-onset pre-eclampsia. *Molecular Human Reproduction* 19, 697–708.

Bourque, D. K., Avila, L., Penaherrera, M., von Dadelszen, P. and Robinson, W. P. (2010). Decreased placental methylation at the H19/IGF2 imprinting control region is associated with normotensive intrauterine growth restriction but not preeclampsia. *Placenta* 31, 197–202.

Bourque, D. K., Penaherrera, M. S., Yuen, R. K., Van Allen, M. I., McFadden, D. E. and Robinson, W. P. (2011). The utility of quantitative methylation assays at imprinted genes for the diagnosis of fetal and placental disorders. *Clinical Genetics* 79, 169–175.

Bouwland-Both, M. I., van Mil, N. H., Stolk, L., Eilers, P. H., Verbiest, M. M., Heijmans, B. T., Tiemeier, H., Hofman, A., Steegers, E. A., Jaddoe, V. W. *et al.* (2013). DNA methylation of IGF2DMR and H19 is associated with fetal and infant growth: the generation R study. *PLoS One* 8, e81731.

Cajaiba, M. M., Witchel, S., Madan-Khetarpal, S., Hoover, J., Hoffner, L., Macpherson, T. and Surti, U. (2011). Prenatal diagnosis of trisomy 6 rescue resulting in paternal UPD6 with novel placental findings. *American Journal of Medical Genetics A* 155A, 1996–2002.

Carter, A. M. (2012). Evolution of placental function in mammals: the molecular basis of gas and nutrient transfer, hormone secretion, and immune responses. *Physiological Review* 92, 1543–1576.

Carter, A. M. and Enders, A. C. (2004). Comparative aspects of trophoblast development and placentation. *Reproductive Biology and Endocrinology* 2, 46.

Cassidy, S. B., Schwartz, S., Miller, J. L. and Driscoll, D. J. (2012). Prader-Willi syndrome. *Genetics in Medicine* 14, 10–26.

Chu, T., Bunce, K., Shaw, P., Shridhar, V., Althouse, A., Hubel, C. and Peters, D. (2014). Comprehensive analysis of preeclampsia-associated DNA methylation in the placenta. *PLoS One* 9, e107318.

Cordeiro, A., Neto, A. P., Carvalho, F., Ramalho, C. and Doria, S. (2014). Relevance of genomic imprinting in intrauterine human growth expression of CDKN1C, H19, IGF2, KCNQ1 and PHLDA2 imprinted genes. *Journal of Assisted Reproduction and Genetics* 31, 1361–1368.

Court, F., Tayama, C., Romanelli, V., Martin Trujillo, A., Iglesias-Platas, I., Okamura, K., Sugahara, N., Simon, C., Moore, H., Harness, J. V. *et al.* (2014). Genome-wide parent-of-origin DNA methylation analysis reveals the intricacies of the human imprintome and suggests a germline methylation independent establishment of imprinting. *Genome Research* 24, 554–569.

Cox, B., Kotlyar, M., Evangelou, A. I., Ignatchenko, V., Ignatchenko, A., Whiteley, K., Jurisica, I., Adamson, S. L., Rossant, J. and Kislinger, T. (2009). Comparative systems biology of human and mouse as a tool to guide the modeling of human placental pathology. *Molecular Systems Biology* 5, 279.

Curley, J. P., Barton, S., Surani, A. and Keverne, E. B. (2004). Coadaptation in mother and infant regulated by a paternally expressed imprinted gene. *Proceedings Biological Science* 271, 1303–1309.

Das, R., Lee, Y. K., Strogantsev, R., Jin, S., Lim, Y. C., Ng, P. Y., Lin, X. M., Chng, K., Yeo, G. S., Ferguson-Smith, A. C. *et al.* (2013). DNMT1 and AIM1 Imprinting in human placenta revealed through a genome-wide screen for allele-specific DNA methylation. *BMC Genomics* 14, 685-2164-14-685.

Diaz, P., Powell, T. L. and Jansson, T. (2014). The role of placental nutrient sensing in maternal-fetal resource allocation. *Biological Reproduction* 91, 82.

Dilworth, M. R. and Sibley, C. P. (2013). Review: Transport across the placenta of mice and women. *Placenta* 34 Suppl, S34–39.

Diplas, A. I., Lambertini, L., Lee, M. J., Sperling, R., Lee, Y. L., Wetmur, J. and Chen, J. (2009). Differential expression of imprinted genes in normal and IUGR human placentas. *Epigenetics* 4, 235–240.

Dudley, O. and Muscatelli, F. (2007). Clinical evidence of intrauterine disturbance in Prader-Willi syndrome, a genetically imprinted neurodevelopmental disorder. *Early Human Development* 83, 471–478.

El-Maarri, O., Seoud, M., Coullin, P., Herbiniaux, U., Oldenburg, J., Rouleau, G. and Slim, R. (2003). Maternal alleles acquiring paternal methylation patterns in biparental complete hydatidiform moles. *Human Molecular Genetics* 12, 1405–1413.

Enders, A. C. and Carter, A. M. (2012). The evolving placenta: convergent evolution of variations in the endotheliochorial relationship. *Placenta* 33, 319–326.

Enders, A. C., Lantz, K. C. and Schlafke, S. (1996). Preference of invasive cytotrophoblast for maternal vessels in early implantation in the macaque. *Acta Anatomica (Basel)* 155, 145–162.

Frost, J. M. and Moore, G. E. (2010). The importance of imprinting in the human placenta. *PLoS Genetics* 6, e1001015.

Girardot, M., Feil, R. and Lleres, D. (2013). Epigenetic deregulation of genomic imprinting in humans: causal mechanisms and clinical implications. *Epigenomics* 5, 715–728.

Guo, L., Choufani, S., Ferreira, J., Smith, A., Chitayat, D., Shuman, C., Uxa, R., Keating, S., Kingdom, J. and Weksberg, R. (2008). Altered gene expression and methylation of the human chromosome 11 imprinted region in small for gestational age (SGA) placentae. *Developmental Biology* 320, 79–91.

Haig, D. and Westoby, M. (1989). Parent-Specific Gene-Expression and the Triploid Endosperm. *The American Naturalist* 134, 147–155.

Hall, J. G. (1990). Genomic imprinting: review and relevance to human diseases. *American Journal of Human Genetics* 46, 857–873.

Hanna, C. W. and Kelsey, G. (2014). The specification of imprints in mammals. *Heredity (Edinb)* 113, 176–183.

Hanna, C. W., McFadden, D. E. and Robinson, W. P. (2013). DNA methylation profiling of placental villi from karyotypically normal miscarriage and recurrent miscarriage. *American Journal of Pathology* 182, 2276–2284.

Hata, K., Okano, M., Lei, H. and Li, E. (2002). Dnmt3L cooperates with the Dnmt3 family of *de novo* DNA methyltransferases to establish maternal imprints in mice. *Development* 129, 1983–1993.

Hayward, B. E., De Vos, M., Talati, N., Abdollahi, M. R., Taylor, G. R., Meyer, E., Williams, D., Maher, E. R., Setna, F., Nazir, K. *et al.* (2009). Genetic and epigenetic analysis of recurrent hydatidiform mole. *Human Mutations* 30, E629–639.

Horsthemke, B. and Wagstaff, J. (2008). Mechanisms of imprinting of the Prader-Willi/Angelman region. *American Journal of Medical Genetics A* 146A, 2041–2052.

Jacob, K. J., Robinson, W. P. and Lefebvre, L. (2013). Beckwith-Wiedemann and Silver-Russell syndromes: opposite developmental imbalances in imprinted regulators of placental function and embryonic growth. *Clinical Genetics* 84, 326–334.

Jia, D., Jurkowska, R. Z., Zhang, X., Jeltsch, A. and Cheng, X. (2007). Structure of Dnmt3a bound to Dnmt3L suggests a model for de novo DNA methylation. *Nature* 449, 248–251.

Jia, R. Z., Zhang, X., Hu, P., Liu, X. M., Hua, X. D., Wang, X. and Ding, H. J. (2012). Screening for differential methylation status in human placenta in preeclampsia using a CpG island plus promoter microarray. *International Journal of Molecular Medicine* 30, 133–141.

Judson, H., Hayward, B. E., Sheridan, E. and Bonthron, D. T. (2002). A global disorder of imprinting in the human female germ line. *Nature* 416, 539–542.

Kagami, M., Matsuoka, K., Nagai, T., Yamanaka, M., Kurosawa, K., Suzumori, N., Sekita, Y., Miyado, M., Matsubara, K., Fuke, T. *et al.* (2012). Paternal uniparental disomy 14 and related disorders: placental gene expression analyses and histological examinations. *Epigenetics* 7, 1142–1150.

Kagami, M., Sekita, Y., Nishimura, G., Irie, M., Kato, F., Okada, M., Yamamori, S., Kishimoto, H., Nakayama, M., Tanaka, Y. *et al.* (2008). Deletions and epimutations affecting the human 14q32.2 imprinted region in individuals with paternal and maternal upd(14)-like phenotypes. *Nature Genetics* 40, 237–242.

Kamiya, M., Judson, H., Okazaki, Y., Kusakabe, M., Muramatsu, M., Takada, S., Takagi, N., Arima, T., Wake, N., Kamimura, K. *et al.* (2000). The cell cycle control gene ZAC/PLAGL1 is imprinted — a strong candidate gene for transient neonatal diabetes. *Human Molecular Genetics* 9, 453–460.

Kanayama, N., Takahashi, K., Matsuura, T., Sugimura, M., Kobayashi, T., Moniwa, N., Tomita, M. and Nakayama, K. (2002). Deficiency in p57Kip2 expression induces preeclampsia-like symptoms in mice. *Molecular Human Reproduction* 8, 1129–1135.

Kaneda, M., Okano, M., Hata, K., Sado, T., Tsujimoto, N., Li, E. and Sasaki, H. (2004). Essential role for *de novo* DNA methyltransferase Dnmt3a in paternal and maternal imprinting. *Nature* 429, 900–903.

Kato, Y., Kaneda, M., Hata, K., Kumaki, K., Hisano, M., Kohara, Y., Okano, M., Li, E., Nozaki, M. and Sasaki, H. (2007). Role of the Dnmt3 family in *de novo* methylation of imprinted and repetitive sequences during male germ cell development in the mouse. *Human Molecular Genetics* 16, 2272–2280.

Keverne, E. B. and Curley, J. P. (2008). Epigenetics, brain evolution and behaviour. *Front Neuroendocrinol* 29, 398–412.

Keverne, E. B., Fundele, R., Narasimha, M., Barton, S. C. and Surani, M. A. (1996). Genomic imprinting and the differential roles of parental genomes in brain development. *Brain Research Developmental Brain Research* 92, 91–100.

Killian, J. K., Byrd, J. C., Jirtle, J. V., Munday, B. L., Stoskopf, M. K., MacDonald, R. G. and Jirtle, R. L. (2000). M6P/IGF2R imprinting evolution in mammals. *Molecular Cell* 5, 707–716.

Kobayashi, H., Suda, C., Abe, T., Kohara, Y., Ikemura, T. and Sasaki, H. (2006). Bisulfite sequencing and dinucleotide content analysis of 15 imprinted mouse differentially methylated regions (DMRs): paternally methylated DMRs contain less CpGs than maternally methylated DMRs. *Cytogenetics Genome Research* 113, 130–137.

Kobayashi, S., Uemura, H., Kohda, T., Nagai, T., Chinen, Y., Naritomi, K., Kinoshita, E. I., Ohashi, H., Imaizumi, K., Tsukahara, M. *et al.* (2001). No evidence of PEG1/MEST gene mutations in Silver-Russell syndrome patients. *American Journal of Medical Genetics* 104, 225–231.

Koukoura, O., Sifakis, S., Zaravinos, A., Apostolidou, S., Jones, A., Hajiioannou, J., Widschwendter, M. and Spandidos, D. A. (2011). Hypomethylation along with increased H19 expression in placentas from pregnancies complicated with fetal growth restriction. *Placenta* 32, 51–57.

Lage, J. M. (1991). Placentomegaly with massive hydrops of placental stem villi, diploid DNA content, and fetal omphaloceles: possible association with Beckwith-Wiedemann syndrome. *Human Pathology* 22, 591–597.

Lakovschek, I. C., Streubel, B. and Ulm, B. (2011). Natural outcome of trisomy 13, trisomy 18, and triploidy after prenatal diagnosis. *American Journal of Medical Genetics A* 155A, 2626–2633.

Lambertini, L., Diplas, A. I., Lee, M. J., Sperling, R., Chen, J. and Wetmur, J. (2008). A sensitive functional assay reveals frequent loss of genomic imprinting in human placenta. *Epigenetics* 3, 261–269.

Lee, H. J., Hore, T. A. and Reik, W. (2014). Reprogramming the methylome: erasing memory and creating diversity. *Cell Stem Cell* 14, 710–719.

Lefebvre, L., Viville, S., Barton, S. C., Ishino, F., Keverne, E. B. and Surani, M. A. (1998). Abnormal maternal behaviour and growth retardation associated with loss of the imprinted gene Mest. *Nature Genetics* 20, 163–169.

Lewis, A., Mitsuya, K., Umlauf, D., Smith, P., Dean, W., Walter, J., Higgins, M., Feil, R. and Reik, W. (2004). Imprinting on distal chromosome 7 in the placenta involves repressive histone methylation independent of DNA methylation. *Nature Genetics* 36, 1291–1295.

Li, X., Ito, M., Zhou, F., Youngson, N., Zuo, X., Leder, P. and Ferguson-Smith, A. C. (2008). A maternal-zygotic effect gene, Zfp57, maintains both maternal and paternal imprints. *Developmental Cell* 15, 547–557.

Lim, A. L., Ng, S., Leow, S. C., Choo, R., Ito, M., Chan, Y. H., Goh, S. K., Tng, E., Kwek, K., Chong, Y. S. *et al.* (2012). Epigenetic state and expression of imprinted genes in umbilical cord correlates with growth parameters in human pregnancy. *Journal of Medical Genetics* 49, 689–697.

Linder, D., McCaw, B. K. and Hecht, F. (1975). Parthenogenic origin of benign ovarian teratomas. *The New England Journal of Medicine* 292, 63–66.

Lionti, T., Reid, S. M., White, S. M. and Rowell, M. M. (2015). A population-based profile of 160 Australians with Prader-Willi syndrome: trends in diagnosis, birth prevalence and birth characteristics. *American Journal of Medical Genetics A* 167A, 371–378.

Liu, Y., Hoyo, C., Murphy, S., Huang, Z., Overcash, F., Thompson, J., Brown, H. and Murtha, A. P. (2013). DNA methylation at imprint regulatory regions in preterm birth and infection. *American Journal of Obstetrics & Gynecology* 208, 395.e1–395.e7.

Liu, Y., Murphy, S. K., Murtha, A. P., Fuemmeler, B. F., Schildkraut, J., Huang, Z., Overcash, F., Kurtzberg, J., Jirtle, R., Iversen, E. S. *et al.* (2012). Depression in pregnancy, infant birth weight and DNA methylation of imprint regulatory elements. *Epigenetics* 7, 735–746.

Lyko, F., Ramsahoye, B. H., Kashevsky, H., Tudor, M., Mastrangelo, M. A., Orr-Weaver, T. L. and Jaenisch, R. (1999). Mammalian (cytosine-5) methyltransferases cause genomic DNA methylation and lethality in Drosophila. *Nature Genetics* 23, 363–366.

Martin, J., Tucker M and Browning, J. C. (2012). Infantile osteoma cutis as a presentation of a GNAS mutation. *Pediatric Dermatology* 29, 483–484.

McFadden, D. E. and Kalousek, D. K. (1991). Two different phenotypes of fetuses with chromosomal triploidy: correlation with parental origin of the extra haploid set. *American Journal of Medical Genetics* 38, 535–538.

McFadden, D. E. and Pantzar, J. T. (1996). Placental pathology of triploidy. *Human Pathology* 27, 1018–1020.

McGrath, J. and Solter, D. (1984). Completion of mouse embryogenesis requires both the maternal and paternal genomes. *Cell* 37, 179–183.

McMinn, J., Wei, M., Schupf, N., Cusmai, J., Johnson, E. B., Smith, A. C., Weksberg, R., Thaker, H. M. and Tycko, B. (2006). Unbalanced placental expression of imprinted genes in human intrauterine growth restriction. *Placenta* 27, 540–549.

Meekins, J. W., Pijnenborg, R., Hanssens, M., McFadyen, I. R. and van Asshe A. (1994). A study of placental bed spiral arteries and trophoblast invasion in normal and severe pre-eclamptic pregnancies. *British Journal of Obstetrics and Gynaecology* 101, 669–674.

Mertz, L. G., Christensen, R., Vogel, I., Hertz, J. M. and Ostergaard, J. R. (2014). Eating behavior, prenatal and postnatal growth in Angelman syndrome. *Research in Developmental Disabilities* 35, 2681–2690.

Messerschmidt, D. M., de Vries, W., Ito, M., Solter, D., Ferguson-Smith, A. and Knowles, B. B. (2012). Trim28 is required for epigenetic stability during mouse oocyte to embryo transition. *Science* 335, 1499–1502.

Monk, D., Arnaud, P., Frost, J., Hills, F. A., Stanier, P., Feil, R. and Moore, G. E. (2009). Reciprocal imprinting of human GRB10 in placental trophoblast and brain: evolutionary conservation of reversed allelic expression. *Human Molecular Genetics* 18, 3066–3074.

Monk, D., Bentley, L., Hitchins, M., Myler, R. A., Clayton-Smith, J., Ismail, S., Price, S. M., Preece, M. A., Stanier, P. and Moore, G. E. (2002). Chromosome 7p disruptions in Silver Russell syndrome: delineating an imprinted candidate gene region. *Human Genetics* 111, 376–387.

Monk, D., Sanches, R., Arnaud, P., Apostolidou, S., Hills, F. A., Abu-Amero, S., Murrell, A., Friess, H., Reik, W., Stanier, P. *et al.* (2006). Imprinting of IGF2 P0 transcript and novel alternatively spliced INS-IGF2 isoforms show differences between mouse and human. *Human Molecular Genetics* 15, 1259–1269.

Moore, T. and Haig, D. (1991). Genomic imprinting in mammalian development: a parental tug-of-war. *Trends Genetics* 7, 45–49.

Murphy, R., Thompson, J. M., Tost, J., Mitchell, E. A. and Auckland Birthweight Collaborative Study Group. (2014). No evidence for copy number and methylation variation in H19 and KCNQ10T1 imprinting control regions in children born small for gestational age. *BMC Medical Genetics* 15, 67–2350–15–67.

Nakamura, T., Arai, Y., Umehara, H., Masuhara, M., Kimura, T., Taniguchi, H., Sekimoto, T., Ikawa, M., Yoneda, Y., Okabe, M. *et al.* (2007). PGC7/Stella protects against DNA demethylation in early embryogenesis. *Nature Cell Biology* 9, 64–71.

Nguyen, N. M. and Slim, R. (2014). Genetics and Epigenetics of Recurrent Hydatidiform Moles: Basic Science and Genetic Counselling. *Current Obstetrics and Gynecology Reports* 3, 55–64.

Noguer-Dance, M., Abu-Amero, S., Al-Khtib, M., Lefevre, A., Coullin, P., Moore, G. E. and Cavaille, J. (2010). The primate-specific microRNA gene cluster (C19MC) is imprinted in the placenta. *Human Molecular Genetics* 19, 3566–3582.

Okae, H., Chiba, H., Hiura, H., Hamada, H., Sato, A., Utsunomiya, T., Kikuchi, H., Yoshida, H., Tanaka, A., Suyama, M. *et al.* (2014). Genome-wide analysis of DNA methylation dynamics during early human development. *PLoS Genetics* 10, e1004868.

Onyango, P., Miller, W., Lehoczky, J., Leung, C. T., Birren, B., Wheelan, S., Dewar, K. and Feinberg, A. P. (2000). Sequence and comparative analysis of the mouse 1-megabase region orthologous to the human 11p15 imprinted domain. *Genome Research* 10, 1697–1710.

Ouyang, Y., Mouillet, J. F., Coyne, C. B. and Sadovsky, Y. (2014). Review: placenta-specific microRNAs in exosomes — good things come in nano-packages. *Placenta* 35, S69–73.

Paradinas, F. J., Sebire, N. J., Fisher, R. A., Rees, H. C., Foskett, M., Seckl, M. J. and Newlands, E. S. (2001). Pseudo-partial moles: placental stem vessel hydrops and the association with Beckwith-Wiedemann syndrome and complete moles. *Histopathology* 39, 447–454.

Parry, D. A., Logan, C. V., Hayward, B. E., Shires, M., Landolsi, H., Diggle, C., Carr, I., Rittore, C., Touitou, I., Philibert, L. *et al.* (2011). Mutations causing familial biparental hydatidiform mole implicate c6orf221 as a possible regulator of genomic imprinting in the human oocyte. *American Journal of Human Genetics* 89, 451–458.

Penaherrera, M. S., Weindler, S., Van Allen, M. I., Yong, S. L., Metzger, D. L., McGillivray, B., Boerkoel, C., Langlois, S. and Robinson, W. P. (2010). Methylation profiling in individuals with Russell-Silver syndrome. *American Journal of Medical Genetics A* 152A, 347–355.

Plagge, A., Gordon, E., Dean, W., Boiani, R., Cinti, S., Peters, J. and Kelsey, G. (2004). The imprinted signaling protein XL alpha s is required for postnatal adaptation to feeding. *Nature Genetics* 36, 818–826.

Plasschaert, R. N. and Bartolomei, M. S. (2014). Genomic imprinting in development, growth, behavior and stem cells. *Development* 141, 1805–1813.

Pliushch, G., Schneider, E., Weise, D., El Hajj, N., Tresch, A., Seidmann, L., Coerdt, W., Muller, A. M., Zechner, U. and Haaf, T. (2010). Extreme methylation values of imprinted genes in human abortions and stillbirths. *American Journal of Pathology* 176, 1084–1090.

Rahimpanah, F. and Smoleniec, J. (2000). Partial mole, triploidy and proteinuric hypertension: two case reports. *Australian and New Zealand Journal of Obstetrics and Gynaecology* 40, 215–218.

Rao, V. V., Schnittger, S. and Hansmann, I. (1991). G protein Gs alpha (GNAS 1), the probable candidate gene for Albright hereditary osteodystrophy, is assigned to human chromosome 20q12-q13.2. *Genomics* 10, 257–261.

Reddy, R., Akoury, E., Phuong Nguyen, N. M., Abdul-Rahman, O. A., Dery, C., Gupta, N., Daley, W. P., Ao, A., Landolsi, H., Ann Fisher, R. *et al.* (2013). Report of four new patients with protein-truncating mutations in C6orf221/KHDC3L and colocalization with NLRP7. *European Journal of Human Genetics* 21, 957–964.

Redline, R. W., Hassold, T. and Zaragoza, M. V. (1998). Prevalence of the partial molar phenotype in triploidy of maternal and paternal origin. *Human Pathology* 29, 505–511.

Renfree, M. B., Ager, E. I., Shaw, G. and Pask, A. J. (2008). Genomic imprinting in marsupial placentation. *Reproduction* 136, 523–531.

Rijhsinghani, A., Yankowitz, J., Strauss, R. A., Kuller, J. A., Patil, S. and Williamson, R. A. (1997). Risk of preeclampsia in second-trimester triploid pregnancies. *Obstetrics & Gynecology* 90, 884–888.

Robinson, W. P., Lauzon, J. L., Innes, A. M., Lim, K., Arsovska, S. and McFadden, D. E. (2007a). Origin and outcome of pregnancies affected by androgenetic/biparental chimerism. *Human Reproduction* 22, 1114–1122.

Robinson, W. P., Slee, J., Smith, N., Murch, A., Watson, S. K., Lam, W. L. and McFadden, D. E. (2007b). Placental mesenchymal dysplasia associated with fetal overgrowth and mosaic deletion of the maternal copy of 11p15.5. *American Journal of Medical Genetics A* 143A, 1752–1759.

Romanelli, V., Belinchon, A., Campos-Barros, A., Heath, K. E., Garcia-Minaur, S., Martinez-Glez, V., Palomo, R., Mercado, G., Gracia, R. and Lapunzina, P. (2009). CDKN1C mutations in HELLP/preeclamptic mothers of Beckwith-Wiedemann Syndrome (BWS) patients. *Placenta* 30, 551–554.

Rossant, J. and Cross, J. C. (2001). Placental development: lessons from mouse mutants. *Nature Review Genetics* 2, 538–548.

Schulz, R., Proudhon, C., Bestor, T. H., Woodfine, K., Lin, C. S., Lin, S. P., Prissette, M., Oakey, R. J. and Bourc'his, D. (2010). The parental non-equivalence of imprinting control regions during mammalian development and evolution. *PLoS Genetics* 6, e1001214.

Seisenberger, S., Andrews, S., Krueger, F., Arand, J., Walter, J., Santos, F., Popp, C., Thienpont, B., Dean, W. and Reik, W. (2012). The dynamics of genome-wide DNA methylation reprogramming in mouse primordial germ cells. *Molecular Cell* 48, 849–862.

Seval, Y., Korgun, E. T. and Demir, R. (2007). Hofbauer cells in early human placenta: possible implications in vasculogenesis and angiogenesis. *Placenta* 28, 841–845.

Shepard, T. H., Fantel, A. G. and Fitzsimmons, J. (1989). Congenital defect rates among spontaneous abortuses: twenty years of monitoring. *Teratology* 39, 325–331.

Sitras, V., Paulssen, R., Leirvik, J., Vartun, A. and Acharya, G. (2009). Placental gene expression profile in intrauterine growth restriction due to placental insufficiency. *Reproductive Science* 16, 701–711.

Smallwood, S. A., Tomizawa, S., Krueger, F., Ruf, N., Carli, N., Segonds-Pichon, A., Sato, S., Hata, K., Andrews, S. R. and Kelsey, G. (2011). Dynamic CpG island methylation landscape in oocytes and preimplantation embryos. *Nature Genetics* 43, 811–814.

Solter D. (1988). Differential imprinting and expression of maternal and paternal genomes. *Annual Review of Genetics* 22, 127–146.

Spencer, H. G. and Clark, A. G. (2014). Non-conflict theories for the evolution of genomic imprinting. *Heredity (Edinb)* 113, 112–118.

Stevens, L. C. and Varnum, D. S. (1974). The development of teratomas from parthenogenetically activated ovarian mouse eggs. *Developmental Biology* 37, 369–380.

St-Pierre, J., Hivert, M. F., Perron, P., Poirier, P., Guay, S. P., Brisson, D. and Bouchard, L. (2012). IGF2 DNA methylation is a modulator of newborn's fetal growth and development. *Epigenetics* 7, 1125–1132.

Surani, M. A., Barton, S. C. and Norris, M. L. (1984). Development of reconstituted mouse eggs suggests imprinting of the genome during gametogenesis. *Nature* 308, 548–550.

Tang, Z., Abrahams, V. M., Mor, G. and Guller, S. (2011). Placental Hofbauer cells and complications of pregnancy. *Annuals of the New York Academy of Scienes* 1221, 103–108.

Turan, N., Ghalwash, M. F., Katari, S., Coutifaris, C., Obradovic, Z. and Sapienza, C. (2012). DNA methylation differences at growth related genes correlate with birth weight: a molecular signature linked to developmental origins of adult disease?. *BMC Medical Genomics* 5, 10-8794-5-10.

van Dijk, M., Drewlo, S. and Oudejans, C. B. (2010). Differential methylation of STOX1 in human placenta. *Epigenetics* 5, 736–742.

van Dijk, M. and Oudejans, C. (2013). (Epi)genetics of pregnancy-associated diseases. *Frontiers in Genetics* 4, 180.

Varmuza, S. and Mann, M. (1994). Genomic imprinting--defusing the ovarian time bomb. *Trends in Genetics* 10, 118–123.

Varmuza, S. and Miri, K. (2015). What does genetics tell us about imprinting and the placenta connection?. *Cellular and Molecular Life Science* 72, 51–72.

Wake, N., Takagi, N. and Sasaki, M. (1978). Androgenesis as a cause of hydatidiform mole. *Journal of the National Cancer Institute* 60, 51–57.

Wan, L. B. and Bartolomei, M. S. (2008). Regulation of imprinting in clusters: noncoding RNAs versus insulators. *Advances in Genetics* 61, 207–223.

Wang, C. M., Dixon, P. H., Decordova, S., Hodges, M. D., Sebire, N. J., Ozalp, S., Fallahian, M., Sensi, A., Ashrafi, F., Repiska, V. *et al.* (2009). Identification of 13 novel NLRP7 mutations in 20 families with recurrent hydatidiform mole; missense mutations cluster in the leucine-rich region. *Journal of Medical Genetics* 46, 569–575.

Wang, Q., Chow, J., Hong, J., Smith, A. F., Moreno, C., Seaby, P., Vrana, P., Miri, K., Tak, J., Chung, E. D. *et al.* (2011). Recent acquisition of imprinting at the rodent Sfmbt2 locus correlates with insertion of a large block of miRNAs. *BMC Genomics* 12, 204-2164-12-204.

Wolf, J. B. and Hager, R. (2006). A maternal-offspring coadaptation theory for the evolution of genomic imprinting. *PLoS Biology* 4, e380.

Yamazawa, K., Kagami, M., Nagai, T., Kondoh, T., Onigata, K., Maeyama, K., Hasegawa, T., Hasegawa, Y., Yamazaki, T., Mizuno, S. *et al.* (2008). Molecular and clinical findings and their correlations in Silver-Russell syndrome: implications for a positive role of IGF2 in growth determination and differential imprinting regulation of the IGF2-H19 domain in bodies and placentas. *Journal of Molecular Medicine (Berl)* 86, 1171–1181.

Yoder, J. A., Soman, N. S., Verdine, G. L. and Bestor, T. H. (1997). DNA (cytosine-5)-methyltransferases in mouse cells and tissues. Studies with a mechanism-based probe. *Journal of Molecular Biology* 270, 385–395.

Yuen, R. K., Jiang, R., Penaherrera, M. S., McFadden, D. E. and Robinson, W. P. (2011), Genome-wide mapping of imprinted differentially methylated regions by DNA methylation profiling of human placentas from triploidies. *Epigenetics Chromatin* 4, 10.

Yuen, R. K., Penaherrera, M. S., von Dadelszen, P., McFadden, D. E. and Robinson, W. P. (2010). DNA methylation profiling of human placentas reveals promoter hypomethylation of multiple genes in early-onset preeclampsia. *European Journal of Human Genetics* 18, 1006–1012.

Zhang, P., Dixon, M., Zucchelli, M., Hambiliki, F., Levkov, L., Hovatta, O. and Kere, J. (2008). Expression analysis of the NLRP gene family suggests a role in human preimplantation development. *PLoS One* 3, e2755.

Genetic and Epigenetic Variation at the *H19* Imprinted Region and Its Effect on Birth Weight

7

Celia MT Greenwood[*,†,‡], *ChangJiang Xu*[*,‡], *Shomoita Alam*[*,‡], *Julia Krushkal*[§], *Ronald M. Adkins*[¶] *and Anna K. Naumova*[||]

[*]*Lady Davis Institute, Jewish General Hospital, Montreal, QC, Canada*
[†]*Department of Oncology, McGill University, Montreal, QC, Canada*
[‡]*Department of Epidemiology, Biostatistics and Occupational Health, McGill University, Montreal, QC, Canada*
[§]*Computational & Systems Biology Branch, Biometric Research Program, National Cancer Institute, NIH, Rockville, MD, USA*
[¶]*Department of Pediatrics, the University of Tennessee Health Science Center, Memphis, Tennessee 38163*
[||]*Departments of Obstetrics and Gynecology and Human Genetics, McGill University and the Research Institute of the McGill University Health Centre, Montreal, QC, Canada*

Imprinted genes influence prenatal growth, and several studies have demonstrated association between genetic polymorphisms in imprinted regions and birth weight. Association between DNA methylation levels at imprinted loci and birth weights has been also detected. Here, we examine the relationship between haplotypes, DNA methylation and birth weight to determine if DNA methylation acts as a mediator between haplotype and phenotype. Analysis of DNA from infant cord blood at the imprinted *H19* locus, in a set of infants of either African American or White ancestry, demonstrated both effects of maternally-inherited haplotypes on birthweight, where one key haplotype was associated with lower birthweight by approximately 210 grams, and also strong effects of DNA methylation levels on birthweight, such that increasing levels of methylation were associated with lower birthweights. Despite the presence of strong cis-associations between the haplotypes and the methylation levels, the genetic and epigenetic effects on birthweight appear to be independent, and the probes involved were not in the imprinting control region.

7.1 The *H19-IGF2* Imprinted Region and Its Role in Human Development

7.1.1 *Genomic imprinting in humans*

Genomic imprinting is an epigenetic mechanism that supports the non-equivalence of the maternal and paternal genomes in mammals and is critical for embryo growth and viability (reviewed in (Barlow and Bartolomei 2014)). About 100 imprinted genes have been discovered to date, and many of them play a role in growth and development (Barlow and Bartolomei 2014). The majority of imprinted genes reside in clusters. Imprinting in these gene clusters depends on DNA elements termed imprinting control regions (ICRs) (Barlow and Bartolomei 2014). ICRs contain direct repeats and CpG islands that gain methylation during gametogenesis, with the majority of the ICRs gaining methylation during oocyte growth (Barlow and Bartolomei 2014). These are termed maternally imprinted regions, the best studied examples of which include the ICRs located within the promoters of paternally expressed genes: potassium voltage-gated channel subfamily Q member 1 opposite strand/antisense transcript (*KCNQ1OT1*) on chromosome 11p15 and small nuclear ribonucleoprotein polypeptide N (*SNRPN*) on chromosome 15q11.2 (Glenn *et al.* 1996, El-Maarri *et al.* 2001, Li *et al.* 2004, Hiura *et al.* 2006). To date, only three ICRs are known to gain methylation during spermatogenesis. These are termed paternally imprinted regions. Paternally imprinted ICRs are the *H19* ICR on chromosome 11p15, the intergenic differentially methylated region (IG DMR) on chromosome 14q32, and the Ras protein specific guanine nucleotide releasing factor 1 (*RAS-GRF1*) on chromosome 15q24.2 (reviewed in (Barlow and Bartolomei 2014)). Differential ICR methylation is maintained post-fertilization and results in parent-of-origin specific DNA methylation of the ICRs in the somatic cells of the embryo, with one parental allele being highly methylated and the other parental allele being completely unmethylated. Some of the imprinted clusters harbor additional differentially methylated regions (DMRs) that also contribute to imprinted expression of genes such as the insulin-like growth factor

2 (*IGF2*), maternally expressed gene 3 (*MEG3*), delta-like 1 homolog (*Drosophila*) (*DLK1*), etc. These DMRs are unmethylated in gametes, but gain parent-of-origin specific methylation during embryonic development (Geuns *et al.* 2003). Abnormal DNA methylation at ICRs and DMRs in the parental germ line or embryos, respectively, or errors in the maintenance of these imprinting marks, results in a range of abnormal phenotypes from early embryo loss to congenital developmental abnormalities (reviewed in (Eggermann *et al.* 2015b)). In humans, abnormal DNA methylation at the imprinted region of chromosome 15 (15q11–q12) is associated with congenital developmental disorders Prader-Willi and Angelman syndromes (reviewed in (Nicholls *et al.* 1998)). Hypermethylation of the *SNRPN* ICR on both parental chromosomes causes Prader-Willi syndrome, which is characterized by endocrine anomalies and short stature; hypomethylation of the same region causes Angelman syndrome, which is characterized by neurodevelopmental delay, mental retardation, seizures and postnatal growth abnormalities, among other secondary characteristics (reviewed in (Nicholls *et al.* 1998)). DNA methylation abnormalities in the imprinted region of chromosome 14q32 (*DLK1-MEG3*) cause Temple syndrome characterized by low birth weight, hypotonia, motor delay, feeding problems and short stature (Ioannides *et al.* 2014, Briggs *et al.* 2016). Imprinting aberrations in the chromosome 20q13.3 imprinted region that harbors the GNAS complex locus result in a number of different endocrine abnormalities, such as pseudohypoparathyroidism type 1a, pseudohypoparathyroidism type 1b, Albright hereditary osteodystrophy, pseudopseudohypoparathyroidism, McCune–Albright syndrome, progressive osseus heteroplasia, polyostotic fibrous dysplasia of bone and pituitary tumors (reviewed in (Mantovani *et al.* 2010)). The specific phenotypes seen in these diseases depend on the gene and allele affected (Mantovani *et al.* 2010, Juppner 2015). Thus, most of the congenital disorders associated with abnormal imprinting in humans are associated with endocrine dysfunction and affect growth and development (reviewed in (Polychronakos and Kukuvitis 2002)).

7.1.2 *The* H19-IGF2 *imprinted region*

The imprinted *H19-IGF2* locus resides in chromosomal region 11p15.5 and harbors two imprinted genes: *IGF2*, encoding an important fetal growth factor and *H19*, a non-coding RNA whose dosage and parent-of-origin dependent expression are critical for embryonic viability (Brannan *et al.* 1990, Ferguson-Smith *et al.* 1991, Kono *et al.* 2004). Expression of these genes depends on the *H19* ICR, which is located upstream of the *H19* gene and is methylated on the paternal chromosome and unmethylated on the maternal chromosome (Zhang *et al.* 1993, Tremblay *et al.* 1995). Several tissue-specific enhancers located downstream of *H19* are shared between *H19* and *IGF2* and are critical for their regulation (Webber *et al.* 1998). The region upstream of the *H19* promoter contains seven CTCCC factor (CTCF) binding sites (Figure 7.1(a)) that define the 3D organization of the chromatin and thereby transcriptional control in the *H19-IGF2* region (Bell and

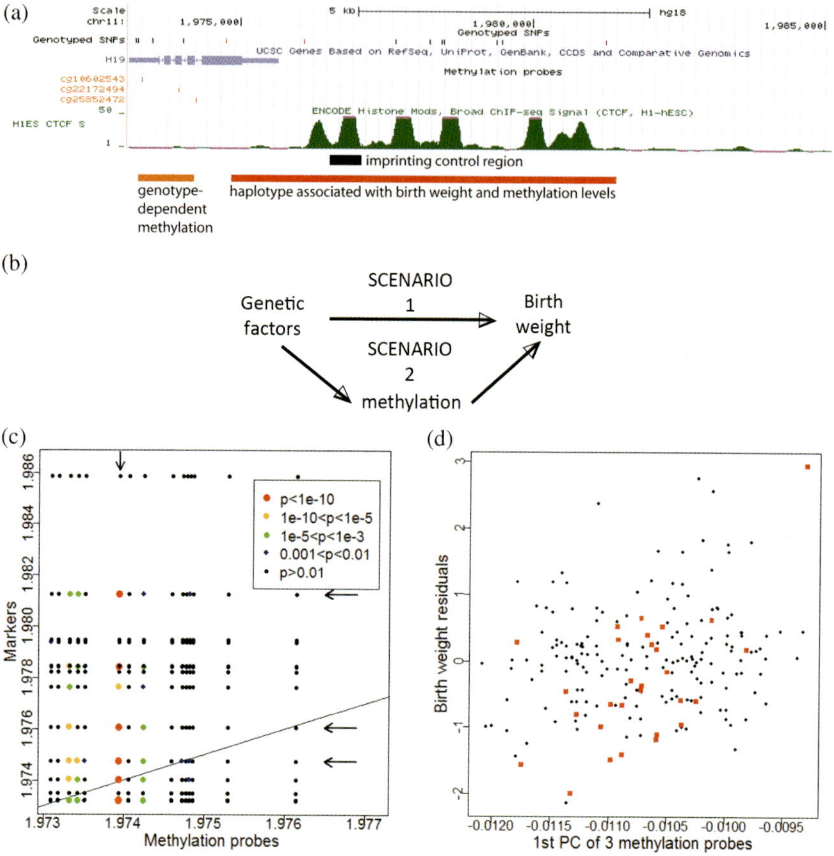

Figure 7.1. (a) The *H19* gene region associated with birth weight (hg18). Genotyped SNPs are shown with black ticks, whereas the three SNPs associated with birth weight are indicated in red. The haplotype spanned by rs2067051, rs2251375 and rs4929984 is indicated by a red bar. Transcriptional and imprinting control of the *H19-IGF2* locus depends upon the ICR containing seven CTCF-binding sites located upstream of the *H19* gene. Position of CTCF binding sites in human embryonic stem cells is shown in green (Mikkelsen *et al.* 2007). The differentially methylated region of the ICR is indicated with a black bar, and the region showing *cis*-meQTL associations is indicated in orange.

Figure 7.1. (*Continued*) To generate this figure we used the UCSC genome browser (http://genome.ucsc.edu). (b) Relationships between genotypes, DNA methylation levels and birth weight. Haplotypes GAC and ACA influence methylation levels, and maternal haplotype GCA is associated with birth weight. Although methylation levels and birth weight show association, the direction of causality is unknown. Since haplotypes GAC and ACA show no direct association with birth weight, the direction indicated by the red arrow may be more likely than the green arrow. (c) *P*-values for the associations between methylation levels and marker genotypes, plotted by CpG probe position (x-axis) and marker position (y-axis). Colors indicate the magnitude of the *p*-values. Probe cg22172494 is indicated by an arrow at the top of the plot. Markers rs2067051, rs2251375 and rs4929984 are indicated by arrows on the right hand side (bottom to top). A diagonal line indicates the same position for CpG probes and SNPs. (d) Association between maternal GCA haplotype and birth weight. The first principal component of the methylation levels at cg22172494, cg10602543 and cg25852472 is plotted against birth weight residuals. Points in red are the infants who received the GCA haplotype from his/her mother.

Felsenfeld 2000, Kanduri *et al.* 2000, Szabo *et al.* 2000). One of the seven CTCF-binding sites overlaps with the differentially methylated region of the ICR, and is referred to as *H19* DMR (Takai *et al.* 2001). CTCF binds the unmethylated maternal allele of the *H19* DMR and promotes formation of a chromatin loop that prevents interaction between the distal enhancers and *IGF2* promoter, but favors interaction between the enhancers and the *H19* promoter (Webber *et al.* 1998, Lopes *et al.* 2003). DNA methylation of the *H19* DMR prevents CTCF binding. Hence, CTCF does not bind the methylated paternal allele of the *H19* DMR. This removes the barriers preventing interaction between *IGF2* and distal enhancers, and activates *IGF2* transcription (Bell and Felsenfeld 2000, Kanduri *et al.* 2000, Szabo *et al.* 2000). Thus, the methylation profile of the *H19* DMR defines the parent-of-origin dependent expression of both *H19* and *IGF2*. The *IGF2* region also harbors secondary differentially methylated regions, i.e. the *IGF2* DMR1 and *IGF2* DMR2 (Sullivan *et al.* 1999). Their methylation patterns are also parent-of-origin dependent, but established during embryonic development (Sullivan *et al.* 1999, Ishihara *et al.* 2000).

The parent-of-origin dependent methylation of the *H19-IGF2* DMRs is essential for normal development. High methylation of both parental *H19* DMR alleles leads to biallelic expression of *IGF2* and causes Beckwith–Wiedemann syndrome (BWS) ((Weksberg *et al.* 1993) and reviewed in (Choufani *et al.* 2013)). BWS includes enhanced fetal growth and is associated with increased risk of Wilms tumor (Choufani *et al.* 2013). Babies with BWS tend to have a very high birth weight due to overgrowth during the second half of gestation (Mussa *et al.*

2016). Their growth rate is also higher than normal during the first years of life, but frequently normalizes later (Weksberg *et al.* 2010). Biallelic hypomethylation of the *H19* DMR is associated with Silver–Russell syndrome (SRS), which is characterized by intrauterine growth retardation and low birth weight. Abnormally low methylation of the *H19* DMR is found in 20–60% of cases of SRS (Eggermann *et al.* 2015a). Both BWS and SRS are extremely rare disorders (Eggermann *et al.* 2015a). However, they illustrate the importance of genomic imprinting of the *H19-IGF2* region and the expression levels of *H19* and *IGF2* for fetal growth.

7.1.3 *Genetic and epigenetic variation at imprinted regions and child development*

Phenotypic variation in humans is largely associated with variation in gene expression levels and regulatory polymorphisms that act in *cis* or *trans*. Imprinted regions contain common genetic variants and, in principle, these common variants may influence expression levels of imprinted genes without affecting their monoallelic expression (Labialle *et al.* 2008). Variation in expression levels of an imprinted gene may have an impact on child growth and development (Hernandez-Valero *et al.* 2013, Murphy *et al.* 2014, Moore *et al.* 2015). Indeed, both genetic and epigenetic variants influencing imprinting of *IGF2* and the insulin-like growth factor 2 receptor (*IGF2R*) have been reported in humans (Giannoukakis *et al.* 1996, Paquette *et al.* 1998, Vafiadis *et al.* 1998, Babak *et al.* 2015) (reviewed in (Naumova and Croteau 2004)). Moreover, several studies detected parent of-origin dependent associations between polymorphisms in imprinted regions and different developmental or physiological traits. Genetic variation in the imprinted *DLK1*, *MAGEL2* and *KCNK9* regions influences the age of menarche (Perry *et al.* 2014). Alleles at the *DLK1* locus also modify the risk of type 1 diabetes in a parent-of-origin dependent fashion (Wallace *et al.* 2010). Birth weight is associated with common alleles in several imprinted region (Kukuvitis *et al.* 2004, Kaku *et al.* 2007, Adkins *et al.* 2010b). The association between rs8191754 at the *IGF2R* and birth weight is unclear. In two studies babies carrying the rs8191754-CC genotype were heavier than heterozygotes (Kaku *et al.* 2007, Adkins *et al.* 2010b), however, in two other studies no such effect was found (Kukuvitis *et al.* 2004, Vidal *et al.* 2013). Lower birth weights were observed when specific alleles at markers flanking the *H19-IGF2* ICR were maternally inherited. The strongest effect was observed for babies who inherited the rs4929984-A allele from their mother (Adkins *et al.* 2010b). Similar findings were reported in another study of birth weight and fetal genotype (Petry *et al.* 2011). Maternal rs6026576 genotype at the *GNAS* locus was also associated with birth weight (Adkins *et al.* 2010a). Genetic variation in the promoter of another imprinted gene, the pleckstrin homology-like domain, family A, member 2 (*PHLDA2*), influenced its expression in the placenta and influenced birth weight (Ishida *et al.* 2012). Another study showed that variation in the expression levels of *PHLDA2* and paternally expressed gene 10 (*PEG10*) were negatively (*PHLDA2*) and positively (*PEG10*) correlated with birth weight (Lim *et al.* 2012). The underlying mechanism has been elucidated in mouse studies where loss of placental expression of *Phlda2* led to increase of the spongiotrophoblast endocrine compartment of the placenta and elevated levels of placental glycogen (Tunster *et al.* 2016). We therefore conclude that genotype-dependent changes in placental *PHLDA2* levels in humans modulate the placental demand on maternal resources and explain the observed effect on birth weight.

DNA methylation plays an important role in the transcriptional control of imprinted genes including the *H19-IGF2* region, and therefore the impact of variation in methylation on birth weight has been assessed. Studies of *H19-IGF2* variation in methylation and its relationship with child development, however, do not always agree on the importance of the effects. Rate of weight gain during the first three months of life was associated with lower DNA methylation at the *IGF2* DMR, but not the *H19* DMR (Bouwland-Both *et al.* 2013). In another study, methylation levels at the *H19-IGF2* locus were negatively associated with head circumference in children, but not birth weight (Huang *et al.* 2012). Thus, the sum of current association data suggests a complex relationship between genotype, DNA methylation at the *H19-IGF2* imprinted region and birth weight.

7.2 Mediation Analysis of Genetic and Epigenetic Effects on Birth Weight

Figure 7.1(b) illustrates that the joint consideration of genotype, methylation and birth weight forms a possible triangle of relationships. As discussed above, several studies have shown relationships between genotypes in the *H19* region and birth weight or growth; likewise, patterns of methylation in the region have shown some associations with birth weight (Adkins *et al.* 2012). All across the genome, strong *cis*-regulation of methylation levels has been clearly demonstrated (e.g. (Bell *et al.* 2011, Liu *et al.* 2013)), and some of these associations lie within our region of interest (Adkins *et al.* 2010b, Bell *et al.* 2011). Therefore, we can ask whether genetic variation in the *H19* region acts directly on birth weight via Scenario 1, i.e. the top arrow in Figure 7.1(b), or whether the genetic effects act through changes in methylation levels (Scenario 2). These kinds of investigations are known in the statistical literature as mediation analyses (Baron and Kenny 1986), where the goal is to gain insight into the possible causal relationships between the vertices of the triangle.

Genotype-birth weight associations were studied in a large group of mothers and infants from Tennessee (Adkins *et al.* 2010b), and for a subset of these infants, methylation levels were measured at several probes near the *H19* ICR. We therefore asked if methylation measures in the *H19* region are mediating or moderating the relationship between *H19* ICR genotypes and birth weight, and if any such relationships vary with African American ancestry.

7.2.1 *Methods*

7.2.1.1 *Study sample and birth weight scores*

All mother–newborn pairs were from Shelby County, Tennessee USA. They were recruited as part of a birth study that later became a longitudinal cohort study of human development from pregnancy to age 3, the Conditions Affecting Neurocognitive Development and Learning in Early Childhood Study (CANDLE). In the second trimester of pregnancy, written informed consent was obtained from all mothers for the collection of their venous and newborn umbilical cord blood during labor and delivery. This study was approved by the institutional review boards of all the participating hospitals. Most infants were born at 38–40 weeks gestation; four infants were born at less than 36 weeks. Standardized birth weight scores were calculated for each newborn (Hamill *et al.* 1979). From this study, mother–child pairs of either White or African American ethnicity were retained. After excluding subjects missing data on parity, maternal age, maternal weight and weight gain during pregnancy, 1,045 mother–child pairs were analyzed. The data overlap extensively with those analyzed in (Adkins *et al.* 2010b), which examined the earliest births from the samples from Shelby County, TN along with the data from two additional US sites.

7.2.1.2 *DNA methylation*

Methylation was measured in leukocytic DNA from infant cord blood using Illumina HumanMethylation27 BeadChip (Illumina, Inc, San Diego, USA) as described in (Adkins *et al.* 2012); methylation data were available for 214 infants where birth weight and SNP genotyping data were also available. The methylation measurements estimate the proportion of chromosomes where the probe's CpG site was methylated. Values have a potential range between 0 and 1.0, where 0 is not methylated at all and 1.0 is completely methylated. The methylation measures for several probes of interest were summarized into a single measure by using principal component analysis.

7.2.1.3 *Genotypes and haplotypes*

SNP genotyping was performed with Taqman (Applied Biosystems, Inc). Direct DNA sequencing and allele-specific PCR were used for rs2251375 (Adkins *et al.* 2010b). Haplotypes in the region of interest near the *H19* ICR were estimated separately for African American and White families, by the software package BEAGLE (Browning and Browning 2007) using the option for a child with one parent. We retained the haplotype estimates for the 214 infants with methylation data. After convergence of the haplotype estimation algorithm, we sampled 200 haplotypes for each (infant, mother) pair to estimate an empirical distribution of haplotypes. We modelled associations with the haplotype distributions formed by the three markers — rs2067051, rs2251375 and rs4929984 — since associations with these markers were reported previously (Adkins *et al.* 2010b). Linear models weighted by the probability of each trio of transmitted and non-transmitted haplotypes were fit under three different genotype coding schemes: (1) counts of minor alleles or haplotypes, (2) a four-category variable (neither parent transmitted the minor allele (MA), mother transmitted MA, father transmitted MA, both parents transmitted MA) and (3) two indicators for maternal or paternal transmission of MA.

7.2.2 *Results*

Reporting on 1,045 mother–infant pairs, Table 7.1 demonstrates the relationships between standardized birth weight scores (already adjusted for sex, gestational age and ethnicity) and measures of maternal age, maternal weight/weight gain, as well as gestational age in weeks. The anticipated relationships were found, i.e. that birth weight increased with maternal age and maternal weight and weight gain. The model estimated no association of sex on birth weight, since the birth weights were already standardized for sex. However, gestational age continued to show a small impact.

After selecting the subset of 214 mature infants for whom methylation data and SNP genotypes are also available, Table 7.2 provides descriptive characteristics. Approximately 55% of these infants were African American and 60% were parous. Figure 7.1(a) and Table 7.3 show the marker and methylation probe locations used in our analysis of the *H19* ICR.

Table 7.1. Covariate effects on infant standardized birth weight Z scores in 1045 White or African American infants for whom covariate data and birth weight were available. The birth weight Z-scores are standardized for gestational age, sex and ethnicity. Covariates for sex, parity, and an indicator for BMI ≥27 were not significant and were removed.

Covariate	Estimate (s.e.)	*p*-value
Gestational Age	0.061 (0.024)	0.0102
Mother's Age	0.033 (0.005)	3.52e-10
Mother's weight gain	0.023 (0.004)	4.90e-09
Mother's BMI	0.031 (0.005)	1.53e-11

Table 7.2. Descriptive statistics of 214 infants for whom methylation data are available.

	N (%)	
# Infants	214 (100%)	
# Female infants	108 (50.5%)	
# Male infants	106 (49.5%)	
Reported ethnicity:		
# African American	118 (55.1%)	
# White	96 (44.9%)	
Number of women previously parous	129 (60.3%)	
	Mean (S.D.)	Median (Range)
Birth weight in grams	3335.21 (471.05)	3339 (1230–4954)
Standardized birth weights (Z values)	0.0276 (0.858)	−0.032 (−2.08, 3.19)
Gestational age in weeks	38.94 (1.30)	39 (32, 41)
Mother's Age	26.58 (5.09)	27 (18, 39)
Mother's BMI	26.89 (7.11)	24.93 (15.38, 54.92)
Mother's weight gain (kg)	14.91(7.35)	14.0 (−7.0, 55.0)
Parity before this birth	1.08 (1.17)	1 (0,5)

Associations between birth weight and three markers in this region (rs2067051, rs2251375 and rs4929984) had been reported in the original analysis of the larger set of infants from which this subset was taken (Adkins *et al.* 2010b). These markers are in linkage disequilibrium with pairwise correlations between 0.55 and 0.75.

Table 7.3. Positions of methylation probes and markers near the *H19* ICR. There are 14 markers and 16 methylation probes considered in this chapter. Markers in bold were used to construct the haplotypes used in analyses.

Marker or probe name	Position (Build 18)	Alleles for SNPs, minor allele first	Mean (standard deviation) for probes	Previous-reporting of cis-meQTLs (Bell *et al.* 2011, Renteria *et al.* 2013)*
cg11492040	1973089		0.941 (0.029)	
cg06197492	1973181		0.908 (0.038)	
rs3741219	1973195	C T		
rs2839703	1973238	G A		
cg10602543	1973323		0.833 (0.043)	
cg23977670	1973424		0.874 (0.031)	Bell *et al.* 2012
rs217727	1973484	T C		
cg11716026	1973513		0.927 (0.031)	
cg22172494	1973938		0.795 (0.080)	rs217727; rs2067051; rs2251375; rs2735469; rs4929984
rs3024270	1974015	C G		
cg26808784	1974059		0.932 (0.104)	
cg25852472[§]	1974240		0.846 (0.045)	rs4929984
cg15269875	1974594		0.943 (0.021)	
cg21167159	1974730		0.902 (0.033)	rs2067051; rs2251375; rs2107425; rs4929984
rs2067051	1974744	A G		
cg13145013	1974781		0.948 (0.018)	
cg15317267	1974823		0.922 (0.034)	
cg04715462	1974863		0.965 (0.024)	
cg07342901	1975300		0.908 (0.060)	
rs2251375	1976072	C A		
cg17769238	1976144		0.629 (0.051)	
cg02657360	1977136		0.562 (0.070)	Bell *et al.* 2012
rs2107425	1977651	T C		
rs2735971	1978225	A G		

(*Continued*)

Table 7.3. (*Continued*)

Marker or probe name	Position (Build 18)	Alleles for SNPs, minor allele first	Mean (standard deviation) for probes	Previous-reporting of cis-meQTLs (Bell *et al.* 2011, Renteria *et al.* 2013)*
rs4930101	1978435	T G		
rs12417375	1978469	A T		
rs2735469	1979380	T C		
rs17658052	1979475	A G		
rs4929984	1981259	A C		
rs7950932	1985854	A G		

* When previous studies (Renteria *et al.* 2013) reported cis-meQTLs with SNPs that are in our data, we report the SNP name. When the SNPs are not in our data, we give the citation.
§One outlier measurement was set to missing.

Therefore, we created a three-marker haplotype formed by rs2067051, rs2251375 and rs4929984 for further analysis. Due to the known differences in allele frequencies between African Americans and Whites at these SNPs, we estimated the haplotype distributions for each infant and mother separately by ethnicity. After estimation using all SNPs in the region, we estimated the individual probability distributions at these three markers by sampling 200 haplotype trios (maternal transmitted haplotype, maternal non-transmitted haplotype and paternal transmitted haplotype) for each mother and infant. Table 7.4 shows the estimated haplotype frequencies for the four most common haplotypes at these markers. Haplotype GAC is much more common in African Americans than Whites and conversely, ACA is more common in Whites.

As stated previously, associations between individual SNPs and birth weight had been previously reported (Adkins *et al.* 2010b), in particular at rs4929984. We separated the haplotype data by parent of origin, and separately estimated the associations of maternally- and paternally-transmitted haplotypes with the birth weight residuals, after adjustment for the covariates in Table 7.1. In the data set of 1,045 infants, Table 7.5 shows that lower birth weights are associated with maternal inheritance of the GCA haplotype (–0.27 standardized units, $p = 0.0031$), and in the subset of 214 infants with methylation data, the effect is even larger: infants receiving the GCA haplotype from their mother are 0.45 standardized units lighter than other infants ($p = 0.014$). The other haplotypes show much weaker trends, and the GCA haplotype result is also stronger than the associations for any single marker.

Table 7.4. Common haplotype frequencies and birth weights for infants receiving these haplotypes from the mother. The top of the table shows the 4 most common haplotype frequencies in African American and White infants. Haplotypes ACA and GAC have very different frequencies between African Americans and Whites. The bottom of the table shows the mean and standard deviation of the standardized birth weight residuals, when this haplotype was maternally transmitted.

	Four most common haplotypes at rs2067051, rs2251375 and rs4929984			
	ACA	GAC	GCA	GCC
Frequency in all 214 infants	0.254	0.467	0.128	0.119
Frequency in African Americans (N = 118)	0.102	0.660	0.174	0.055
Frequency in Whites (N = 96)	0.442	0.230	0.071	0.197
Birth weight, all infants	0.091 (0.780)	0.037 (0.790)	−0.390 (0.824)	0.097 (0.577)
Birth weight, African Americans	0.065 (0.567)	−0.020 (0.771)	−0.379 (0.666)	0.173 (0.535)
Birth weight, Whites	0.097 (0.829)	0.261 (0.839)	−0.427 (1.327)	0.074 (0.599)

We note that the maternally non-transmitted haplotype did not appear to have an effect on birth weight and furthermore, that if birthweight was not adjusted for maternal weight or weight gain, similar associations were found. Hence, these results imply an effect of *H19* infant genotypes on birth weight, rather than any maternal environment effects, in a parent-of-origin dependent manner.

In some models in Table 7.5, African ancestry was slightly associated with birth weight; however, the associations were inconsistent and small. Furthermore, we found no evidence that haplotype associations with birth weight varied by ancestry.

Figure 7.1(c) shows p-values for associations between the methylation proportion at each probe and SNP minor allele counts near *H19*. Several strong associations (p<1e-10) were seen with individual SNPs and methylation at cg22172494, cg10602543 and cg25852472. These probes were highly methylated, and strongly correlated ($r > 0.55$); therefore, we summarized methylation data at these three probes by their first principal component (which we refer to as PCmeth). PCmeth demonstrated very strong associations with our two haplotypes of interest and with marker rs4929984 (Table 7.5); furthermore, PCmeth was

Table 7.5. Associations of haplotypes with birthweight and methylation levels. From the eight haplotypes formed by rs2067051, rs2251375 and rs4929984, associations of two haplotypes are shown, with residual birth weight and with the first principal component of methylation levels across three probes: cg22172494, cg10602543 and cg25852472.

			Residual birthweight		Methylation PC	
Haplotype	N infants	Variable	Slope (s.e.)	*p*-value	Slope (s.e.)	*p*-value
ACA	1045	Maternal	−0.032 (0.067)	0.64	—	
		Paternal	0.119 (0.064)	0.063	—	
		African	−0.095 (0.055)	0.087	—	
	214	Maternal	0.038 (0.142)	0.79	−0.558 (0.137)	5.9e-5
		Paternal	0.210 (0.136)	0.12	−0.377 (0.132)	0.0046
		African	−0.079 (0.131)	0.55	0.758 (0.127)	6.6e-9
GAC	1045	Maternal	0.083 (0.054)	0.13	—	
		Paternal	−0.098 (0.054)	0.070	—	
		African	0.111 (0.056)	0.047	—	
	214	Maternal	0.147 (0.128)	0.25	0.366 (0.127)	0.0041
		Paternal	−0.165 (0.121)	0.17	0.315 (0.121)	0.0094
		African	−0.176 (0.137)	0.20	0.782 (0.137)	2.6e-8
GCA	1045	Maternal	−0.268 (0.090)	0.0031	—	
		Paternal	0.064 (0.086)	0.46	—	
		African	−0.080 (0.056)	0.15	—	
	214	Maternal	−0.446 (0.180)	0.014	−0.039 (0.185)	0.83
		Paternal	0.045 (0.157)	0.78	0.062 (0.163)	0.70
		African	−0.236 (0.114)	0.27	1.075 (0.117)	8.0e-18
GCC	1045	Maternal	0.096 (0.080)	0.23	—	
		Paternal	−0.016 (0.089)	0.86	—	
		African	−0.103 (0.054)	0.055	—	
	214	Maternal	0.043 (0.171)	0.80	0.304 (0.173)	0.79
		Paternal	0.076 (0.189)	0.69	−0.029 (0.191)	0.88
		African	−0.149 (0.118)	0.21	1.13 (0.120)	1.1e-18

strongly associated with birth weight (p = 0.0008, slope –0.189, s.e. 0.056) such that higher levels of methylation were associated with lower birth weights. It is also interesting to note the extremely strong difference in methylation levels associated with African ancestry in Table 7.5; the levels of methylation are higher by approximately 0.8 of a standard deviation among African Americans.

Returning to the possible triangle of relationships illustrated in Figure 7.1(b), we have demonstrated two strong haplotype-methylation associations (Figure 7.1(c) and Table 7.5), and a strong relationship between PCmeth and birthweight, i.e. associations are convincing along the Scenario 2 pathway for haplotypes ACA and GAC. For Scenario 1, we have observed an association between the GCA haplotype, when maternally transmitted, and birth weight (Table 7.5), which is not affected by ancestry, despite the very different haplotype frequencies in African Americans.

One might therefore ask whether there is evidence for genetic factors acting through methylation levels on birth weight. Using principles of mediation analysis (Baron and Kenny 1986), we estimated the relationship between birth weight and, simultaneously, PCmeth and each of our haplotypes of interest, also adjusting for African ancestry. If the methylation levels were mediating the relationship between SNPs and birth weight, any SNP associations should disappear. Table 7.6 shows the analyses of residual birth weight including covariates for parent-of-origin derived haplotypes and PCmeth. Methylation remains strongly associated with lower residual birth weights for all of the four haplotypes. Also, the maternally-inherited copy of the GCA haplotypes continues to show a strong association with lower birth weight (effect: –0.462, s.e. 0.176, p = 0.0090), suggesting that mediation is not occurring.

7.3 Discussion

DNA methylation is an important regulatory mechanism controlling gene expression and may depend upon genetic factors ((Fang *et al.* 2012), reviewed in (Meaburn *et al.* 2010, Tycko 2010)). Given that the *H19/IGF2* locus and its parent-of-origin methylation are implicated in prenatal growth control (Weksberg *et al.* 1993, Pidsley *et al.* 2012), we might expect genetic effects on birth weight to act only through the methylation levels at this locus (Figure 7.1(b)). We have shown strong effects of the SNPs and haplotypes on methylation levels and furthermore that infants receiving the GCA haplotype from their mothers tend to be lighter by 0.46 standardized scores; this corresponds to approximately

Table 7.6. Estimates from linear model explaining residual birth weight as a function of both SNP or haplotype effects and methylation data (N = 214). Methylation data are summarized into a single principal component.

SNP data	Covariate	Slope (s.e.)	*p*-value
ACA haplotype	Maternal	–0.069 (0.144)	0.63
	Paternal	0.151 (0.135)	0.27
	PCmeth	–0.204 (0.072)	0.0049
	African	0.085 (0.143)	0.55
GAC haplotype	Maternal	0.252 (0.128)	0.050
	Paternal	–0.096 (0.120)	0.42
	PCmeth	–0.224 (0.071)	0.0017
	African	–0.007 (0.146)	0.96
GCA haplotype	Maternal	**–0.462 (0.176)**	**0.0090**
	Paternal	0.058 (0.153)	0.70
	PCmeth	–0.215 (0.068)	0.0016
	African	0.110 (0.136)	0.42
GCC haplotype	Maternal	0.071 (0.171)	0.68
	Paternal	0.047 (0.186)	0.80
	PCmeth	–0.210 (0.069)	0.0027
	African	0.084 (0.142)	0.55

210 grams, a significant effect. However, this parent-of-origin genetic effect appeared unchanged in an analysis of mediation, suggesting that the genetic effect on birth weight, captured by the maternal GCA haplotype, is not acting through altered methylation levels at the *H19* locus.

Figure 7.1(d) shows that the relationship between the maternally-transmitted GCA haplotype and birth weight is not affected by the strong meQTL relationships with the ACA or GAC haplotypes. The *cis*-meQTL probes used in our analyses are located near a weak enhancer element at the *H19* locus. In fact, given the lack of marginal association between ACA or GCA haplotypes and birth weight, perhaps other factors affecting birth weight are altering methylation levels instead of the reverse (Figure 7.1(b)). Nevertheless, the maternal GCA haplotype appears to act on birth weight through a mechanism independent of methylation of the *H19* CpG probes. The model does not exclude the possibility that haplotype-birth weight association may still be dependent on methylation of other CpG loci, e.g. via associations with *IGF2* DMR2 region methylation reported by (Tobi *et al.* 2012).

In whole blood DNA methylation in adults, Bell *et al* (Bell *et al.* 2012) also identified *cis*-meQTLs at five of our probes (Table 7.3), some of which were subsequently confirmed in a study of adolescent twins (Renteria *et al.* 2013). In contrast, the Dutch Hunger Winter Families Study (Tobi *et al.* 2012) did not find association between *H19* SNPs and whole blood methylation of CpG probes in the *H19* locus; however, they found an association with DNA methylation at the DMR2 region of *IGF2*; hence, associations between *H19* SNPs and DNA methylation in newborn umbilical cord blood may change throughout the life course.

Since the haplotype frequencies varied substantially by ethnicity, we included African ancestry as a covariate in all models. Methylation levels at our probes of interest were substantially higher, by approximately 0.8 standard deviations, in African Americans. However, any effects of African ancestry on birth weight (which was previously standardized for ethnicity) were small and inconsistent. This provides further indirect support for the conclusion that the genetic effects are not acting through methylation at this locus.

Genetic effects on methylation levels often act through *cis*-regulation. The methylation data in this study were obtained from the Illumina 27K array, which estimates average methylation levels at a genomic position without the ability to separate the signal by chromosome. Approaches such as bisulphite sequencing (Laird 2010) lead to estimates of methylation levels that are intrinsically linked to DNA sequence, thereby enabling a direct assessment of allelic-specific methylation levels (Shoemaker *et al.* 2010), however these technologies have only recently started to be used in large, high-throughput studies (e.g. (Busche *et al.* 2015)). If we had been able to measure allele-specific methylation in this region, it may have been easier to explore why the GAC and ACA haplotypes showed such strong associations with methylation levels but not birth weight, whereas the GCA haplotype was associated with birthweight and not methylation. Additionally, most assays, including the Illumina 27K array used in this study, do not distinguish between levels of 5-methylcytosine (5-mC) and 5-hydroxymethylcytosine (5-hmC). Interestingly, a recent study of birth weight distinguishing the two modifications found an association with 5-hmC levels at the *H19* locus in placental tissue (Piyasena *et al.* 2015). It remains to be investigated whether the levels of DNA methylation, hydroxymethylation, or a combination of both at the *H19* locus in umbilical cord blood represent a biologically functional association with birth weight.

In the 214 infants analyzed in our study, the association between birth weight and the maternal GCA haplotype was stronger than the associations with any single marker. This suggests that neither SNP rs4929984 nor the other markers defining the haplotype are likely to be causal variants. However, one or more closely-linked genetic variants may influence birth weight. Furthermore, cg22172494 is located in exon 3 of the *H19* gene within a putative weak enhancer element.

In summary, our analysis indicates that the association between the GCA maternal haplotype and birth weight is not accounted for by the level of methylation at cg22172494, cg10602543 or cg25852472. This implies that the GCA haplotype is associated with an element that, when maternally inherited, tends to lead to lower birth weights — this could be a genetic variant or methylation elsewhere (e.g. (Tobi *et al.* 2012)). Since we examined cord blood DNA, we cannot exclude the possibility that DNA methylation in developing fetal cells may be both dependent upon this haplotype and have an effect on fetal growth. Nevertheless, we conclude that the genetic effect of the GCA haplotype on birth weight is independent of methylation levels in this enhancer element; therefore, other molecular mechanisms are in action here.

Acknowledgments

CG and CX were supported by MITACS, a program of the Networks of Centres of Excellence, Canada. AN was supported by a grant from NSERC. This project was supported by grants HD060713 and HD055462 from the US National Institute of Child Health and Human Development. Additional funding was provided by the Clinical and Translational Science Institute, the Center for Integrative and Translational Genomics, the Office of Research of the University of Tennessee Health Science Center (UTHSC), and the Urban Child Institute (Memphis, TN). We acknowledge our late colleague, Grant Somes (UTHSC), who initiated the longitudinal birth cohort study (CANDLE) from which these samples were derived and participated in early statistical analyses of the *H19* genetic data and helped recognize their association with birth weight variation. We also thank UTHSC staff members Frances Tylavsky for her leadership of the continued CANDLE study, and Jeanette Peeples, Joycelynn Butler, Priyanka Jani, and Yanhua Qu, who provided laboratory and analytical assistance to this project.

This work was performed while Ronald Adkins was employed at The University of Tennessee Health Science Center. The opinions expressed in this article are the authors' own and do not reflect the view of the National Institutes of Health, the Department of Health and Human Services, or the United States government.

References

Adkins, R. M., Krushkal, J., Magann, E. F., Klauser, C. K., Morrison, J. C. *et al.* (2010a). Association of maternally inherited GNAS alleles with African-American male birth weight. *International Journal of Pediatric Obesity* 5, 177–184.

Adkins, R. M., Somes, G., Morrison, J. C., Hill, J. B., Watson, E. M. *et al.* (2010b). Association of birth weight with polymorphisms in the IGF2, H19, and IGF2R genes. *Pediatric Research* 68, 429–434.

Adkins, R. M., Tylavsky, F. A. and Krushkal, J. (2012). Newborn umbilical cord blood DNA methylation and gene expression levels exhibit limited association with birth weight. *Chemistry & Biodiversity* 9, 888–899.

Babak, T., DeVeale, B., Tsang, E. K., Zhou, Y., Li, X. *et al.* (2015). Genetic conflict reflected in tissue-specific maps of genomic imprinting in human and mouse. *Natural Genetics* 47, 544–549.

Barlow, D. P. and Bartolomei, M. S. (2014). Genomic imprinting in mammals. *Cold Spring Harbor Perspectives in Biology* 6 pii: a018382.

Baron, R. M. and Kenny, D. A. (1986). The moderator-mediator variable distinction in social psychological research: conceptual, strategic, and statistical considerations. *Journal of Personal and Social Psychology* 51, 1173–1182.

Bell, A. C. and Felsenfeld, G. (2000). Methylation of a CTCF-dependent boundary controls imprinted expression of the Igf2 gene. *Nature* 405, 482–485.

Bell, J. T., Pai, A. A., Pickrell, J. K., Gaffney, D. J., Pique-Regi, R. *et al.* (2011). DNA methylation patterns associate with genetic and gene expression variation in HapMap cell lines. *Genome Biology* 12, R10.

Bell, J. T., Tsai, P. C., Yang, T. P., Pidsley, R., Nisbet, J. *et al.* (2012). Epigenome-wide scans identify differentially methylated regions for age and age-related phenotypes in a healthy ageing population. *PLoS Genetics* 8, e1002629.

Bouwland-Both, M. I., van Mil, N. H., Stolk, L., Eilers, P. H., Verbiest, M. M. *et al.* (2013). DNA methylation of IGF2DMR and H19 is associated with fetal and infant growth: the generation R study. *PLoS One* 8, e81731.

Brannan, C. I., Dees, E. C., Ingram, R. S. and Tilghman, S. M. (1990). The product of the H19 gene may function as an RNA. *Molecular Cell Biology* 10, 28–36.

Briggs, T. A., Lokulo-Sodipe, K., Chandler, K. E., Mackay, D. J. and Temple, I. K. (2016). Temple syndrome as a result of isolated hypomethylation of the 14q32 imprinted DLK1/MEG3 region. *The American Journal of Medical Genetics A* 170, 170–175.

Browning, S. R. and Browning, B. L. (2007). Rapid and accurate haplotype phasing and missing-data inference for whole-genome association studies by use of localized haplotype clustering. *The American Journal of Human Genetics* 81, 1084–1097.

Busche, S., Shao, X., Caron, M., Kwan, T., Allum, F. *et al.* (2015). Population whole-genome bisulfite sequencing across two tissues highlights the environment as the principal source of human methylome variation. *Genome Biology* 16, 290.

Choufani, S., Shuman, C. and Weksberg, R. (2013). Molecular findings in Beckwith-Wiedemann syndrome. *American Journal of Medical Genetics C Seminar Medical Genetics* 163c, 131–140.

Eggermann, T., Brioude, F., Russo, S., Lombardi, M. P., Bliek, J. *et al.* (2015a). Prenatal molecular testing for Beckwith-Wiedemann and Silver-Russell syndromes: a challenge for molecular analysis and genetic counseling. *European Journal of Human Genetics* 24, 784–793.

Eggermann, T., Perez de Nanclares, G., Maher, E. R., Temple, I. K., Tumer, Z. *et al.* (2015b). Imprinting disorders: a group of congenital disorders with overlapping patterns of molecular changes affecting imprinted loci. *Clinical Epigenetics* 7, 123.

El-Maarri, O., Buiting, K., Peery, E. G., Kroisel, P. M., Balaban, B. *et al.* (2001). Maternal methylation imprints on human chromosome 15 are established during or after fertilization. *Natural Genetics* 27, 341–344.

Fang, F., Hodges, E., Molaro, A., Dean, M., Hannon, G. J. *et al.* (2012). Genomic landscape of human allele-specific DNA methylation. *Proceedings of the National Academy of Sciences of the United States of America* 109, 7332–7337.

Ferguson-Smith, A. C., Cattanach, B. M., Barton, S. C., Beechey, C. V. and Surani, M. A. (1991). Embryological and molecular investigations of parental imprinting on mouse chromosome 7. *Nature* 351, 667–670.

Geuns, E., De Rycke, M., Van Steirteghem, A. and Liebaers, I. (2003). Methylation imprints of the imprint control region of the SNRPN-gene in human gametes and preimplantation embryos. *Human Molecular Genetics* 12, 2873–2879.

Giannoukakis, N., Deal, C., Paquette, J., Kukuvitis, A. and Polychronakos, C. (1996). Polymorphic functional imprinting of the human IGF2 gene among individuals, in blood cells, is associated with H19 expression. *Biochemical and Biophysical Research Communications* 220, 1014–1019.

Glenn, C. C., Saitoh, S., Jong, M. T., Filbrandt, M. M., Surti, U. *et al.* (1996). Gene structure, DNA methylation, and imprinted expression of the human SNRPN gene. *The American Journal of Human Genetics* 58, 335–346.

Hamill, P. V., Drizd, T. A., Johnson, C. L., Reed, R. B., Roche, A. F. *et al.* (1979). Physical growth: National Center for Health Statistics percentiles. *The American Journal of Clinical Nutrition* 32, 607–629.

Hernandez-Valero, M. A., Rother, J., Gorlov, I., Frazier, M. and Gorlova, O. (2013). Interplay between polymorphisms and methylation in the H19/IGF2 gene region may contribute to obesity in Mexican-American children. *Journal of Developmental Origins of Health Disease* 4, 499–506.

Hiura, H., Obata, Y., Komiyama, J., Shirai, M. and Kono, T. (2006). Oocyte growth-dependent progression of maternal imprinting in mice. *Genes Cells* 11, 353–361.

Huang, R. C., Galati, J. C., Burrows, S., Beilin, L. J., Li, X. *et al.* (2012). DNA methylation of the IGF2/H19 imprinting control region and adiposity distribution in young adults. *Clinical Epigenetics* 4, 21.

Ioannides, Y., Lokulo-Sodipe, K., Mackay, D. J., Davies, J. H. and Temple, I. K. (2014). Temple syndrome: improving the recognition of an underdiagnosed chromosome 14 imprinting disorder: an analysis of 51 published cases. *Journal of Medical Genetics* 51, 495–501.

Ishida, M., Monk, D., Duncan, A. J., Abu-Amero, S., Chong, J. *et al.* (2012). Maternal inheritance of a promoter variant in the imprinted PHLDA2 gene significantly increases birth weight. *The American Journal of Human Genetics* 90, 715–719.

Ishihara, K., Hatano, N., Furuumi, H., Kato, R., Iwaki, T. *et al.* (2000). Comparative genomic sequencing identifies novel tissue-specific enhancers and sequence elements for methylation-sensitive factors implicated in Igf2/H19 imprinting. *Genome Research* 10, 664–671.

Juppner, H. (2015). Genetic and epigenetic defects at the GNAS locus cause different forms of pseudohypoparathyroidism. *Annales d'Endocrinologie* (Paris) 76, 92–97.

Kaku, K., Osada, H., Seki, K. and Sekiya, S. (2007). Insulin-like growth factor 2 (IGF2) and IGF2 receptor gene variants are associated with fetal growth. *Acta Paediatrica* 96, 363–367.

Kanduri, C., Pant, V., Loukinov, D., Pugacheva, E., Qi, C. F. *et al.* (2000). Functional association of CTCF with the insulator upstream of the H19 gene is parent of origin-specific and methylation-sensitive. *Current Biology* 10, 853–856.

Kono, T., Obata, Y., Wu, Q., Niwa, K., Ono, Y. *et al.* (2004). Birth of parthenogenetic mice that can develop to adulthood. *Nature* 428, 860–864.

Kukuvitis, A., Georgiou, I., Syrrou, M., Andronikou, S., Dickerman, Z. *et al.* (2004). Lack of association of birth size with polymorphisms of two imprinted genes, IGF2R and GRB10. *Journal of Pediatric Endocrinology and Metabolism* 17, 1215–1220.

Labialle, S., Yang, L., Ruan, X., Villemain, A., Schmidt, J. V. *et al.* (2008). Coordinated diurnal regulation of genes from the Dlk1-Dio3 imprinted domain: implications for regulation of clusters of non-paralogous genes. *Human Molecular Genetics* 17, 15–26.

Laird, P. W. (2010). Principles and challenges of genomewide DNA methylation analysis. *Nature Reviews Genetics* 11, 191–203.

Li, J. Y., Lees-Murdock, D. J., Xu, G. L. and Walsh, C. P. (2004). Timing of establishment of paternal methylation imprints in the mouse. *Genomics* 84, 952–960.

Lim, A. L., Ng, S., Leow, S. C., Choo, R., Ito, M. *et al.* (2012). Epigenetic state and expression of imprinted genes in umbilical cord correlates with growth parameters in human pregnancy. Journal of Medical Genetics 49, 689–697.

Liu, Y., Aryee, M. J., Padyukov, L., Fallin, M. D., Hesselberg, E. *et al.* (2013). Epigenome-wide association data implicate DNA methylation as an intermediary of genetic risk in rheumatoid arthritis. *Nature Biotechnology* 31, 142–147.

Lopes, S., Lewis, A., Hajkova, P., Dean, W., Oswald, J. *et al.* (2003). Epigenetic modifications in an imprinting cluster are controlled by a hierarchy of DMRs suggesting long-range chromatin interactions. *Human Molecular Genetics* 12, 295–305.

Mantovani, G., Lania, A. G. and Spada, A. (2010). GNAS imprinting and pituitary tumors. *Molecular Cell Endocrinology* 326, 15–18.

Meaburn, E. L., Schalkwyk, L. C. and Mill, J. (2010). Allele-specific methylation in the human genome: implications for genetic studies of complex disease. *Epigenetics: Official Journal of the DNA Methylation Society* 5, 578–582.

Mikkelsen, T. S., Ku, M., Jaffe, D. B., Issac, B., Lieberman, E. *et al.* (2007). Genome-wide maps of chromatin state in pluripotent and lineage-committed cells. *Nature* 448, 553–560.

Moore, G. E., Ishida, M., Demetriou, C., Al-Olabi, L., Leon, L. J. *et al.* (2015). The role and interaction of imprinted genes in human fetal growth. *Philosophical Transactions of the Royal Society of London B Biological Science* 370, 20140074.

Murphy, R., Thompson, J. M., Tost, J. and Mitchell, E. A. (2014). No evidence for copy number and methylation variation in H19 and KCNQ10T1 imprinting control regions in children born small for gestational age. *BMC Medical Genetics* 15, 67.

Mussa, A., Russo, S., De Crescenzo, A., Freschi, A., Calzari, L. *et al.* (2016). Fetal growth patterns in Beckwith-Wiedemann syndrome. *Clinical Genetics* 90, 21–27.

Naumova, A. K. and Croteau, S. (2004). Mechanisms of Epigenetic Variation: Polymorphic Imprinting. *Current Genomics* 5, 417–429.

Nicholls, R. D., Saitoh, S. and Horsthemke, B. (1998). Imprinting in Prader-Willi and Angelman syndromes. *Trends in Genetics* 14, 194–200.

Paquette, J., Giannoukakis, N., Polychronakos, C., Vafiadis, P. and Deal, C. (1998). The INS 5' variable number of tandem repeats is associated with IGF2 expression in humans. *Journal of Biological Chemistry* 273, 14158–14164.

Perry, J. R., Day, F., Elks, C. E., Sulem, P., Thompson, D. J. *et al.* (2014). Parent-of-origin-specific allelic associations among 106 genomic loci for age at menarche. *Nature* 514, 92–97.

Petry, C. J., Seear, R. V., Wingate, D. L., Acerini, C. L., Ong, K. K. *et al.* (2011). Maternally transmitted foetal H19 variants and associations with birth weight. Human Genetics 130, 663–670.

Pidsley, R., Dempster, E., Troakes, C., Al-Sarraj, S. and Mill, J. (2012). Epigenetic and genetic variation at the IGF2/H19 imprinting control region on 11p15.5 is associated with cerebellum weight. *Epigenetics: Official Journal of the DNA Methylation Society* 7, 155–163.

Piyasena, C., Reynolds, R. M., Khulan, B., Seckl, J. R., Menon, G. *et al.* (2015). Placental 5-methylcytosine and 5-hydroxymethylcytosine patterns associate with size at birth. *Epigenetics* 10, 692–697.

Polychronakos, C. and Kukuvitis, A. (2002). Parental genomic imprinting in endocrinopathies. *European Journal of Endocrinology* 147, 561–569.

Renteria, M. E., Coolen, M. W., Statham, A. L., Choi, R. S., Qu, W. *et al.* (2013). GWAS of DNA methylation variation within imprinting control regions suggests parent-of-origin association. *Twin Research and Human Genetics* 16, 767–781.

Shoemaker, R., Deng, J., Wang, W. and Zhang, K. (2010). Allele-specific methylation is prevalent and is contributed by CpG-SNPs in the human genome. *Genome Research* 20, 883–889.

Sullivan, M. J., Taniguchi, T., Jhee, A., Kerr, N. and Reeve, A. E. (1999). Relaxation of IGF2 imprinting in Wilms tumours associated with specific changes in IGF2 methylation. *Oncogene* 18, 7527–7534.

Szabo, P., Tang, S. H., Rentsendorj, A., Pfeifer, G. P. and Mann, J. R. (2000). Maternal-specific footprints at putative CTCF sites in the H19 imprinting control region give evidence for insulator function. *Current Biology* 10, 607–610.

Takai, D., Gonzales, F. A., Tsai, Y. C., Thayer, M. J. and Jones, P. A. (2001). Large scale mapping of methylcytosines in CTCF-binding sites in the human H19 promoter and aberrant hypomethylation in human bladder cancer. *Human Molecular Genetics* 10, 2619–2626.

Tobi, E. W., Slagboom, P. E., van Dongen, J., Kremer, D., Stein, A. D. *et al.* (2012). Prenatal Famine and Genetic Variation Are Independently and Additively Associated with DNA Methylation at Regulatory Loci within IGF2/H19. *PLoS One* 7, e37933.

Tremblay, K. D., Saam, J. R., Ingram, R. S., Tilghman, S. M. and Bartolomei, M. S. (1995). A paternal-specific methylation imprint marks the alleles of the mouse H19 gene. *Nature Genetics* 9, 407–413.

Tunster, S. J., Creeth, H. D. and John, R. M. (2016). The imprinted Phlda2 gene modulates a major endocrine compartment of the placenta to regulate placental demands for maternal resources. *Developmental Biology* 409, 251–260.

Tycko, B. (2010). Allele-specific DNA methylation: beyond imprinting. *Human Molecular Genetics* 19, R210–220.

Vafiadis, P., Bennett, S. T., Todd, J. A., Grabs, R. and Polychronakos, C. (1998). Divergence between genetic determinants of IGF2 transcription levels in leukocytes and of IDDM2-encoded susceptibility to type 1 diabetes. *Journal of Clinical Endocrinology and Metabolism* 83, 2933–2939.

Vidal, A. C., Overcash, F., Murphy, S. K., Murtha, A. P., Schildkraut, J. M. *et al.* (2013). Associations between birth and one year anthropometric measurements and IGF2 and IGF2R genetic variants in African American and Caucasian American infants. *Journal of Pediatrics Genetics* 2, 119–127.

Wallace, C., Smyth, D. J., Maisuria-Armer, M., Walker, N. M., Todd, J. A. *et al.* (2010). The imprinted DLK1-MEG3 gene region on chromosome 14q32.2 alters susceptibility to type 1 diabetes. *Nature Genetics* 42, 68–71.

Webber, A. L., Ingram, R. S., Levorse, J. M. and Tilghman, S. M. (1998). Location of enhancers is essential for the imprinting of H19 and Igf2 genes. *Nature* 391, 711–715.

Weksberg, R., Shen, D. R., Fei, Y. L., Song, Q. L. and Squire, J. (1993). Disruption of insulin-like growth factor 2 imprinting in Beckwith-Wiedemann syndrome. *Nature Genetics* 5, 143–150.

Weksberg, R., Shuman, C. and Beckwith, J. B. (2010). Beckwith-Wiedemann syndrome. *European Journal of Human Genetics* 18, 8–14.

Zhang, Y., Shields, T., Crenshaw, T., Hao, Y., Moulton, T. *et al.* (1993). Imprinting of human H19: allele-specific CpG methylation, loss of the active allele in Wilms tumor, and potential for somatic allele switching. *The American Journal of Human Genetics* 53, 113–124.

Disturbances in Folate Metabolism and Their Impact on Development

8

Renata H. Bahous, Karen E. Christensen* and Rima Rozen*,†*

**Department of Human Genetics, McGill University, The Research Institute of the McGill University Health Centre, Montreal, Quebec, Canada*
†*Department of Pediatrics, McGill University, The Research Institute of the McGill University Health Centre, Montreal, Quebec, Canada*

Folate is an important vitamin that is well known in the prevention of neural tube defects. It is required for the synthesis of nucleotides and S-adenosylmethionine (SAM), a ubiquitous methyl donor. SAM is an important molecule in epigenetic processes that rely on methylation of DNA, histones and other proteins. Decreased production of SAM or high levels of homocysteine and its metabolite S-adenosylhomocysteine, intermediates in the methylation cycle, can inhibit methylation reactions. Common genetic polymorphisms in genes in the folate pathway have been shown to affect the methylation cycle and increase the levels of homocysteine. Maternal nutrition before and during pregnancy has attracted a great deal of interest in recent years due its role in embryonic development and its potential long-term effect on offspring and adult health. One possible mechanism for the disturbances in phenotypes is a change in the epigenetic signature, which can affect expression levels of key genes in a given pathway. Folate intake, including dietary deficiency and excessive supplementation, has also been shown to alter DNA methylation and affect offspring phenotype. In this chapter, we will review one-carbon folate metabolism and describe how disturbances in the pathway, genetic and/or dietary, affect epigenetic processes, developmental outcomes and adult health.

8.1 Introduction

Maternal nutrition, periconceptionally and during pregnancy, has been extensively studied to determine its effect on the development of offspring. Considerable attention has been focused on the role of folate in development since the demonstration that inadequate folate intake is a risk factor for neural tube defects (NTD) (reviewed in Hoffbrand and Weir 2001). "Folate" is a general term that refers to all derivatives of folate (including all oxidation states, polyglutamate tail lengths and the synthetic folic acid). Folate is particularly important during embryonic development because it is required to transfer one-carbon units in several key biosynthetic pathways. These pathways include the methionine/methylation cycle (which includes the de-toxification of homocysteine, a non-protein amino acid), the metabolism of choline (an alternate source of one-carbon units) and nucleotide synthesis (Figure 8.1). Folate cannot be synthesized by humans and must be acquired through the diet. In contrast, choline can be synthesized *de novo* or acquired from the diet (Caudill 2009).

Folate is measured in plasma/serum, which reflects recent intake, or in red blood cells (RBC), which is indicative of long-term consumption. The cutoff values for folate deficiency are 10 nmol/L for serum folate and 340 nmol/L for RBC folate (Berry *et al.* 2010). To reduce the risk of NTD, the World Health Organization recommends that RBC folate should be 906 nmol/L in women of reproductive age (Cordero *et al.* 2015). To meet this requirement, supplementation with 0.4 mg folic acid per day is recommended (CDC 1992). The recommended upper limit for folic acid intake is 1 mg/day. However, there are concerns that a small but significant proportion of the population is exceeding this upper limit due to increased use of high dose supplements (Bailey *et al.* 2010, Yeung *et al.* 2011). Several reports suggest that over-consumption of folic acid can have negative consequences for health, including promotion of cancer or masking of vitamin B_{12} deficiency, and may alter epigenetic marks (reviewed in Smith *et al.* 2008, reviewed in Crider *et al.* 2011).

Since almost half of pregnancies are unplanned, mandatory supplementation of grains and cereals with folic acid was adopted in Canada and the US in the late 1990s to increase folate intake by 0.1–0.2 mg/day to help prevent NTD (Berry *et al.* 2010). Approximately 50 countries worldwide have implemented folate fortification programs (EUROCAT); however there are notable exceptions, such as

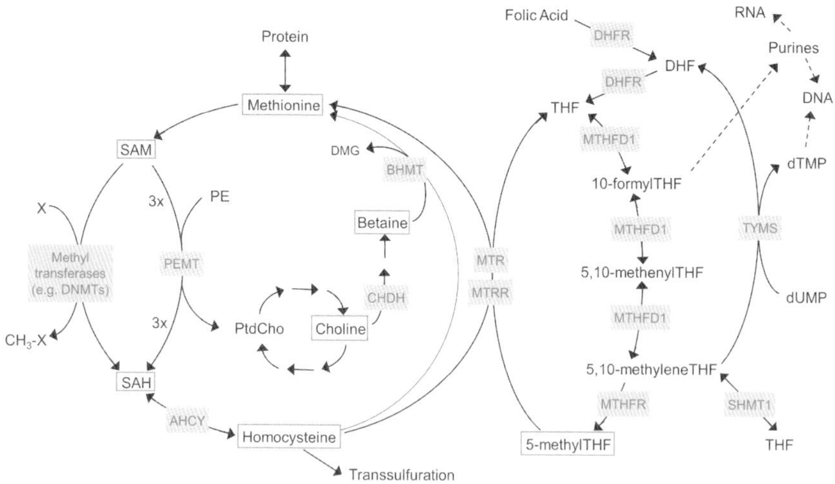

Figure 8.1. One-carbon folate metabolism. Enzymes are shown in solid boxes, metabolites of particular importance are shown in open boxes. Most naturally occurring food folate enters the pathway as 5-methylTHF. Folic acid, the synthetic folate commonly used in supplements, must first be reduced by DHFR and enters the pathway as THF.
Abbreviations: AHCY: S-adenosylhomocysteine hydrolase; BHMT: betaine-homocysteine methyltransferase; CHDH: choline dehydrogenase; DHF: dihydrofolate; DHFR: DHF reductase; DMG: dimethylglycine; DNMT: DNA methyltransferase; MTHFD1: methyleneTHF dehydrogenase-methenylTHF cyclohydrolase-formylTHF synthetase; MTHFR: methylTHF reductase; MTR: methionine synthase; MTRR: MTR reductase: PE: phosphatidylethanolamine; PEMT: PE methyltransferase; PtdCho: phosphatidylcholine; SAH: S-adenosylhomocysteine; SAM: S-adenosylmethionine; SHMT1: serine hydroxy-methyltransferase; THF: tetrahydrofolate; TYMS: thymidylate synthase.

the UK, Ireland and New Zealand. In spite of the controversy, these programs have been successful, with a reduction of NTD incidence (by 19–55%) in the US and Canada. In contrast, the rates of NTD decline are not as significant in countries without fortification (reviewed in Crider *et al.* 2011)

The major folate in blood is 5-methyltetrahydrofolate (5-methylTHF), which is generated by methylenetetrahydrofolate reductase (MTHFR). This enzyme is the gateway to the methylation cycle. 5-MethylTHF is required for the remethylation of homocysteine to methionine. Methionine can then be used to synthesize the ubiquitous methyl donor S-adenosylmethionine (SAM) that is used in a host of methylation reactions including methylation of DNA, RNA and histones and the synthesis of phospholipids and neurotransmitters. A summary of other relevant enzymes in folate metabolism is provided in Figure 8.1 and in the section below.

Since folate is essential for nucleotide synthesis and methylation reactions, it is of particular importance during development, when cells are rapidly proliferating. Methylation of DNA and histones is necessary for the establishment of epigenetic profiles and the regulation and timing of gene expression, which is essential for proper development (reviewed in Xu and Sinclair 2015). Therefore, dietary disturbances in one-carbon folate metabolism can affect development. In addition to NTD, low folate intake has been associated with other adverse pregnancy outcomes including congenital heart defects, intrauterine growth retardation and premature birth (Czeizel 1996, Copp and Greene 2013, Furness *et al.* 2013).

There are many single nucleotide polymorphisms (SNPs) in genes of folate metabolism that may affect the risk for disease (reviewed in Christensen and Rozen, 2010). Mouse models have been created for SNPs in several of these genes, such as mice with deficiencies of *Mthfr*, methionine synthase reductase (*Mtrr*) and methylenetetrahydrofolate dehydrogenase (NADP-dependent)-methenyltetrahydrofolate cyclohydrolase-formyltetrahydrofolate synthetase (*Mthfd1*) (Chen *et al.* 2005, Deng *et al.* 2008, Christensen *et al.* 2013), which have been used to investigate the *in vivo* impact of these genetic disruptions and their potential interaction with dietary folate intake.

Some of the polymorphisms in genes of folate metabolism may affect blood levels of folate and related metabolites (Fredriksen *et al.* 2007, reviewed in Yetley and Johnson, 2011). For example, individuals homozygous for *MTHFR* 677C>T (677TT) (rs1801133) have a mild MTHFR deficiency that is associated with reduced folate and increased homocysteine levels in serum (Fredriksen *et al.* 2007). Gene-nutrient interactions between the SNP and dietary folate intake may modulate their effects on biomarkers or clinical outcomes. For example, the association between *MTHFR* 677TT and elevated homocysteine may be overcome by maintaining higher blood folate (Jacques *et al.* 1996); this phenomenon is likely due to stabilization of the mutant protein by folate (Yamada *et al.* 2001). The *MTHFD1* 1958G>A (rs2236225) variant protein is also stabilized by elevated folate *in vitro* (Christensen *et al.* 2009), and may therefore exhibit similar gene-nutrient interactions, but studies on the interaction *in vivo* are limited (Ivanov *et al.* 2009). Genetic polymorphisms in folate-dependent enzymes are associated with risk for many of the developmental disorders discussed below.

When folate supply is limited or there are genetic blocks in folate-dependent enzymes, the distribution of one-carbon folate forms may be altered. Consequently, the supply of folate for methylation reactions may be impaired. DNA methyltransferases must compete for SAM with other SAM-dependent pathways. The consumption of SAM for methylation reactions in epigenetic events (methylation of DNA, methylation of histones) is actually quite minimal quantitatively in comparison to major SAM consuming pathways such as phosphatidylcholine and creatine synthesis. Therefore the supply of one-carbon units for epigenetic reactions may be compromised depending on cellular needs for one-carbons in other pathways. The literature is sparse regarding the competition for one-carbon units between different reactions (e.g. between epigenetics or nucleotide synthesis), or between different cell types or developmental stages. An understanding of folate-dependent pathways and their intersection with choline/phosphatidylcholine metabolism is important to address the mechanisms by which disturbances in one-carbon metabolism contribute to developmental anomalies. This is particularly important when trying to determine cause and effect of folate-related developmental disturbances or methylation-related epigenetic phenomena, since a wide range of metabolic alterations may accompany the methylation changes.

This chapter will briefly review the critical pathways in folate metabolism, describe potential mechanisms that can lead to disease, and discuss several important developmental disorders that are influenced by folate status.

8.2 Folate Metabolism and Regulation

The term "one-carbon folate metabolism" refers to the metabolic pathways that depend on the transfer and use of one-carbon units carried by folates (Figure 8.1). The enzymes that attach one-carbon units to tetrahydrofolate (THF) and interconvert the oxidation states of those carbons are central to one-carbon metabolism (reviewed in Tibbetts and Appling 2010). One-carbon folates are used directly for nucleotide synthesis and indirectly, via SAM, for methylation. When folic acid enters these pathways, it must first be reduced to THF by dihydrofolate reductase (DHFR). One-carbon units are then added to THF to form 5,10-methyleneTHF or 10-formylTHF, both of which are used to synthesize nucleotides that can later be incorporated into DNA and RNA. 10-FormylTHF provides carbons for the synthesis of purines or is reduced to 5,10-methyleneTHF, which is used by thymidylate synthase (TYMS) to methylate deoxyuridine monophosphate (dUMP) to generate thymidine monophosphate (dTMP) (reviewed in Tibbetts and Appling, 2010).

5,10-MethyleneTHF can also be reduced to 5-methylTHF by MTHFR. This step is crucial in folate metabolism as 5-methylTHF is the primary form of folate in the circulation. Furthermore, 5-methylTHF supports the methylation cycle as it is required for the remethylation of homocysteine to methionine, a reaction catalyzed by methionine synthase (MTR), which needs methionine synthase reductase (MTRR) for MTR reactivation. Methionine is in turn used to synthesize SAM. When SAM donates a methyl group, S-adenosylhomocysteine (SAH) is produced which is then broken down to homocysteine and adenosine by SAH hydrolase (AHCY). In some tissues, such as the liver, homocysteine is degraded via the transsulfuration pathway; in most tissues it is remethylated to methionine, completing the methylation cycle.

The folate pathways and choline metabolism intersect at the methylation cycle because choline metabolites are used as an alternate folate-independent pathway for homocysteine remethylation (reviewed in Caudill, 2009). Choline is oxidized to betaine by the enzyme choline dehydrogenase (CHDH). Betaine can then donate a methyl group to homocysteine to synthesize methionine via betaine hydroxymethyltransferase (BHMT) in the liver and kidney. Choline is an important molecule that can be used to synthesize phosphatidylcholine (PtdCho), a phospholipid essential for cell membrane integrity (reviewed in Vance and Vance, 2009). In the liver, PtdCho can also be synthesized by phosphatidylethanolamine methyltransferase (PEMT), which converts phosphatidylethanolamine to PtdCho using three molecules of SAM. PEMT is the major consumer of SAM in the liver (Jacobs *et al.* 2005), a second example of the connection between choline and folate metabolism. In the nervous system, choline is used to synthesize the neurotransmitter acetylcholine, as well as to synthesize PtdCho for neuronal membranes (reviewed in Li and Vance 2008). The intersection of choline and folate metabolism can have consequences for liver health when folate metabolism is disrupted by dietary or genetic deficiencies (reviewed in da Silva *et al.* 2014) since choline utilization for methylation may be increased to compensate for the folate disturbance. Furthermore, the folate disturbance may have a domino effect on brain development because of the importance of choline in that organ (reviewed in Zeisel 2006).

Given the importance of the metabolites produced by one-carbon metabolism, it is not surprising that the enzymes in the folate pathway are tightly regulated. The patterns of expression of the enzymes are tissue and temporally specific (Pickell *et al.* 2011). For example, the major organ involved in one-carbon folate metabolism is the liver, which expresses the full complement of enzymes in these pathways. In contrast, organs such as the brain are lacking betaine-dependent remethylation and transsulfuration pathways for the disposal of homocysteine. The expression of other enzymes in the pathways is limited to specific stages of embryonic development. MTHFD2, which is essential for the supply of one-carbon units for purine synthesis, is expressed primarily in embryos and tumor tissues (reviewed in Tibbetts and Appling 2010). Folate-dependent enzymes are often regulated by product inhibition, such as inhibition of many members of the methyltransferase family by their product SAH or MTHFR inhibition by SAM. These proteins may also regulate their own expression. For instance, TYMS and DHFR can bind to their own mRNA and repress translation (reviewed in Stover 2009).

MTHFR is an example of a number of regulatory mechanisms. Its expression is tissue specific, with higher expression in testes and intermediate expression in liver, brain and kidney. MTHFR expression is also influenced by folic acid intake. In mice, hepatic MTHFR increased when the folic acid content of the diet was low, and decreased when the folic acid content was high (Christensen *et al.* 2010, Christensen *et al.* 2015). In addition to the afore-mentioned feedback inhibition by SAM, MTHFR activity is modulated by phosphorylation (Yamada *et al.* 2005). The relative amount of the phosphorylated form in the liver, which is less active than the non-phosphorylated form, appears to depend on the amount of folate in the diet (Christensen *et al.* 2015). The combined effect of decreased expression and increased phosphorylation when folate intake is high appears to cause a pseudo-MTHFR deficiency that reduces SAM and 5-methylTHF levels in liver and leads to liver damage (Christensen *et al.* 2015).

8.3 Folate Disruptions and Development: Possible Mechanisms

Folate metabolism is a biological pathway well known for its importance in embryonic development. However, disturbances in folate metabolism are also implicated in diseases such as cancer (reviewed in Giovannucci 2002), neuropsychiatric disorders (reviewed in Gilbody *et al.* 2007) and cardiovascular disease (reviewed in Refsum *et al.* 1998). Collective efforts in these fields have helped to elucidate the mechanisms by which disturbances in folate metabolism alter cellular function and contribute to developmental anomalies. Most of the mechanisms that have been examined to date have been related to low folate intake and/or the effects of genetic variation that alter one-carbon fluxes. However, with the recent concerns about over-supplementation, the biochemical effects of high folate intake have come under greater scrutiny.

8.3.1 *Altered methylation*

Reduced SAM availability during folate deficiency can have several consequences. Newly synthesized DNA requires methylation following replication for stability and for preservation of epigenetic marks. Methylation of DNA can determine levels of gene expression; it is generally assumed that reduced DNA methylation corresponds to increased gene expression (reviewed in Attwood *et al.* 2002). Promoters often have low methylation levels to enable gene expression, whereas transposable elements, including Long Terminal Repeats, which can be active retroviruses, are heavily methylated to be silenced (reviewed in Smith and Meissner 2013). Methylation status can be measured at different levels: genome-wide (global methylation) or gene level (local methylation). Several studies, mainly in animal and/or cancer models, have examined the effects of folate deficiency, which is often associated with global hypomethylation. DNA hypomethylation can result in increased expression of proto-oncogenes, aberrant chromosome condensation or segregation, or DNA damage by deamination of unmethylated cytosines to uracil with resultant double strand breaks (reviewed in Crott and Mason 2005). For example, low folate may lead to reduced p53 promoter methylation (Kim *et al.* 1996). Paradoxically, folate deficiency has also been associated with hypermethylation of some genes in specific cancers (van den Donk *et al.* 2007). In colorectal cancer, hypermethylation and reduced expression of a gene involved in vitamin A metabolism, *Bcmo1*, was reported in mice fed low folate diets, and linked to tumorigenesis (Leclerc *et al.* 2013). Low RBC folate was a strong risk factor in women with HPV infection that progressed to cervical cancer, with hypermethylation of tumor suppressor genes such as *DAPK*, *CDH1* and *HIC1* (Flatley *et al.* 2009).

Adequate supplies of SAM are crucial during early embryonic development. In the early zygote stage, there is an almost complete erasure of DNA methylation genome-wide (excluding regions such as repeat elements, transposable elements and imprinted genes) followed by the reacquisition of DNA methylation marks (reviewed in Smith and Meissner 2013). A second wave of demethylation/remethylation occurs later in development, in the primordial germ cells of the developing gonads (Hajkova *et al.* 2002). The methylation patterns are acquired in a sex-specific manner resulting in the differential methylation of maternally and paternally imprinted genes in the germ cells (Kobayashi *et al.* 2013). In somatic cells, DNA methylation is also important for X-chromosome inactivation (McGraw *et al.* 2013) and for cells to progress from the pluripotent state to specific cell lineages (reviewed in Smith and Meissner 2013). The establishment of proper epigenetic marks is crucial during embryogenesis, both to regulate gene expression in the developing embryo, and to transmit these expression patterns to future offspring (reviewed in Smith and Meissner 2013, Xu and Sinclair, 2015). The window in which DNA methylation is reacquired is particularly sensitive to SAM levels. Therefore alterations of proper DNA methylation due to impaired folate metabolism will influence the embryo and possibly subsequent generations, a process referred to as transgenerational epigenetics. Murine *Mtrr* deficiency in maternal grandparents led to birth defects in the grandprogeny due to epigenetic changes that were transmitted through the germline, and to changes in the uterine environment

(Padmanabhan *et al.* 2013). In another study, adult mice fed methyl rich diets had changes in CpG methylation which were transmitted from generation to generation, emphasizing that epigenetic variation can be heritable (Li *et al.* 2011).

DNA methylation and histone modifications are interdependent and work together to ensure proper gene expression and chromatin structure. Histone modification can direct DNA methylation and conversely, DNA methylation can help maintain histone modifications during replication (reviewed in Cedar and Bergman 2009). Therefore, although reduced SAM levels may affect the availability of methyl groups for histone and DNA methylation, altered DNA methylation itself may have consequences for histone modification (Cedar and Bergman 2009).

DNA methylation is catalyzed by a family of DNA methyltransferases (DNMTs). DNMTs and other methyltransferases are very sensitive to the levels of SAM and SAH. Each methyltransferase has a specific affinity for SAM and therefore these enzymes may have different sensitivities to SAM availability. Furthermore, reduced remethylation of homocysteine to methionine may lead to elevated homocysteine. This in turn may cause SAH levels to rise because the kinetics of SAH hydrolase, the enzyme that produces homocysteine from SAH, favor SAH production when homocysteine concentrations are high. SAH accumulation may result in the feedback inhibition of methyltransferases in general; the degree of inhibition is dependent on the affinity of each methyltransferase for SAH. Importantly, the degree of SAH accumulation and subsequent hypomethylation is tissue specific. The tissue specificity may relate to the presence of alternative, folate-independent pathways to reduce the levels of homocysteine, such as the transsulfuration pathway, which ultimately degrades homocysteine or the presence of the choline-betaine pathway, which provides an alternate one-carbon source for homocysteine remethylation. In tissues where excess homocysteine can be removed via transsulfuration, the reduction in SAH concentrations may prevent methyltransferase inhibition (reviewed in James *et al.* 2002).

The impact of high methyl donor intake on methylation and offspring phenotype was demonstrated in viable yellow agouti (A^{vy}) mice. The predominant coat color in the offspring of the mice shifted from yellow in control mice to brown (pseudoagouti) in the litters of mothers fed a methyl donor rich diet (high in folic acid, vitamin B_{12}, choline and betaine) (Waterland and Jirtle 2003). This phenomenon was due to increased methylation of the intracisternal A particle retrotransposon in the A^{vy} allele that consequently reduced expression of the protein that causes yellow coat color.

8.3.2 *Decreased DNA synthesis and repair*

Disturbed folate metabolism may lead to DNA damage through uracil misincorporation and chromosome breaks. Reduced methyleneTHF folate availability, due to dietary deficiency or gene variants, may result in reduced conversion of dUMP to dTMP. This can result in the accumulation of dUMP that can be misincorporated into DNA (reviewed in Duthie *et al.* 2002). Removal of the misincorporated uracil to repair the DNA may result in double strand breaks. The reduced availability of thymidine also impairs DNA repair, which eventually leads to an accumulation of DNA damage and cell death. This process has been demonstrated in many systems including Chinese hamster ovary cells cultured in folate-depleted media (James *et al.* 1994) and in the livers of rats fed low folate diets (James *et al.* 1997). Decreased availability of folates can also reduce purine synthesis, which may limit cell proliferation. For example, impaired purine synthesis and reduced white blood cell counts were observed in a mouse model for a variant in the MTHFD1 synthetase domain (Christensen *et al.* 2013).

8.3.3 *Hyperhomocysteinemia*

Disturbed folate metabolism can result in elevated levels of plasma homocysteine, (hyperhomocysteinemia, HHcy) and/or urine homocysteine (as seen in homocystinuria, an inborn error of metabolism). HHcy has been implicated as a risk factor for many disorders including cardiovascular disease (reviewed in Refsum *et al.* 1998), stroke (Casas *et al.* 2005), neuropsychiatric diseases (e.g. Alzheimer) (Seshadri *et al.* 2002), and pregnancy complications such as hypertension (Hernández-Díaz *et al.* 2002) and pre-term birth (Dhobale *et al.* 2011). A number of possible mechanisms of action of homocysteine in disease have been investigated. HHcy may be detrimental to the cell by inducing oxidative stress, increasing ER stress, creating a pro-apoptotic environment, altering immune responses or modifying cellular proteins (Tyagi *et al.* 2005, reviewed in Lubos *et al.* 2007, da Cunha *et al.* 2012). These effects of homocysteine may lead to methylation changes independent of a direct effect on levels of SAM or SAH.

8.3.4 *Disturbances in choline metabolism*

Folate and choline metabolism are interconnected because the choline metabolite betaine can also be used to remethylate homocysteine via BHMT. Although this folate-independent pathway may be beneficial for maintaining methylation reactions when folate-dependent remethylation is disturbed (by genetic or dietary deficiencies), this interconnection leads to changes in choline metabolism when there is a problem in folate metabolism. A disturbance in choline metabolism can lead to fatty liver (Christensen *et al.* 2010) or delayed development, as choline is required to synthesize the phospholipid phosphatidylcholine (reviewed in Vance and Vance, 2009). The Kennedy pathway is the predominant pathway for the synthesis of phosphatidylcholine from choline, through phosphocholine and CDP-choline. In mouse models where genes in this pathway have been disrupted, embryos die *in utero* or are born with limb defects (reviewed in Vance and Vance 2009). Phosphatidylcholine can also be synthesized from phosphatidylethanolamine by PEMT. *Pemt*$^{-/-}$ mice have increased neurogenesis but decreased neuronal differentiation in the hippocampus (Zhu and Zeisel 2004). Choline is also a precursor of acetylcholine, an important neurotransmitter. Adequate choline supplies are necessary for brain development in the offspring, particularly in the hippocampus (reviewed in Zeisel 2006, 2011) and choline metabolism is disturbed in the brain of *Mthfr*$^{-/-}$ mice (Jadavji *et al.* 2012). High folate intake in mice, which leads to pseudo-MTHFR deficiency as mentioned above, resulted in reduced levels of 5-methylTHF, SAM and PtdCho; the decreased PtdCho contributed to altered lipid metabolism and liver disease (Christensen *et al.* 2015).

The role of folic acid and one-carbon units in many important cellular and epigenetic processes underscores the influence of disruptions in one-carbon metabolism on a host of clinical disorders including cancer, cardiovascular disease and neurodegenerative diseases. In this chapter we will concentrate on disorders related to development: birth defects, neuropsychiatric disorders and diseases with developmental origins that manifest in adult life.

8.4 Developmental Defects

8.4.1 *Neural tube defects (NTD)*

Neural tube defects (NTD) include the failure of the neural tube to close in the spine (spina bifida) or the brain (anencephaly/exencephaly) (reviewed in Copp and Greene 2013). Low folate intake was first discussed as a possible cause of NTD in the 1960s by Hibbard (Hibbard 1964). It is currently believed that folate supplementation can prevent 50–75% of NTD (reviewed in Blom *et al.* 2006). The Medical Research Council trial in 1991 demonstrated that periconceptional folate supplementation (4 mg/day) reduced the risk of recurrent NTD by 75% (MRC Vitamin Research Group, 1991). A year later, a Hungarian randomized controlled trial clearly demonstrated the protective effect of 0.8 mg/day of folic acid periconceptionally to prevent NTD (reviewed in Czeizel and Dudas 1992). Subsequently,

Daly *et al.* demonstrated that 0.4 mg/day of folic acid was sufficient to increase RBC folate concentrations and reduce NTD risk (reviewed in Daly and Scott 1998). The neural tube starts forming at day 14 in human embryos and is closed by day 28, before most pregnancies are discovered (reviewed in Copp and Greene 2013). Consequently many countries have instituted folic acid fortification programs to ensure that maternal folate intakes are adequate to prevent NTD at this early stage (reviewed in Berry *et al.* 2010).

The genetic factors that contribute to NTD risk are not fully identified. The *MTHFR* 677C>T (rs1801133) polymorphism was the first recognized genetic risk factor for NTD in many populations, including those of Ireland, China, Canada and the US (van der Put *et al.* 1995, reviewed in Copp and Greene 2010). The *MTHFD1* 1958G>A polymorphism (rs2236225) has been associated with NTD, particularly in Ireland (Parle-McDermott *et al.* 2006). The effects of gene–gene interactions in folate metabolism have been investigated in several studies. For example, the combination of the *MTHFR* 677C>T and *MTRR* 66A>G (rs1801394) variants may increase the infant's risk of NTD and the combination of maternal *MTRR* 66A>G and fetal *MTHFR* 677C>T increased the risk of NTD compared with that of the *MTHFR* variant alone (Relton *et al.* 2004). A recent meta-analysis determined that maternal *MTHFR* 677C>T (50 studies) and *MTRR* 66A>G (11 studies) significantly increased NTD risk in their offspring (Yadav *et al.* 2014).

Several animal models have been used to study the causes of NTD, including chicken, zebrafish, mice and rats. NTD are not usually observed in knockouts of genes in the folate pathways. In many of these models, homozygous null embryos die *in utero*; although disrupted neural tube closure is observed in these embryos, they are accompanied by a number of other severe defects (Piedrahita *et al.* 1999, Christensen *et al.* 2013, Momb *et al.* 2013). In the SHMT1 knockout mouse, NTD are observed, but only in the embryos of mothers fed folate-deficient diets (Beaudin *et al.* 2011). Currently, there are over 200 mouse models of NTD, but very few of the implicated genes have been shown to be relevant for human NTD (reviewed in Harris and Juriloff 2010).

The process of neurulation, wherein the neural plate folds and becomes the neural tube, requires cells to migrate and establish interactions to form a closed structure in a short time span. NTD can arise due to disturbed rates of cell proliferation or differentiation (reviewed in Copp and Greene 2013). For example, if the cell proliferation process is too slow, there will be fewer cells available at the time of closure, thus leading to an open neural tube. In the splotch mouse (*Pax3*-deficient) model for folate-responsive NTD, rescue of the NTD phenotype by folic acid was attributed to increased proliferation potential of the neural crest cells which form the neural tube (Ichi *et al.* 2010).

Recent efforts have focused on identification of epigenetic changes that might contribute to NTD risk. Folic acid may affect the expression of micro RNAs (miRNAs) that are associated with chromatin structure and methylation of histone 3 lysine 27 (H3K27). In the splotch mouse, folic acid reduced miRNA expression, particularly miR-138, miR-148a, miR-185 which in turn normalized the chromatin structure by reducing methylation of H3K27. This allowed for proper gene expression and neuronal differentiation (Ichi *et al.* 2010). As previously mentioned, retrotransposons such as LINE elements are heavily methylated to control their expression. In one study, the methylation of LINE elements in neural tissues was lower in human fetal NTD cases than in controls, and the decreased LINE methylation correlated with increased risk of NTD (Wang *et al.* 2010). There was also a trend toward lower plasma folate concentrations in the cases (Wang *et al.* 2010). CpGs in the homeobox (*HOX*) *HOXA* or *HOXB* clusters were hypomethylated in leukocytes of patients with myelomeningocele (a subset of NTD) compared to controls (Rochtus *et al.* 2014). In another study, the methylation at specific CpGs of the reduced folate carrier (*RFC1*) gene was found to be sensitive to homocysteine levels, which was used as proxy for poor folate nutrition. Moreover, there were differences in methylation of *RFC1* between healthy fetuses and NTD cases with the *RFC1* 80A>G variant (rs1051266) (Farkas *et al.* 2013).

8.4.2 *Intrauterine growth restriction (IUGR) and low birth weight*

Low birth weight is commonly associated with poor nutrition, and as such, is most often found in the developing world or in groups with low socio-economic status. Low folate and high homocysteine were associated with an increased risk of IUGR in an Australian cohort at 20 weeks of gestation (Furness *et al.* 2013). In a recent multi-ethnic American study, women whose RBC folate levels were in the lowest quartile had children with lower birth weights than those with higher levels (Hoyo *et al.* 2014). However, overconsumption of folic acid may carry its own risks; a recent Dutch study suggested that excessively high folate intake was associated with reduced embryo size (van Uitert *et al.* 2014). Epigenetic mechanisms may contribute to these risks. Maternal folate intake has been reported to alter the methylation of nine imprinted genes (Hoyo *et al.* 2014); changes in methylation of these genes and others may affect infant birth weight (Turan *et al.* 2012).

SNPs in folate-related genes have also been associated with IUGR/low birth weight and associated disorders such as placental abnormalities and preeclampsia. In a study of over 5,000 Norwegian mothers, women with the *MTHFR* 677C>T polymorphism had an increased risk of restricted uterine growth and placental abruption (Nurk *et al.* 2004). Embryos from MTHFR-deficient mice fed low folate diets were smaller, and had an increased incidence of developmental delays and placental abnormalities (Pickell *et al.* 2009). Mice with reduced MTHFD1 expression fed folate- or choline-deficient diets had smaller embryos (Beaudin *et al.* 2012).

8.4.3 *Congenital heart defects (CHD)*

Congential heart defects (CHD) is the general term for a variety of malformations of the heart and its connection to the major blood vessels, such as septation defects, outflow tract defects, and more complex syndromes such as Tetralogy of Fallot. The etiology of CHD is not fully understood. Early evidence for a protective role of folate in preventing CHD came from the Hungarian randomized controlled trial on the effect of folic acid supplementation on incidence of NTD and other birth defects. Offspring of women taking folate supplements had significantly fewer CHD, specifically ventricular septal defects (reviewed in Czeizel and Dudas 1992, Czeizel 1996). Folate fortification programs and voluntary supplementation may have reduced the incidence of CHD in more recent years (Ionescu-Ittu *et al.* 2009).

The association of variants in folate metabolism with CHD risk, and the modification of SNP-associated risk by folic acid have been investigated. The findings of these studies differ based on the ethnic background and nutritional environment of the population (van Beynum *et al.* 2006, Brandalize *et al.* 2009, Zhu *et al.* 2012, Xuan *et al.* 2014, Gong *et al.* 2012, Christensen *et al.* 2009). Furthermore, increased risk may be associated with the maternal and/or the embryonic genotype. The *MTHFR* 677C>T variant has been associated with increased risk for CHD in a number of studies (van Beynum *et al.* 2006, Brandalize *et al.* 2009). In a Hispanic cohort, the combination of maternal *MTHFR* 677TT genotype and low maternal folate intake increased the risk of CHD in offspring five-fold (Zhu *et al.* 2012). A recent meta-analysis of 35 case-control studies of different ethnic backgrounds showed that the maternal *MTHFR* 677TT genotype significantly increased CHD risk (Xuan *et al.* 2014). Other SNPs in folate pathway genes may also affect CHD risk. In a Chinese population, the *RFC1* 80A > G variant in the child was reported to be associated with increased CHD risk, particularly for conotruncal defects (malformation of the cardiac outflow tract) (Gong *et al.* 2012). In a pre-fortification Canadian cohort, the *MTHFD1* 1958 AA genotype in the child increased the risk of CHD (Christensen *et al.* 2009) and in the Hispanic population of California, infant and maternal *MTHFD1* rs11627387, an intronic SNP, was associated with increased CHD risk, particularly in mothers who did not use folic acid supplements (Zhu *et al.* 2012).

The association between a gene variant and a phenotype, particularly a complex trait, may be affected by diet. Since it is often difficult to obtain nutrient data in human studies at the critical time point in pregnancy, animal models may be more informative in this regard. Two mouse models for SNPs in folate metabolism have been examined for their impact on heart defects: mouse models with MTHFR or MTRR deficiency. In the $Mthfr^{+/-}$ mouse model, maternal MTHFR deficiency increased the proportion of embryos with heart defects in their litters compared with $Mthfr^{+/+}$ mothers, whereas embryonic genotype did not appear to influence CHD incidence (Li *et al.* 2005). In contrast to the $Mthfr^{+/-}$ model, both maternal and embryonic MTRR deficiency contributed to increased risk of CHD (Deng *et al.* 2008). A low folate diet also increased incidence of CHD in $Mthfr^{+/+}$ mothers, which may be due to reduced proliferation of cells in the myocardium (Li and Rozen 2006). High folate intake in pregnant mice also increased the incidence of CHD in embryos, regardless of maternal or embryonic genotype (Mikael *et al.* 2013).

Recent studies suggest that epigenetic changes may contribute to CHD risk. The incidence of CHD was increased in murine embryos whose maternal grandparents were MTRR-deficient (Padmanabhan *et al.* 2013). The authors suggested that the severity of the phenotype was influenced by epigenetic disturbances because DNA methylation was increased at certain CpGs in the placenta and that the level of methylation correlated with the defects. In a recent human study, some transcriptional regulatory regions were hypermethylated in fetal heart tissue of embryos with isolated or syndromic CHD (Serra-Juhé *et al.* 2014).

8.4.4 *Neuropsychiatric diseases*

Homocystinuria, an inborn error of metabolism with severe deficiencies of enzymes in folate metabolism, is characterized by neurologic complications in early childhood. These include brain atrophy, seizures, delayed motor development and mental retardation (reviewed in Whitehead 2006). Dietary folate deficiency has been associated with memory impairment, depression and psychosis (Coppen and Bolander-Gouaille 2005). In addition to providing nucleotides for neurogenesis, folates are required for the synthesis of neurotransmitters, particularly monoamines such as dopamine, serotonin and epinephrine (reviewed in Miller 2008). Furthermore, choline is a precursor of acetylcholine, another important neurotransmitter (reviewed in Zeisel 2006). In rats, folate intake has been reported to alter the expression of neurotrophic genes, which belong to a class of growth factors and are essential for promoting neuronal survival and differentiation, such as brain derived neurotrophic factor (BDNF) (Sable *et al.* 2011). Importantly, reduced levels of neurotrophins have been linked to psychiatric disease; for example, reduced BDNF levels in prefrontal cortex and hippocampus have been reported in patients with schizophrenia (reviewed in Ikegame *et al.* 2013).

Recent attention has focused on the potential connection between one-carbon folate metabolism and autism spectrum disorder. Some studies have suggested that the use of prenatal vitamins in early pregnancy reduced the risk of autism (reviewed in Lyall *et al.* 2014). A recent study by the NUTRIMENTHE consortium, examining up to 4,000 children, suggests that low maternal folate can negatively affect offspring behavior; children exposed to low folate *in utero* were more likely to have mood instability, obsession, irritability and inattention (reviewed in Anjos *et al.* 2013). Maternal folate levels were found to correlate positively with cognition scores in three-year-olds (Villamor *et al.* 2012).

Variants in genes of folate metabolism may also modify risk for neuropsychiatric disorders. One of the most prominent examples is the association between the *MTHFR* 677C>T variant and schizophrenia risk (reviewed in Gilbody *et al.* 2007). The etiology of schizophrenia and the factors that increase the risk for this disease are not fully understood, although there appears to be a strong genetic basis for schizophrenia susceptibility (reviewed in Mitchell *et al.* 2014). In association studies, the *MTHFR* 677C>T variant is one of the SNPs that is often significantly associated with schizophrenia risk (reviewed in Mitchell *et al.* 2014). Moreover, schizophrenia patients have been frequently shown to have low folate levels that may contribute to the severity of the disease and drug response, with improvement of patients following treatment with 5-methyltetrahydrofolate (5-methylTHF), the product of MTHFR (Ramaekers *et al.* 2014). More recently, it has been suggested that the levels of 5-methylTHF in cerebral spinal fluid may be lower in patients due to the presence of folate receptor auto-antibodies (Ramaekers *et al.* 2014). MTHFR variants have also been associated with increased risk for other neuropsychiatric disorders, including major depression and bipolar disorder (Peerbooms *et al.* 2011).

Animal models have shed some light on the mechanisms that lead to disturbed brain function. $Mthfr^{-/-}$ mice, a model for severe MTHFR deficiency, have smaller cerebella with reduced proliferation of granular cells and disruption of Purkinje neurons (Chen *et al.* 2005). *Mthfr* mice on a different background have short-term memory impairment, reduced hippocampal volume, increased apoptosis in the hippocampus and increased expression of choline acetyltransferase (ChAT), the enzyme that synthesizes acetycholine from choline. Increased ChAT expression is an indication of an attempt to upregulate acetylcholine synthesis because MTHFR deficiency enhances choline-/betaine-dependent methylation and disrupts choline pools (Jadavji *et al.* 2012). Memory impairment has also been observed in $Mtrr^{gt/gt}$ mice, along with increased ChAT expression and decreased mRNA levels of acetylcholinesterase, the enzyme that breaks down acetylcholine (Jadavji *et al.* 2014).

Low folate diets during pregnancy result in altered development of the hippocampus in offspring, with reduced cell differentiation and mitosis. This phenotype could be partially rescued by supplementing the diet with choline (Craciunescu *et al.* 2010). In a different study, short-term memory impairment was observed in 3-week-old pups whose mothers were MTHFR deficient ($Mthfr^{+/-}$) or if they were fed folate- or choline-deficient diets (Jadavji *et al.* 2015); maternal hyperhomocysteinemia and increased apoptosis in offspring hippocampus was a common feature in these conditions. An increasing number of studies are examining gene-specific methylation changes. When pregnant rats were fed folate-deficient diets, the offspring were found to have reduced expression of synapsins, which are essential for neuroplasticity (Pourié *et al.* 2015). The decreased synapsin was associated with reduced expression of estrogen receptor alpha (ER-α); methylation of *ER*-α was in turn decreased (Pourié *et al.* 2015).

The effect of high folic acid intake on brain development and methylation has recently been investigated. A recent study demonstrated that children whose mothers consumed high amounts of folic acid periconceptionally (over 5 mg/day) had delayed psychomotor development (Valera-Gran *et al.* 2014). In mice, the methylation of several genes expressed in the cerebellum of offspring changed due to high maternal folate intake (Barua *et al.* 2015). These changes, which were associated with altered gene expression, included genes essential for neurodevelopment and genes associated with autism spectrum disorder. The observed effects on methylation and expression were significantly affected by offspring sex (Barua *et al.* 2015).

8.4.5 *Developmental origin of late onset disease*

The Developmental Origin of Health and Disease (DOHaD) hypothesis (Barker and Osmond 1986, Barker *et al.* 1989) states that prenatal nutrition affects not only embryonic development and the incidence of birth defects, but also susceptibility to chronic disorders later in life such as metabolic syndrome, diabetes and cardiovascular disease. Impact of *in utero* exposure to nutritional deficiencies and their lasting effects on offspring has been studied in a cohort of adults born during the Dutch famine of World War II. Adults whose mothers were exposed to the famine in early pregnancy had earlier onset of coronary artery disease (Roseboom *et al.* 2000). A second cohort of adults, who were also exposed prenatally to the famine, had changes in the methylation of *IGF2* (Lumey *et al.* 2012), an imprinted gene essential for normal development and insulin production (Heijmans *et al.* 2008). The timing of the nutritional deficiency may be important. The offspring of women affected by the famine late in pregnancy had lower birth weights than unexposed children, whereas children whose mothers were exposed in early pregnancy had higher weights (Roseboom *et al.* 2000). Birth weight has been used as a proxy to assess *in utero* environment and may be associated with risk for chronic diseases later in life. Low birth weight may be associated with hypertension in adulthood, which can in turn increase the risk for stroke and coronary heart disease (Curhan *et al.* 1996, reviewed in Langley-Evans 2015). Although these types of findings are often consequences of deficiencies of multiple nutrients, changes in methylation certainly allude to the potential involvement of folate or its partner nutrients. *IGF2* is often studied because it is an imprinted gene with important roles in offspring birth weight and diabetes (St-Pierre *et al.* 2012). The conclusions of human studies in the field have been mixed; some report that *IGF2* methylation in the offspring is unaffected by maternal folate intake (Hoyo *et al.* 2011), while others have reported increased *IGF2* methylation and associated lower birth weights in the offspring of mothers consuming folate supplements (Steegers-Theunissen *et al.* 2009). High maternal folate intake has also been reported to increase adiposity and insulin resistance in six-year-old children (Yajnik *et al.* 2008). The development of type 2 diabetes mellitus has a potential epigenetic etiology that could be due to one-carbon metabolism (reviewed in Finer *et al.* 2014).

Work in animal models has been useful in understanding the impact of the *in utero* environment on adult health, because of the elimination of many confounders in human studies. In sheep, the combination of maternal B_{12}, folate and methionine deficiency was reported to increase body weight, insulin resistance and blood pressure in adult offspring (Sinclair *et al.* 2007). Furthermore, DNA methylation patterns in fetal liver were changed due to maternal diet. Offspring of rats fed a protein-restricted diet had increased blood pressure, which could be prevented if the maternal low protein diet was supplemented with folate (Torrens *et al.* 2006). In another study, offspring of rats fed a methyl-deficient diet (low in folate, choline and methionine) periconceptionally were heavier and were more glucose intolerant and insulin resistant than controls (Maloney *et al.* 2011). At the other extreme, the offspring of rats over-supplemented with a folate-containing multivitamin mix were also heavier and insulin resistant (Szeto *et al.* 2008). These types of studies highlight the importance of an upper limit for supplementation.

Interestingly, most of the findings were in male offspring, which suggests that there is a sex-specific response in offspring to maternal diet. A paternal diet deficient in folate may also affect offspring health. When male mice were fed a folate-deficient diet, changes were detected in the sperm methylome, particularly in genes linked to diabetes, such as *Aff3* and *Nkx2-2* (Lambrot *et al.* 2013).

A recent study using 90 inbred mouse strains suggested that DNA methylation levels in liver were regulated by genetics, and were associated with such traits as osteoporosis, diabetes, and heart disease. The authors found that variation in *Mtrr* affected liver methylation patterns throughout the genome (Orozco *et al.* 2015).

The afore-mentioned conditions increase the risk of early mortality and morbidity. The findings from these studies will aid in setting an optimal folate and nutrient intake during pregnancy. In turn, this could reduce the incidence of diabetes, hypertension and obesity, not only in the adult offspring but potentially in their offspring (F2), as suggested by hypotheses related to transgenerational effects on health and disease (Padmanabhan *et al.* 2013).

8.5 Conclusion

Folate is an important vitamin required for normal cell function and proper embryonic development. Low or high dietary folate and genetic variation in one-carbon metabolism can affect embryonic development through a number of cellular disturbances. Low folate intake may interact with genetic variants and contribute to the etiology of the multifactorial disorders discussed in this chapter. Alternatively, higher folate intake might overcome some of the deleterious effects of genetic variants, as long as the increased intake is modest. Unfortunately, nutrient levels are not always measured in the investigation of candidate SNPs or genome-wide association studies, and therefore gene–nutrient interactions can be difficult to assess. Large cohort sizes are required to properly assess the effects of these interactions as well as gene–gene interactions on the risk of multifactorial disorders. Nonetheless, it is clear that while adequate folate intake from food sources and moderate supplementation may be beneficial, more investigation is required to establish the safe upper limit for optimal health.

Disrupted folate metabolism may affect DNA synthesis and repair, de-toxification of homocysteine, phospholipid synthesis and DNA methylation. This multiplicity of effects makes it difficult to assess the individual impact of folate-related epigenetic changes, because a wide range of metabolic changes will accompany altered methylation. Nonetheless, DNA methylation is particularly important because a transient disruption of folate metabolism during embryogenesis may change epigenetic signatures and affect not only early development, but may also have consequences for adult health and possibly future offspring. As the methodologies for measuring methylation changes continue to evolve, future studies should have increased reliability and may be able to elucidate more effectively the epigenetic effects associated with folate disturbances. Identifying epigenetic changes that are related to disturbances in folate metabolism could eventually aid in developing markers for prenatal screening or early diagnosis of childhood or later onset disorders. The potential for early diagnosis and/or early intervention with methyl donor manipulation represents an exciting opportunity in developmental and adult-onset disorders.

References

Anjos, T., Altmäe, S., Emmett, P., Tiemeier, H., Closa-Monasterolo, R., Luque, V., Wiseman, S., Pérez-García, M., Lattka, E. and Demmelmair, H. (2013). Nutrition and neurodevelopment in children: focus on NUTRIMENTHE project. *European Journal of Nutrition* 52, 1825–1842.

Attwood, J., Yung, R. and Richardson, B. (2002). DNA methylation and the regulation of gene transcription. *Cellular and Molecular Life Sciences* 59, 241–257.

Bailey, R. L., Mills, J. L., Yetley, E. A., Gahche, J. J., Pfeiffer, C. M., Dwyer, J. T., Dodd, K. W., Sempos, C. T., Betz, J. M. and Picciano, M. F. (2010). Unmetabolized serum folic acid and its relation to folic acid intake from diet and supplements in a nationally representative sample of adults aged > or =60 y in the United States. *American Journal of Clinical Nutrition* 92, 383–389.

Barker, D. J. and Osmond, C. (1986). Infant mortality, childhood nutrition, and ischaemic heart disease in England and Wales. *Lancet* 327, 1077–1081.

Barker, D. J., Osmond, C., Winter, P., Margetts, B. and Simmonds, S. (1989). Weight in infancy and death from ischaemic heart disease. *Lancet* 334, 577–580.

Barua, S., Kuizon, S., Chadman, K. K., Brown, W. T. and Junaid, M. A. (2015). Microarray analysis reveals higher gestational folic acid alters expression of genes in the cerebellum of mice offspring — a pilot study. *Brain Science* 5, 14–31.

Beaudin, A. E., Abarinov, E. V., Noden, D. M., Perry, C. A., Chu, S., Stabler, S. P., Allen, R. H. and Stover, P. J. (2011). *Shmt1* and *de novo* thymidylate biosynthesis underlie folate-responsive neural tube defects in mice. *American Journal of Clinical Nutrition* 93, 789–798.

Beaudin, A. E., Perry, C. A., Stabler, S. P., Allen, R. H. and Stover, P. J. (2012). Maternal *Mthfd1* disruption impairs fetal growth but does not cause neural tube defects in mice. *American Journal of Clinical Nutrition* 95, 882–891.

Blom, H. J., Shaw, G. M., den Heijer, M. and Finnell, R. H. (2006). Neural tube defects and folate: case far from closed. *Nature Reviews Neuroscience* 7, 724–731.

Berry, R. J., Bailey, L., Mulinare, J., Bower, C. and Dary, O. (2010). Fortification of flour with folic acid. *Food Nutrition Bulletin* 31, 22S–35S.

Brandalize, A. P. C., Bandinelli, E., dos Santos, P. A., Roisenberg, I. and Schüler-Faccini, L. (2009). Evaluation of C677T and A1298C polymorphisms of the MTHFR gene as maternal risk factors for Down syndrome and congenital heart defects. *American Journal of Medical Genetics A* 149, 2080–2087.

Casas, J. P., Bautista, L. E., Smeeth, L., Sharma, P. and Hingorani, A. D. (2005). Homocysteine and stroke: evidence on a causal link from mendelian randomisation. *Lancet* 365, 224–232.

Caudill, M. A. (2009). Folate and choline interrelationships: metabolic and potential health implications. In *Folate in Health and Disease*, L. B. Bailey, ed., 449–465.

CDC. (1992). Recommendations for the use of folic acid to reduce the number of cases of spina bifida and other neural tube defects. *MMWR Recommendations and Reports* 41, 1–7.

Cedar, H. and Bergman, Y. (2009). Linking DNA methylation and histone modification: patterns and paradigms. *Nature Review Genetics* 10, 295–304.

Chen, Z., Schwahn, B. C., Wu, Q., He, X. and Rozen, R. (2005). Postnatal cerebellar defects in mice deficient in methylenetetrahydrofolate reductase. *International Journal of Developmental Neuroscience* 23, 465–474.

Christensen, K. E., Rohlicek, C. V., Andelfinger, G. U., Michaud, J., Bigras, J. L., Richter, A., MacKenzie, R. E. and Rozen, R. (2009). The MTHFD1 p. Arg653Gln variant alters enzyme function and increases risk for congenital heart defects. *Human Mutation* 30, 212–220.

Christensen, K. E. and Rozen, R. (2010). Genetic variation: effect on folate metabolism and health. In *Folate in Health and Disease*, L. B. Bailey, ed. (Boca Raton, FL: CRC Press), 75–110.

Christensen, K. E., Wu, Q., Wang, X., Deng, L., Caudill, M. A. and Rozen, R. (2010). Steatosis in mice is associated with gender, folate intake, and expression of genes of one-carbon metabolism. *Journal of Nutrition* 140, 1736–1741.

Christensen, K., Deng, L., Leung, K., Arning, E., Bottiglieri, T., Malysheva, O., Caudill, M., Krupenko, N., Greene, N. and Jerome-Majewska, L. (2013). A novel mouse model for genetic variation in 10-formyltetrahydrofolate synthetase exhibits disturbed purine synthesis with impacts on pregnancy and embryonic development. *Human Molecular Genetics* 22, 3705–3719.

Christensen, K. E., Mikael, L. G., Leung, K.-Y., Lévesque, N., Deng, L., Wu, Q., Malysheva, O. V., Best, A., Caudill, M. A. and Greene, N. D. (2015). High folic acid consumption leads to pseudo-MTHFR deficiency, altered lipid metabolism, and liver injury in mice. *The American Journal of Clinical Nutrition* 101, 646–658.

Copp, A. J. and Greene, N. D. (2010). Genetics and development of neural tube defects. *Journal of Pathology* 220, 217–230.

Copp, A. J. and Greene, N. D. (2013). Neural tube defects—disorders of neurulation and related embryonic processes. *Wiley Interdisciplinary Review of Developmental Biology* 2, 213–227.

Coppen, A. and Bolander-Gouaille, C. (2005). Treatment of depression: time to consider folic acid and vitamin B_{12}. *Journal of Psychopharmacology* 19, 59–65.

Cordero, A. M., Crider, K. S., Rogers, L. M., Cannon, M. J. and Berry, R. J. (2015). Optimal serum and red blood cell folate concentrations in women of reproductive age for prevention of neural tube defects: world health organization guidelines. *MMWR Morbidity and Mortality Weekly Report* 64, 421–423.

Craciunescu, C. N., Johnson, A. R. and Zeisel, S. H. (2010). Dietary choline reverses some, but not all, effects of folate deficiency on neurogenesis and apoptosis in fetal mouse brain. *Journal of Nutrition* 140, 1162–1166.

Crider, K. S., Bailey, L. B. and Berry, R. J. (2011). Folic acid food fortification—its history, effect, concerns, and future directions. *Nutrients* 3, 370–384.

Crott, J. W. and Mason, J. B. (2005). *MTHFR* polymorphisms and colorectal neoplasia. In *MTHFR Polymorphisms and Disease*, P. M. Ueland, and R. Rozen, eds. (Georgetown, TX: Landes Bioscience), 179–196.

Curhan, G. C., Chertow, G. M., Willett, W. C., Spiegelman, D., Colditz, G. A., Manson, J. E., Speizer, F. E. and Stampfer, M. J. (1996). Birth weight and adult hypertension and obesity in women. *Circulation* 94, 1310–1315.

Czeizel, A. E. and Dudas, I. (1992). Prevention of the first occurrence of neural-tube defects by periconceptional vitamin supplementation. *The New England Journal of Medicine* 327, 1832–1835.

Czeizel, A. E. (1996). Reduction of urinary tract and cardiovascular defects by periconceptional multivitamin supplementation. *The American Journal of Medical Genetics* 62, 179–183.

da Cunha, A. A., Ferreira, A. G., Loureiro, S. O., da Cunha, M. J., Schmitz, F., Netto, C. A. and Wyse, A. T. (2012). Chronic hyperhomocysteinemia increases inflammatory markers in hippocampus and serum of rats. *Neurochemical Research* 37, 1660–1669.

da Silva, R. P., Kelly, K. B., Al Rajabi, A. and Jacobs, R. L. (2014). Novel insights on interactions between folate and lipid metabolism. *Biofactors* 40, 277–283.

Daly, S. and Scott, J. M. (1998). The prevention of neural tube defects. *Current Opinion in Obstetrics and Gynecology* 10, 85–89.

Deng, L., Elmore, C. L., Lawrance, A. K., Matthews, R. G. and Rozen, R. (2008). Methionine synthase reductase deficiency results in adverse reproductive outcomes and congenital heart defects in mice. *Molecular Genetics and Metabolism* 94, 336–342.

Dhobale, M., Chavan, P., Kulkarni, A., Mehendale, S., Pisal, H. and Joshi, S. (2011). Reduced folate, increased vitamin B (12) and homocysteine concentrations in women delivering preterm. *Annals of Nutrition and Metabolism* 61, 7–14.

Duthie, S. J., Narayanan, S., Brand, G. M., Pirie, L. and Grant, G. (2002). Impact of folate deficiency on DNA stability. *Journal of Nutrition* 132, 2444S–2449S.

EUROCAT. Folic Acid Mandatory Fortification <http://www.eurocat-network.eu/pagecontent.aspx?pageid=161>. Accessed June 6th 2015.

Farkas, S. A., Böttiger, A. K., Isaksson, H. S., Finnell, R. H., Ren, A. and Nilsson, T. K. (2013). Epigenetic alterations in folate transport genes in placental tissue from fetuses with neural tube defects and in leukocytes from subjects with hyperhomocysteinemia. *Epigenetics* 8, 303–316.

Finer, S., Saravanan, P., Hitman, G. and Yajnik, C. (2014). The role of the one-carbon cycle in the developmental origins of Type 2 diabetes and obesity. *Diabetic Medicine* 31, 263–272.

Flatley, J. E., McNeir, K., Balasubramani, L., Tidy, J., Stuart, E. L., Young, T. A. and Powers, H. J. (2009). Folate status and aberrant DNA methylation are associated with HPV infection and cervical pathogenesis. *Cancer Epidemiology Biomarkers and Prevention* 18, 2782–2789.

Fredriksen, Å., Meyer, K., Ueland, P. M., Vollset, S. E., Grotmol, T. and Schneede, J. (2007). Large-scale population-based metabolic phenotyping of thirteen genetic polymorphisms related to one-carbon metabolism. *Human Mutation* 28, 856–865.

Furness, D., Fenech, M., Dekker, G., Khong, T. Y., Roberts, C. and Hague, W. (2013). Folate, vitamin B_{12}, vitamin B_6 and homocysteine: impact on pregnancy outcome. *Maternal & Child Nutrition* 9, 155–166.

Gilbody, S., Lewis, S. and Lightfoot, T. (2007). Methylenetetrahydrofolate reductase (MTHFR) genetic polymorphisms and psychiatric disorders: a HuGE review. *American Journal of Epidemiology* 165, 1–13.

Giovannucci, E. (2002). Epidemiologic studies of folate and colorectal neoplasia: a review. *Journal of Nutrition* 132, 2350S–2355S.

Gong, D., Gu, H., Zhang, Y. J., Gong, J., Nie, Y., Wang, J., Zhang, H., Liu, R., Hu, S. and Zhang, H. (2012). Methylenetetrahydrofolate reductase C677T and reduced folate carrier 80 G> A polymorphisms are associated with an increased risk of conotruncal heart defects. *Clinical Chemistry and Laboratory Medicine* 50, 1455–1461.

Hajkova, P., Erhardt, S., Lane, N., Haaf, T., El-Maarri, O., Reik, W., Walter, J. and Surani, M. A. (2002). Epigenetic reprogramming in mouse primordial germ cells. *Mechanisms of Development* 117, 15–23.

Harris, M. J. and Juriloff, D. M. (2010). An update to the list of mouse mutants with neural tube closure defects and advances toward a complete genetic perspective of neural tube closure. *Birth Defects Research Part A: Clinical and Molecular Teratology* 88, 653–669.

Heijmans, B. T., Tobi, E. W., Stein, A. D., Putter, H., Blauw, G. J., Susser, E. S., Slagboom, P. E. and Lumey, L. (2008). Persistent epigenetic differences associated with prenatal exposure to famine in humans. *Proceedings of the National Academy of Sciences* 105, 17046–17049.

Hernández-Díaz, S., Werler, M. M., Louik, C. and Mitchell, A. A. (2002). Risk of gestational hypertension in relation to folic acid supplementation during pregnancy. *American Journal of Epidemiology Sciences* 156, 806–812.

Hibbard, B. M. (1964). The role of folic acid in pregnancy with particular reference to anemia, abruption and abortion. *The Journal of Obstetrics and Gynecology of the British Commonwealth* 7, 529–542.

Hoffbrand, A. and Weir, D. (2001). The history of folic acid. *British Journal of Haematology* 113, 579–589.

Hoyo, C., Daltveit, A. K., Iversen, E., Benjamin-Neelon, S. E., Fuemmeler, B., Schildkraut, J., Murtha, A. P., Overcash, F., Vidal, A. C. and Wang, F. (2014). Erythrocyte folate concentrations, CpG methylation at genomically imprinted domains, and birth weight in a multiethnic newborn cohort. *Epigenetics* 9, 1120–1130.

Hoyo, C., Murtha, A. P., Schildkraut, J. M., Jirtle, R. L., Demark-Wahnefried, W., Forman, M. R., Iversen, E. S., Kurtzberg, J., Overcash, F. and Huang, Z. (2011). Methylation variation at IGF2 differentially methylated regions and maternal folic acid use before and during pregnancy. *Epigenetics* 6, 928–936.

Ichi, S., Costa, F. F., Bischof, J. M., Nakazaki, H., Shen, Y.-W., Boshnjaku, V., Sharma, S., Mania-Farnell, B., McLone, D. G. and Tomita, T. (2010). Folic acid remodels chromatin on *Hes1* and *Neurog2* promoters during caudal neural tube development. *Journal of Biological Chemistry* 285, 36922–36932.

Ikegame, T., Bundo, M., Murata, Y., Kasai, K., Kato, T. and Iwamoto, K. (2013). DNA methylation of the BDNF gene and its relevance to psychiatric disorders. *Journal of Human Genetics* 58, 434–438.

Ionescu-Ittu, R., Marelli, A. J., Mackie, A. S. and Pilote, L. (2009). Prevalence of severe congenital heart disease after folic acid fortification of grain products: time trend analysis in Quebec, Canada. *BMJ* 338.

Ivanov, A., Nash-Barboza, S., Hinkis, S. and Caudill, M. A. (2009). Genetic variants in phosphatidylethanolamine N-methyltransferase and methylenetetrahydrofolate dehydrogenase influence biomarkers of choline metabolism when folate intake is restricted. *Journal of the American Dietetic Association* 109, 313–318.

Jacobs, R. L., Stead, L. M., Devlin, C., Tabas, I., Brosnan, M. E., Brosnan, J. T. and Vance, D. E. (2005). Physiological regulation of phospholipid methylation alters plasma homocysteine in mice. *Journal of Biological Chemistry* 280, 28299–28305.

Jacques, P. F., Bostom, A. G., Williams, R. R., Ellison, R. C., Eckfeldt, J. H., Rosenberg, I. H., Selhub, J. and Rozen, R. (1996). Relation Between Folate Status, a Common Mutation in Methylenetetrahydrofolate Reductase, and Plasma Homocysteine Concentrations. *Circulation* 93, 7–9.

Jadavji, N. M., Deng, L., Leclerc, D., Malysheva, O., Bedell, B. J., Caudill, M. A. and Rozen, R. (2012). Severe methylenetetrahydrofolate reductase deficiency in mice results in behavioral anomalies with morphological and biochemical changes in hippocampus. *Molecular Genetics and Metabolism* 106, 149–159.

Jadavji, N. M., Bahous, R. H., Deng, L., Malysheva, O., Grand'maison, M., Bedell, B. J., Caudill, M. A. and Rozen, R. (2014). Mouse model for deficiency of methionine synthase reductase exhibits short-term memory impairment and disturbances in brain choline metabolism. *Biochemical Journal* 461, 205–212.

Jadavji, N., Deng, L., Malysheva, O., Caudill, M. and Rozen, R. (2015). MTHFR deficiency or reduced intake of folate or choline in pregnant mice results in impaired short-term memory and increased apoptosis in the hippocampus of wild-type offspring. *Neuroscience* 300, 1–9.

James, S. J., Basnakian, A. G. and Miller, B. J. (1994). *In vitro* folate deficiency induces deoxynucleotide pool imbalance, apoptosis, and mutagenesis in Chinese hamster ovary cells. *Cancer Research* 54, 5075–5080.

James, S. J., Melnyk, S., Pogribna, M., Pogribny, I. P. and Caudill, M. A. (2002). Elevation in S-adenosylhomocysteine and DNA hypomethylation: potential epigenetic mechanism for homocysteine-related pathology. *Journal of Nutrition* 132, 2361S-2366S.

James, S. J., Miller, B. J., Basnakian, A. G., Pogribny, I. P., Pogribna, M. and Muskhelishvili, L. (1997). Apoptosis and proliferation under conditions of deoxynucleotide pool imbalance in liver of folate/methyl deficient rats. *Carcinogenesis* 18, 287–293.

Kim, Y.-I., Pogribny, I. P., Salomon, R. N., Choi, S. W., Smith, D. E., James, S. J. and Mason, J. B. (1996). Exon-specific DNA hypomethylation of the p53 gene of rat colon induced by dimethylhydrazine. Modulation by dietary folate. *American Journal of Pathology* 149, 1129.

Kobayashi, H., Sakurai, T., Miura, F., Imai, M., Mochiduki, K., Yanagisawa, E., Sakashita, A., Wakai, T., Suzuki, Y. and Ito, T. (2013). High-resolution DNA methylome analysis of primordial germ cells identifies gender-specific reprogramming in mice. *Genome Research* 23, 616–627.

Lambrot, R., Xu, C., Saint-Phar, S., Chountalos, G., Cohen, T., Paquet, M., Suderman, M., Hallett, M. and Kimmins, S. (2013). Low paternal dietary folate alters the mouse sperm epigenome and is associated with negative pregnancy outcomes. *Nature Communications* 4, 2889.

Langley-Evans, S. (2015). Nutrition in early life and the programming of adult disease: a review. *Journal of Human Nutrition and Dietetics* 28, 1–14.

Leclerc, D., Cao, Y., Deng, L., Mikael, L. G., Wu, Q. and Rozen, R. (2013). Differential gene expression and methylation in the retinoid/PPARA pathway and of tumor suppressors may modify intestinal tumorigenesis induced by low folate in mice. *Molecular Nutrition & Food Research* 57, 686–697.

Li, C. C., Cropley, J. E., Cowley, M. J., Preiss, T., Martin, D. I. and Suter, C. M. (2011). A sustained dietary change increases epigenetic variation in isogenic mice. *PLoS Genet* 7, e1001380.

Li, D., Pickell, L., Liu, Y., Wu, Q., Cohn, J. S. and Rozen, R. (2005). Maternal methylenetetrahydrofolate reductase deficiency and low dietary folate lead to adverse reproductive outcomes and congenital heart defects in mice. *American Journal of Clinical Nutrition* 82, 188–195.

Li, D. and Rozen, R. (2006). Maternal folate deficiency affects proliferation, but not apoptosis, in embryonic mouse heart. *Journal of Nutrition* 136, 1774–1778.

Li, Z. and Vance, D. E. (2008). Thematic Review Series: Glycerolipids. Phosphatidylcholine and choline homeostasis. *Journal of Lipid Research* 49, 1187–1194.

Lubos, E., Loscalzo, J. and Handy, D. E. (2007). Homocysteine and glutathione peroxidase-1. *Antioxidants & Redox Signaling* 9, 1923–1940.

Lumey, L., Terry, M. B., Delgado-Cruzata, L., Liao, Y., Wang, Q., Susser, E., McKeague, I. and Santella, R. M. (2012). Adult global DNA methylation in relation to pre-natal nutrition. *International Journal of Epidemiology* 41, 116–123.

Lyall, K., Schmidt, R. J. and Hertz-Picciotto, I. (2014). Maternal lifestyle and environmental risk factors for autism spectrum disorders. *International Journal of Epidemiology* 43, 443–464.

Maloney, C. A., Hay, S. M., Young, L. E., Sinclair, K. D. and Rees, W. D. (2011). A methyl-deficient diet fed to rat dams during the peri-conception period programs glucose homeostasis in adult male but not female offspring. *Journal of Nutrition* 141, 95–100.

McGraw, S., Oakes, C. C., Martel, J., Cirio, M. C., de Zeeuw, P., Mak, W., Plass, C., Bartolomei, M. S., Chaillet, J. R. and Trasler, J. M. (2013). Loss of DNMT1o disrupts imprinted X chromosome inactivation and accentuates placental defects in females. *PLoS Genetics* 9, e1003873.

Mikael, L. G., Deng, L., Paul, L., Selhub, J. and Rozen, R. (2013). Moderately high intake of folic acid has a negative impact on mouse embryonic development. *Birth Defects Research Part A: Clinical and Molecular Teratology* 97, 47–52.

Miller, A. L. (2008). The methylation, neurotransmitter, and antioxidant connections between folate and depression. *Alternative Medicine Review* 13, 216–226.

Mitchell, E. S., Conus, N. and Kaput, J. (2014). B vitamin polymorphisms and behavior: Evidence of associations with neurodevelopment, depression, schizophrenia, bipolar disorder and cognitive decline. *Neuroscience & Biobehavioral Reviews* 47, 307–320.

Momb, J., Lewandowski, J. P., Bryant, J. D., Fitch, R., Surman, D. R., Vokes, S. A. and Appling, D. R. (2013). Deletion of Mthfd1l causes embryonic lethality and neural tube and craniofacial defects in mice. *Proceedings of the National Academy of Sciences USA* 110, 549–554.

MRC Vitamin Research Group (1991). Prevention of neural tube defects: results of the Medical Research Council Vitamin Study. *Lancet* 338, 131–137.

Nurk, E., Tell, G. S., Refsum, H., Ueland, P. M. and Vollset, S. E. (2004). Associations between maternal methylenetetrahydrofolate reductase polymorphisms and adverse outcomes of pregnancy: the Hordaland Homocysteine Study. *American Journal of Medicine* 117, 26–31.

Orozco, L. D., Morselli, M., Rubbi, L., Guo, W., Go, J., Shi, H., Lopez, D., Furlotte, N. A., Bennett, B. J., Farber, C. R., *et al.* (2015). Epigenome-wide association of liver methylation patterns and complex metabolic traits in mice. *Cell Metabolism* 21, 905–917.

Padmanabhan, N., Jia, D., Geary-Joo, C., Wu, X., Ferguson-Smith, A. C., Fung, E., Bieda, M. C., Snyder, F. F., Gravel, R. A. and Cross, J. C. (2013). Mutation in folate metabolism causes epigenetic instability and transgenerational effects on development. *Cell* 155, 81–93.

Parle-McDermott, A., Kirke, P. N., Mills, J. L., Molloy, A. M., Cox, C., O'Leary, V. B., Pangilinan, F., Conley, M., Cleary, L., Brody, L. C., *et al.* (2006). Confirmation of the R653Q polymorphism of the trifunctional C1-synthase enzyme as a maternal risk for neural tube defects in the Irish population. *European Journal of Human Genetics* 14, 768–772.

Peerbooms, O. L., van Os, J., Drukker, M., Kenis, G., Hoogveld, L., De Hert, M., Delespaul, P., van Winkel, R. and Rutten, B. P. (2011). Meta-analysis of MTHFR gene variants in schizophrenia, bipolar disorder and unipolar depressive disorder: evidence for a common genetic vulnerability? *Brain, Behaviour and Immunity* 25, 1530–1543.

Pickell, L., Li, D., Brown, K., Mikael, L. G., Wang, X. L., Wu, Q., Luo, L., Jerome-Majewska, L. and Rozen, R. (2009). Methylenetetrahydrofolate reductase deficiency and low dietary folate increase embryonic delay and placental abnormalities in mice. *Birth Defects Research Part A: Clinical and Molecular Teratology* 85, 531–541.

Pickell, L., Wu, Q., Wang, X.-L., Leclerc, D., Friedman, H., Peterson, A. C. and Rozen, R. (2011). Targeted insertion of two Mthfr promoters in mice reveals temporal-and tissue-specific regulation. *Mammalian Genome* 22, 635–647.

Piedrahita, J. A., Oetama, B., Bennett, G. D., van Waes, J., Kamen, B. A., Richardson, J., Lacey, S. W., Anderson, R. G. and Finnell, R. H. (1999). Mice lacking the folic acid-binding protein Folbp1 are defective in early embryonic development. *Nature Genetics* 23, 228–232.

Pourié, G., Martin, N., Bossenmeyer-Pourié, C., Akchiche, N., Guéant-Rodriguez, R. M., Geoffroy, A., Jeannesson, E., Chehadeh, S. E. H., Mimoun, K. and Brachet, P. (2015). Folate-and vitamin B_{12}–deficient diet during gestation and lactation alters cerebellar synapsin expression via impaired influence of estrogen nuclear receptor α. *The FASEB Journal* 9, 3713–3725.

Ramaekers, V., Thöny, B., Sequeira, J., Ansseau, M., Philippe, P., Boemer, F., Bours, V. and Quadros, E. (2014). Folinic acid treatment for schizophrenia associated with folate receptor autoantibodies. *Molecular Genetics and Metabolism* 113, 307–314.

Refsum, H., Ueland, P. M., Nygård, O. and Vollset, S. E. (1998). Homocysteine and cardiovascular disease. *Annual Review of Medicine* 49, 31–62.

Relton, C., Wilding, C., Pearce, M., Laffling, A., Jonas, P., Lynch, S., Tawn, E. and Burn, J. (2004). Gene–gene interaction in folate-related genes and risk of neural tube defects in a UK population. *Journal of Medical Genetics* 41, 256–260.

Rochtus, A., Izzi, B., Vangeel, E., Louwette, S., Wittevrongel, C., Lambrechts, D., Moreau, Y., Winand, R., Verpoorten, C. and Jansen, K. (2014). DNA methylation analysis of Homeobox genes implicates *HOXB7* hypomethylation as risk factor for neural tube defects. *Epigenetics*, 10, 92–101.

Roseboom, T. J., van der Meulen, J. H., Osmond, C., Barker, D. J., Ravelli, A. C., Schroeder-Tanka, J. M., van Montfrans, G. A., Michels, R. P. and Bleker, O. P. (2000). Coronary heart disease after prenatal exposure to the Dutch famine, 1944–1945. *Heart* 84, 595–598.

Sable, P., Dangat, K., Kale, A. and Joshi, S. (2011). Altered brain neurotrophins at birth: consequence of imbalance in maternal folic acid and vitamin B 12 metabolism. *Neuroscience* 190, 127–134.

Serra-Juhé, C., Cuscó, I., Homs, A., Flores, R., Torán, N. and Pérez-Jurado, L. A. (2014). DNA methylation abnormalities in congenital heart disease. *Epigenetics* 10, 167–177.

Seshadri, S., Beiser, A., Selhub, J., Jacques, P. F., Rosenberg, I. H., D'Agostino, R. B., Wilson, P. W. and Wolf, P. A. (2002). Plasma homocysteine as a risk factor for dementia and Alzheimer's disease. *The New England Journal of Medicine* 346, 476–483.

Sinclair, K. D., Allegrucci, C., Singh, R., Gardner, D. S., Sebastian, S., Bispham, J., Thurston, A., Huntley, J. F., Rees, W. D. and Maloney, C. A. (2007). DNA methylation, insulin resistance, and blood pressure in offspring determined by maternal periconceptional B vitamin and methionine status. *Proceedings of the National Academy of Sciences of the United States of America* 104, 19351–19356.

Smith, A. D., Kim, Y. I. and Refsum, H. (2008). Is folic acid good for everyone? *American Journal of Clinical Nutrition* 87, 517–533.

Smith, Z. D. and Meissner, A. (2013). DNA methylation: roles in mammalian development. *Nature Reviews Genetics* 14, 204–220.

St-Pierre, J., Hivert, M.-F., Perron, P., Poirier, P., Guay, S.-P., Brisson, D. and Bouchard, L. (2012). *IGF2* DNA methylation is a modulator of newborn's fetal growth and development. *Epigenetics* 7, 1125–1132.

Steegers-Theunissen, R. P., Obermann-Borst, S. A., Kremer, D., Lindemans, J., Siebel, C., Steegers, E. A., Slagboom, P. E. and Heijmans, B. T. (2009). Periconceptional maternal folic acid use of 400 μg per day is related to increased methylation of the IGF2 gene in the very young child. *PloS One* 4, e7845.

Stover, P. J. (2009). Folate Biochemical Pathways and Their Regulation. In *Folate in Health and Disease* L.B. Bailey, ed, 49–74

Szeto, I. M., Aziz, A., Das, P. J., Taha, A. Y., Okubo, N., Reza-Lopez, S., Giacca, A. and Anderson, G. H. (2008). High multivitamin intake by Wistar rats during pregnancy results in increased food intake and components of the metabolic syndrome in male offspring. *The American Journal of Physiology — Regulatory Integrative Comparitive Physiology* 295, R575-R582.

Tibbetts, A. S. and Appling, D. R. (2010). Compartmentalization of Mammalian folate mediated one-carbon metabolism. *Annual Review of Nutrition* 30, 57–81

Torrens, C., Brawley, L., Anthony, F. W., Dance, C. S., Dunn, R., Jackson, A. A., Poston, L. and Hanson, M. A. (2006). Folate supplementation during pregnancy improves offspring cardiovascular dysfunction induced by protein restriction. *Hypertension* 47, 982–987.

Turan, N., Ghalwash, M. F., Katari, S., Coutifaris, C., Obradovic, Z. and Sapienza, C. (2012). DNA methylation differences at growth related genes correlate with birth weight: a molecular signature linked to developmental origins of adult disease? *BMC Medical Genomics* 5, 10.

Tyagi, N., Sedoris, K. C., Steed, M., Ovechkin, A. V., Moshal, K. S. and Tyagi, S. C. (2005). Mechanisms of homocysteine-induced oxidative stress. *American Journal of Physiology — Heart and Circulatory Physiology* 289, H2649–H2656.

Valera-Gran, D., de la Hera, M. G., Navarrete-Muñoz, E. M., Fernandez-Somoano, A., Tardón, A., Julvez, J., Forns, J., Lertxundi, N., Ibarluzea, J. M. and Murcia, M.

(2014). Folic acid supplements during pregnancy and child psychomotor development after the first year of life. *JAMA Pediatrics* 168, e142611-e142611.

van Beynum, I. M., Kapusta, L., den Heijer, M., Vermeulen, S. H., Kouwenberg, M., Daniels, O. and Blom, H. J. (2006). Maternal *MTHFR* 677C>T is a risk factor for congenital heart defects: effect modification by periconceptional folate supplementation. *European Heart Journal* 27, 981–987.

van den Donk, M., van Engeland, M., Pellis, L., Witteman, B. J., Kok, F. J., Keijer, J. and Kampman, E. (2007). Dietary folate intake in combination with *MTHFR* C677T genotype and promoter methylation of tumor suppressor and DNA repair genes in sporadic colorectal adenomas. *Cancer Epidemiology, Biomarkers & Prevention* 16, 327–333.

van der Put, N. M., Steegers-Theunissen, R. P., Frosst, P., Trijbels, F. J., Eskes, T. K., van den Heuvel, L. P., Mariman, E. C., den Heyer, M., Rozen, R. and Blom, H. J. (1995). Mutated methylenetetrahydrofolate reductase as a risk factor for spina bifida. *Lancet* 346, 1070–1071.

van Uitert, E., van Ginkel, S., Willemsen, S., Lindemans, J., Koning, A., Eilers, P., Exalto, N., Laven, J., Steegers, E. and Steegers-Theunissen, R. (2014). An optimal periconception maternal folate status for embryonic size: the Rotterdam Predict study. *BJOG* 7, 821–829.

Vance, D. E. and Vance, J. E. (2009). Physiological consequences of disruption of mammalian phospholipid biosynthetic genes. *Journal of Lipid Research* 50, S132–S137.

Villamor, E., Rifas-Shiman, S. L., Gillman, M. W. and Oken, E. (2012). Maternal intake of methyl-donor nutrients and child cognition at 3 years of age. *Paediatric and Perinatal Epidemiology* 26, 328–335.

Wang, L., Wang, F., Guan, J., Le, J., Wu, L., Zou, J., Zhao, H., Pei, L., Zheng, X. and Zhang, T. (2010). Relation between hypomethylation of long interspersed nucleotide elements and risk of neural tube defects. *The American Journal of Clinical Nutrition* 91, 1359–1367.

Waterland, R. A. and Jirtle, R. L. (2003). Transposable elements: targets for early nutritional effects on epigenetic gene regulation. *Molecular and Cellular Biology* 23, 5293–5300.

Whitehead, V. M. (2006). Acquired and inherited disorders of cobalamin and folate in children. *British Journal of Haematology* 134, 125–136.

Xu, J. and Sinclair, K. (2015). One-carbon metabolism and epigenetic regulation of embryo development. *Reproduction, Fertility and Development* 27, 667–676.

Xuan, C., Li, H., Zhao, J.-X., Wang, H.-W., Wang, Y., Ning, C.-P., Liu, Z., Zhang, B.-B., He, G.-W. and Lun, L.-M. (2014). Association Between MTHFR Polymorphisms and Congenital Heart Disease: A Meta-analysis based on 9,329 cases and 15,076 controls. *Scientific Reports* 4, 7311.

Yadav, U., Kumar, P., Yadav, S. K., Mishra, O. P. and Rai, V. (2014). Polymorphisms in folate metabolism genes as maternal risk factor for neural tube defects: an updated meta-analysis. *Metabolic Brain Disease* 30, 7–24.

Yajnik, C., Deshpande, S., Jackson, A., Refsum, H., Rao, S., Fisher, D., Bhat, D., Naik, S., Coyaji, K. and Joglekar, C. (2008). Vitamin B_{12} and folate concentrations during pregnancy and insulin resistance in the offspring: the Pune Maternal Nutrition Study. *Diabetologia* 51, 29–38.

Yamada, K., Chen, Z., Rozen, R. and Matthews, R. G. (2001). Effects of common polymorphisms on the properties of recombinant human methylenetetrahydrofolate reductase. *Proceedings of the National Academy of Sciences of the United States of America* 98, 14853–14858.

Yamada, K., Strahler, J. R., Andrews, P. C. and Matthews, R. G. (2005). Regulation of human methylenetetrahydrofolate reductase by phosphorylation. *Proceedings of the National Academy of Sciences of the United States of America* 102, 10454–10459.

Yetley, E. A. and Johnson, C. L. (2011). Folate and vitamin B-12 biomarkers in NHANES: history of their measurement and use. *The American Journal of Clinical Nutrition* 94, 322S-331S.

Yeung, L. F., Cogswell, M. E., Carriquiry, A. L., Bailey, L. B., Pfeiffer, C. M. and Berry, R. J. (2011). Contributions of enriched cereal-grain products, ready-to-eat cereals, and supplements to folic acid and vitamin B-12 usual intake and folate and vitamin B-12 status in US children: National Health and Nutrition Examination Survey (NHANES), 2003–2006. *The American Journal of Clinical Nutrition* 93, 172–185.

Zeisel, S. H. (2006). The fetal origins of memory: the role of dietary choline in optimal brain development. *Journal of Pediatrics* 149, S131–S136.

Zeisel, S. H. (2011). The supply of choline is important for fetal progenitor cells. *Seminars in Cell and Developmental Biology* 22, 624–628.

Zhu, H., Yang, W., Lu, W., Etheredge, A. J., Lammer, E. J., Finnell, R. H., Carmichael, S. L. and Shaw, G. M. (2012). Gene variants in the folate-mediated one-carbon metabolism (FOCM) pathway as risk factors for conotruncal heart defects. *The American Journal of Medical & Genetics A* 158, 1124–1134.

Zhu, X. and Zeisel, S. H. (2004). Deletion of the Pemt gene increases progenitor cell mitosis, DNA and protein methylation and decreases calretinin expression in embryonic day 17 mouse hippocampus. *Brain Research Molecular Brain Research* 149, 121–129.

Early Life Experiences and Their Influence on Epigenetic Marks and Human Brain Development

9

Anke Hoffmann and Dietmar Spengler

Max Planck Institute of Psychiatry,
Translational Research in Psychiatry,
Kraepelinstr. 2-10, 80804, Munich, Germany

Early life adversity (ELA) frequently associates with maternal depression, anxiety, or trauma during pregnancy and is an important risk factor for the development of psychiatric disease. All of these maternal states manifest deregulation of the hypothalamic-pituitary-adrenal (HPA) axis and can elicit epigenetic changes that are lastingly inscribed into the fetus's genetic blueprint presetting future interactions with the environment. The HPA axis mediates the stress response and epigenetic programming of HPA-axis regulatory genes offers a versatile mechanism by which ELA gets under the skin. The glucocorticoid receptor (*NR3C1*) plays an important role in terminating the stress response and can be at the same time epigenetically programmed by stress. Here, we discuss recent evidence on the impact of maternal depression or trauma, but also of mundane stress, on epigenetic programing of *NR3C1* and its association with infant HPA-axis activity and neurobehavioral performance. Furthermore, integrate analysis of placental glucocorticoid signaling pathways may enhance insight into stress dynamics in utero. Overall, epigenetic bookmarking of fetal stress systems may have far-reaching implications for mental well-being of generations to come.

9.1 Early Life Adversity is a Major Risk Factor for Mental Disease

The individual, societal, and economic burden of mental disorders grows steadily (Murray and Lopez 1996) despite continuous improvements in pharmacological and psychological interventions over the last decades (Wittchen *et al.* 2011). An increasing number of epidemiological and experimental studies have aimed to identify the genetic and environmental factors underlying the onset, course and treatment response of mental disease. Although the relevance of the quality of early life (in utero, perinatal, and in infancy) for later physical health is well known (Gluckman *et al.* 2008, 2009), its impact on the development of mental disorders has been recognized only more recently (Kessler *et al.* 1997). Exposure to ELA is one of the strongest risk factors for psychiatric disease (Heim and Binder 2012) concerning mood and affective disorders (particularly major depressive disorders [MDD]) (Cougle *et al.* 2010, Kendler *et al.* 1999, Teicher *et al.* 2002), schizophrenia, borderline personal disorders (BPD) (Green *et al.* 2010, McLaughlin *et al.* 2010), and posttraumatic stress disorders (PTSD) (Lang *et al.* 2008). ELA refers to parental maladjustment (mental illness, substance abuse, violence and criminality), maltreatment (sexual abuse, physical abuse or neglect), interpersonal loss (parental death or divorce, separation from parents or caregivers), life threatening childhood physical illness in the respondent or severe childhood family financial distress (Green *et al.* 2010). For 2012, approximately 686,000 victims of abuse and neglect were reported, corresponding to 9.2 victims per 1,000 children across the US population (NCANDS 2013). The victimization rate was highest for children younger than 1 year (21.9 per 1,000 children) and decreased with age. Until today most studies on the influence of ELA have focused on childhood trauma, in particular sexual (9.3%) and physical (18.3%) abuse. In contrast, neglect during early childhood receives much less publicity although it is the most common form of maltreatment, accounting for more than three quarters (78.3%) of all cases.

Similarly, maternal stress-related diseases, particularly MDD or PTSD, are largely overlooked though they represent major risk factors for the developing fetus and newborn. MDD and anxiety disorders are nearly twice as prevalent in women as in men (Murray and Lopez 1996, Wittchen *et al.* 2011) and manifest in pregnancy with an estimated prevalence of 7–18% for depressive (Bennett *et al.* 2004) and 8.5% for generalized anxiety disorders (Ross and McLean 2006). Depressed women are more likely to deliver prematurely (Beydoun and Saftlas 2008) or to deliver growth-restricted neonates (Wadhwa *et al.* 2001); moreover, children of depressed mothers show impaired mental, emotional and motor development as infants (Nancy Aaron Jones 1998) and are at higher risk of developing depression (Silberg *et al.* 2010), anxiety (Gerardin *et al.* 2011) or violent behavior (Hay *et al.* 2010).

Up to 8% of women in the US are treated with antidepressants during pregnancy (Andrade *et al.* 2008, Oberlander *et al.* 2008a), which have been reported to confer an increased risk for poor birth outcomes as inferred from lower birth weight (Oberlander *et al.* 2006, 2008b), lower Apgar scores (Ross *et al.* 2013) or antidepressant withdrawal symptoms after birth (Moses-Kolko *et al.* 2005). In any case, maternal mood and anxiety disorders are a major threat for the developing fetus and newborn by influencing their immediate and long-term psychological and physiological development (Green *et al.* 2010, McLaughlin *et al.* 2010).

In this review we will discuss evidence for the particular vulnerability of the fetal and newborn brain and how adverse experiences can couple to enduring epigenetic effects on neuronal phenotypes, and ultimately behavior. Epigenetic mechanisms (reviewed in (Allis *et al.* 2015, Armstrong 2013)) are increasingly thought to be involved in the interactions between dynamically changing environments and the preset genetic blueprint. As a result, epigenetic marking of genes involved in distinct neuronal functions can contribute to an organism's capability to mount an adaptive response (reviewed in (Hoffmann and Spengler 2014, Patchev *et al.* 2014, Szyf and Bick 2013)). In the present context, we will refer mainly to humans despite obvious limitations on the availability of (fetal) brain tissues for epigenetic analysis.

9.2 Windows of Sensitivity

Developmental (psycho-) biologists have been intrigued since a long time by the 'Nature versus Nurture' debate in terms of individual differences in life course trajectories (Sameroff 2010) and relatedly, by the effects of early lifetime events on the prevalence, severity, and course of mood, affective, and cognitive disorders (Heim and Binder 2012, Hunter and McEwen 2013, Lupien *et al.* 2009).

The basic architecture of the human brain develops through an ongoing process that begins before birth and extends into adulthood (Stiles and Jernigan 2010). The rudimentary structures of the central nervous system are established by the end of the embryonic period and develop steadily through the end of gestation. This period is characterized by rapid growth and elaboration of both cortical and subcortical structures, including the formation of major fiber pathways. These gross morphological changes of the prenatal neural system are accompanied by the production of neurons from human embryonic day 42 (E42) onward until completion of major parts of neurogenesis by midgestation. Following migration to their final destination, neurons form rudimentary neural networks and complete fiber pathways. It is important to note that brain development continues for an extended period postnatally (Reiss *et al.* 1996) whereby structural changes underpin changes in functional reorganization regulating cognition, emotion, and behavior among other higher functions. Levels of connectivity throughout the developing brain exceed by far those of adults (Innocenti and Price 2005) and are gradually pruned back via competitive processes that are influenced by an organism's experiences. These processes underlie the well-known plasticity and capacity for adaptations characteristic of early brain development.

Early life experiences — for better or for worse — are essential for brain development and make up a time of great opportunities and great vulnerabilities (Shonkoff 2003). Although the brain communicates with the environment from the 'womb to the tomb', there are periods in life during which the brain is more sensitive to the effects of experience. Such 'sensitive periods' associate with the formation of specific circuits that underpin specific abilities like vision and hearing, language acquisition, and higher cognitive functions. This steady genes-environment dialog can also elicit 'predictive' changes in phenotypes (Bateson *et al.* 2014) (also known as 'programming') that allow organisms to adapt more easily to changing environments by resetting physiological thresholds and set points. In this respect, growing evidence suggests that epigenetic mechanisms are important to maintain brain plasticity and to inscribe at the molecular level lasting, or even lifetime, programing of phenotypes (Hoffmann and Spengler 2014, Murgatroyd *et al.* 2010).

To serve as a basis for this chapter, we will represent a snapshot of epigenetic mechanisms before asking which physiological systems can transmit the effects of adverse early life experiences via epigenetic marking.

9.3 A Snapshot of Molecular Epigenetic Mechanisms

The term epigenetics dates back to Conrad Waddington who referred to this term in context of different cellular and organismal phenotypes resulting from the same genotype during development (Waddington CH 2014). On the other side, Arthur Riggs defined epigenetics as inheritable changes in gene expression controlling cell fate decisions and cellular phenotypes irrespective of changes in DNA sequence (Russo *et al.* 1996). Perceptions of molecular epigenetics have undergone major revisions over time due to important advancements in understanding the role of DNA methylation, posttranslational modifications of core histones, nucleosome positioning and non-coding RNAs (ncRNA), among others (reviewed in

(Allis *et al.* 2015, Armstrong 2013)). Particularly two type of processes — DNA methylation and histone modifications — are core concepts of molecular epigenetics and have elicited broad interest since they are thought to inscribe enduringly epigenetic states into the genome and will be therefore briefly surveyed in this manuscript.

9.3.1 *DNA modifications*

Although DNA methylation was originally discovered in 1925 (Johnson and Coghill 1925) and subsequently analyzed in the 1950s (Wyatt 1950), its functional role in gene expression was only evidenced by studies on X-chromosome inactivation (Riggs 1975) and cancer (Holliday 1979). DNA methylation denotes the addition of a methyl group to the fifth carbon of cytosine (5mC) and occurs in somatic cells primarily in the context of palindromic CpG dinucleotides (reviewed in (Schübeler 2015)). Additionally, non-CpG methylation (CpH, H = A, T, C), which predominates in plants, has been more recently also discovered in embryonic stem cells (ESCs) (Xie *et al.* 2012) and the developing mammalian nervous system although its relevance remains still unresolved (Lister *et al.* 2013).

DNA methylation is carried out by a family of conserved DNA methyltransferases (DNMT) comprising DNMT1, DNMT3A and DNMT3B and the enzymatically inactive homologue DNMT3L (reviewed in (Jeltsch and Jurkowska 2014)). DNMT1, together with its obligate partner UHRF1 (ubiquitin-like plant homeodomain and RING finger domain), recognizes predominantly hemimethylated DNA during replication and maintains the methylation pattern of the parental strand. Conversely, DNMT3A and DNMT3B catalyze the *de novo* transfer of methyl groups to unmethylated DNA. This function is supported by assembly with DNMT3L, which enhances catalytic activities through its scaffolding function (reviewed in (Jeltsch and Jurkowska 2014)) (Figure 9.1).

CpG sites are in general depleted in the genome, except from punctuated stretches of DNA termed CpG islands (CGI), where CpG content is high. Approximately 70% of all annotated promoters associate with a CGI and stay in general methylation free (Saxonov *et al.* 2006), whereas CpGs outside of CGIs are mostly methylated (Gibney and Nolan 2010). DNA methylation of promoter CGIs has been commonly thought to associate with gene repression, particularly during cancer development. Yet, recent studies have evidenced that the popular view of DNA methylation as an all-purpose repressive mark needs to be revised and applies only to allele-specific methylation of promoter CGIs on the inactive copy of the female X chromosome, to imprinting control regions, and germ cell-expressed genes in somatic cells (Auclair *et al.* 2014).

In fact, genome-wide methylation analyses show that the effects of DNA methylation depend strongly on the underlying sequence, genomic location and transcriptional status and associate with both gene activation and repression (Lister *et al.* 2009, 2013, Maunakea *et al.* 2010, Weber *et al.* 2007). Moreover, the concept that DNA methylation represents a fairly stable or even irreversible, covalent modification established during development has to be revised following the transformative discovery of active demethylation (reviewed in (Wu and Zhang 2014)). Initially, oxidation-independent mechanisms were suggested to catalyze active demethylation; however, these biochemical reactions perform poorly under physiological conditions. Instead, the family of ten–eleven translocation (Tet) enzymes seem to efficiently catalyze the oxidation of 5mC into 5-hydroxymethylcytosine (5hmC) (Ito *et al.* 2010, Tahiliani *et al.* 2009) and into further oxidized derivatives [5-formylcytosine (5fC) followed by 5-carboxycytosine (5caC)] (He *et al.* 2011, Ito *et al.* 2011). These products interfere with the binding and/or activity of maintenance DNMT complexes and support replication-dependent dilution of oxidized 5mC during preimplantation and primordial germ development. Alternatively, oxidized cytosines can be also directly removed by the DNA repair machinery in embryonic stem cells (He *et al.* 2011, Maiti and Drohat 2011) (Figure 9.1).

Additionally, oxidized cytosines may disrupt interactions with proteins recognizing 5mC and thus counteract their various functions (Spruijt *et al.* 2013). In accord with this hypothesis, 5hmC is enriched at specific regulatory regions throughout the genome pointing to a more active role in gene regulation (Lister *et al.* 2013).

Collectively, these findings suggest that the methylome undergoes dynamic changes in terms of 5mC distribution, cyclic cascades consisting of cytosine methylation, iterative oxidation, replication dependent dilution and/or DNA base repair, and recruitment of proteins recognizing these modifications across the genome (Figure 9.1).

9.3.2 *Histone modifications*

Histone proteins and associated factors package genomic DNA into a highly condensed form, termed chromatin, within the cell nucleus (Misteli 2007). Hereby, 146 bp of DNA are wrapped around a histone octamer consisting of two copies of each core histone (i.e. H2A, H2B, H3 and H4) to form the nucleosome — the building block of chromatin. Nucleosomes are further arrayed into higher order

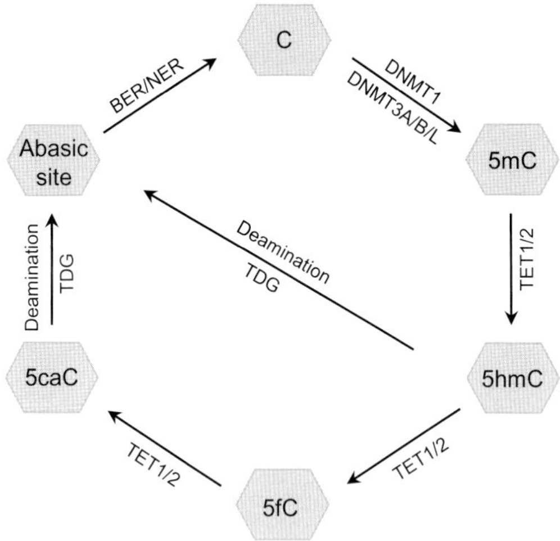

Figure 9.1. Iterative cycles of DNA methylation in mammalian cells. DNA maintenance methyltransferase 1 (DNMT1) and *de novo* DNMT3A/DNMT3B catalyze the transfer of a methyl group to the nucleotide cytosine (C) at the 5th carbon. Catalytically inactive DNMT3L enhances the activity of *de novo* DNMTs due to its scaffolding function. Active demethylation of 5-methylcytosine (5mC) results from iterative oxidation by ten-eleven translocation proteins (TET1/2), which catalyze the conversion to 5-hydroxymethylcytosine (5hmC), then 5-formylcytosine (5fC), and lastly 5-carboxylcytosine (5caC). These cytosine analogs are thought to inhibit binding and activity of the maintenance methylation machinery in mitotically active cells and to passively dilute oxidized 5mC nucleotides. Alternatively, deamination of 5caC, but also of 5hmC, to thymine enables excision by thymine DNA glycosylase (TDG). Subsequently, mismatched bases are repaired by the base excision and/or nucleotide excision repair machinery (BER/NER).

structures with the help of linker histones (i.e. H1) and other non-histone proteins and are in this conformation largely inaccessible to the transcription machinery (reviewed in (Li *et al.* 2007)). Nonetheless, various enzymatic complexes can remodel histone-DNA structures (e.g. ATP-dependent mobilization of core histones relative to the DNA) to expose regulatory sequences and initiate transcription subsequently (Cairns 2007).

The core histones' free amino-terminal tails protrude from the nucleosome surface and can be subject to various posttranslational modifications (PTMs) at specific amino acid residues including lysine acetylation, lysine and arginine methylation, serine phosphorylation and covalent binding of ubiquitin among others (reviewed in (Kouzarides 2007)). Histone marks are established by lysine methyltransferases and histone acetyltransferases and are counteracted by lysine demethylases and histone deacetylases. Histone modifications frequently occur in a combinatorial pattern assimilating cross-regulatory and well-ordered relationships between single PTMs. Different groups of chromatin-binding proteins are capable to decipher such chromatin patterns by virtue of specialized domains (e.g. bromodomains and chromodomains) and to modify these marks on their own or following the recruitment of additional effector proteins (Kouzarides 2007). Therefore, nucleosomes show dynamic changes in histone marking that favor or disfavor chromatin accessibility and transcription.

Relatedly, a number of DNA-binding proteins can recognize and bind to 5mC (e.g. methyl-CpG-binding proteins, Kaiso, Kaiso-like and Set and Ring Associated (SRA) domain proteins) and serve as a platform for various effector proteins that regulate chromatin and DNA methylation marks (Fournier *et al.* 2012). As a result, DNA methylation and chromatin marks mutually interact with each other in deposition and maintenance.

Taken together, chromatin controls the transcriptional machinery's access to regulatory DNA sequences and integrates dynamically intrinsic and environmental signals across the genome.

Whilst this snapshot is intended to provide a guidance for the sections to follow, readers more interested in this field are referred to a number of recent reviews and monographs (Allis *et al.* 2015, Armstrong 2013, Eccleston *et al.* 2007, Goldberg *et al.* 2007, Tollefsbol 2011).

9.4 The Stress System Mediates the Effects of ELA

Foregoing sections raise the intriguing question through which physiological system(s) ELA is transduced into enduring epigenetic marking of genes underlying mental well-being and behavior.

The subjective state of sensing potential or actual threats in the environment is commonly referred to as 'stress' and causes the release of molecules acting on specific neuronal populations orchestrating unique downstream effects such as rise in blood pressure, increase in heart rate and elevated levels of stress hormones (Fink 2007, de Kloet *et al.* 2005). Together, these responses constitute the 'stress response' and endow organisms with the capability to adapt to challenges through immediate changes as wells as the modification of future behaviors (Lupien *et al.* 2009).

A central driver of the mammalian stress response is the HPA axis, which integrates a broad scope of physiological and psychological stressors (de Kloet *et al.* 2005, Lupien *et al.* 2009). Parvocellular neurons of the hypothalamic paraventricular nucleus (PVN) produce and release two neuropeptides, namely corticotropin-releasing hormone (CRH) and arginine vasopressin (AVP), which coordinate the behavioral and metabolic responses to stress (Figure 9.2(a)). Following binding to membrane bound receptors at the anterior pituitary, they synergistically stimulate pro-opiomelanocortin (*POMC*) gene expression and secretion of its posttranslational product adrenocorticotropin (ACTH) (reviewed in (Charmandari *et al.* 2012, Chrousos 2009)). Subsequently, ACTH acts on the adrenal cortex to enhance the secretion of cortisol (in humans) and corticosterone (in humans, rats and mice). These lipophilic hormones readily enter the brain to bind to nuclear glucocorticoid and mineralocorticoid receptors (GR and MR encoded by *NR3C1* and *NR3C2*, respectively). These receptors represent closely related ligand-gated transcription factors that are highly coexpressed in neurons of the limbic brain, with high levels of GR being additionally present in the PVN and anterior pituitary (Figure 9.2(a)). The MR, requiring lower glucocorticoid concentrations for activation, is thought to initiate the stress response, whereas GR, dependent on higher glucocorticoid concentrations, curtails, and, ultimately, terminates the stress response (reviewed in (Fink 2007, de Kloet *et al.* 2005)).

It is important to note that the HPA axis orchestrates the response to stress and can be at the same time enduringly programmed by stress. Activity-dependent plasticity allows the integration of environmental signals into developmental trajectories underlying the perception and subsequent responses to various stressors. During sensitive windows of prenatal and postnatal mammalian development, stress can elicit permanent structural and regulatory changes predisposing to stress-related diseases in later life. For example, in the hippocampus and medial prefrontal cortex, chronic stress causes shrinkage and reduced branching of dendrites with an accompanying reduction in synaptic input (McEwen 2007). These changes are commonly thought to underpin structural plasticity. In contrast, we focus here on regulatory, potentially reversible, changes in gene expression through epigenetic marking, which we propose to underlie molecular plasticity and to control HPA-axis activity on the long run. Accordingly, we will discuss next current evidence for a relationship between ELA and DNA methylation in humans during the prenatal and perinatal period.

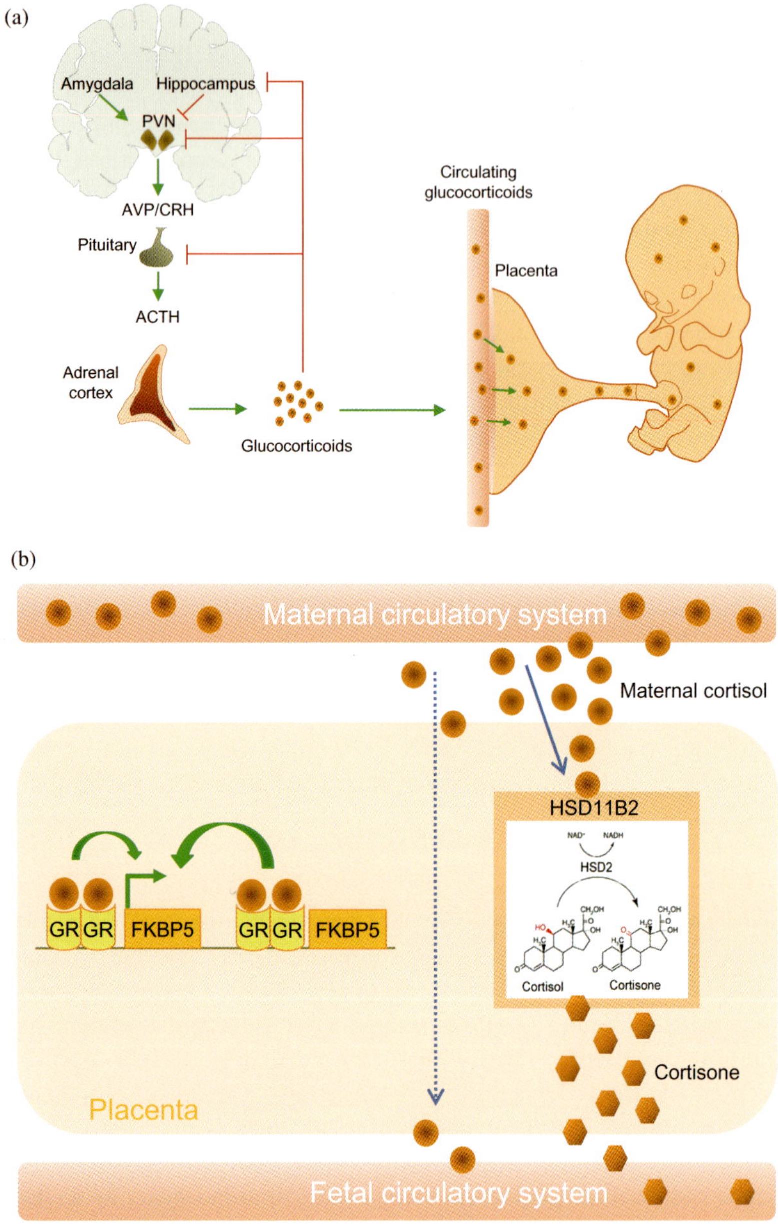

Figure 9.2. Maternal HPA-axis activity and fetal glucocorticoid exposure. (a) The two neuropeptides arginine vasopressin (AVP) and corticotropin-releasing hormone (CRH) are produced in the parvocellular neurons of the hypothalamic paraventricular nucleus (PVN)

Figure 9.2. (*Continued*) and are released in the hypophyseal-portal system in response to various stressors. Following binding to G-protein coupled receptors at the anterior pituitary; they synergistically stimulate the production of pro-opiomelanocortin (POMC) and secretion of its post-translational product adrenocorticotropin (ACTH). In turn, ACTH acts on the adrenal cortex to enhance the synthesis and secretion of glucocorticoids. These lipophilic hormones readily pass into the brain to bind nuclear mineralocorticoid and glucocorticoid receptors, which are ligand-gated transcription factors with an important role in the assessment and termination of the stress response, respectively. The glucocorticoid cortisol can as well cross the placenta and reach the fetus to trigger transcriptional activation of placental and fetal glucocorticoid receptors. (b) The placenta forms a physiological interface between the maternal and fetal circulatory system. Maternal derived cortisol is converted by the enzyme 11-β hydroxysteroid dehydrogenase type 2 (HSD11B2) into the oxidized inactive metabolite cortisone (inset). Therefore, this enzyme is thought to shield the fetus against harmful concentrations of maternal glucocorticoids. Still, excessive and/or prolonged surges in cortisol due to severe maternal trauma or depression can compromise the placental barrier and expose the developing fetus to an increased glucocorticoid load. As a result, developing programs can be adjusted leading to various adaptations, partly via epigenetic mechanisms, which confer an increased risk for later stress-related diseases. Excess glucocorticoid exposure may affect as well the placenta via activation of local glucocorticoid receptors (GR) that may trigger, or even maintain, programming of various placenta functions across gestation. The co-chaperone FKBP5 plays an important role in limiting GR activity. On the other side, following cortisol binding, GR dimers (yellow symbols) bind at the upstream promoter and intronic regions of *FKBP5* to confer transcriptional activation (green arrowhead). In turn, enhanced *FKBP5* transcription will translate into increased levels of FKBP5 protein attenuating GR function. Together, this regulatory process constitutes an important intracellular negative feedback loop to curtail GR function. Changes in DNA methylation at regulatory regions of both the *NR3C1* or *FKBP5* loci can underpin lasting changes in gene transcription. These can tilt the balance between GR and FKBP5 expression and result in an overall altered placental GR function during pregnancy.

9.5 Epigenetic Memories of Maternal Care

9.5.1 *NR3C1's* modular promoter structure is well conserved

A series of candidate gene studies have focused on the GR due to its critical role in restraining HPA-axis activity. Human *NR3C1* maps on chromosome 5q31-q21 and contains multiple first exons (Steiger *et al.* 2013, Turner and Muller 2005). Among these, four exons are located in the distal promoter region (A_{1-3} and I) and ten (D, J, E, B, F, G, C_{1-3} and H) in the proximal composite promoter region (Figure 9.3). The 5'untranslated region comprises 14 exon 1 splice variants all of which harbor unique splice donor sites and share a common exon 2 splice acceptor

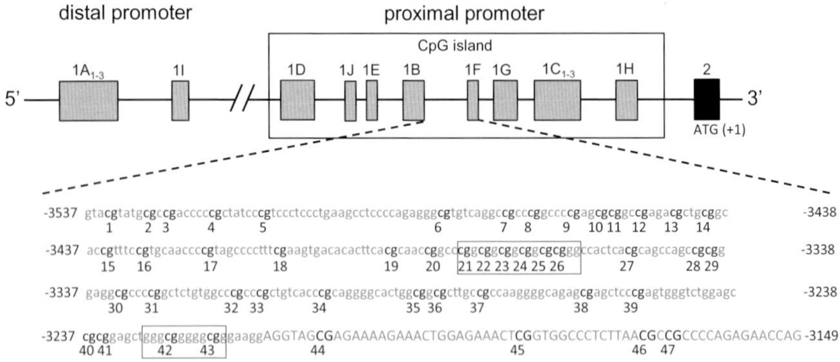

Figure 9.3. Structure of human NR3C1 distal and proximal promoters. Numbered open boxes show the different untranslated exons in 5'-3' orientation. The distal and proximal promoters localize 30 kb and 5 kb upstream of the start codon (ATG) in exon 2 (black box) and contain four (A_{1-3} and I) or ten (D, J, E, B, F, G, C_{1-3}, and H) alternative exons, respectively. The proximal promoter is straddled by a high CpG frequency island (grey frame). The enlargement below depicts the sequence of the exon 1F promoter (lower case) and exon 1F (upper case) with 47 CpG residues numbered. Canonical (solid-line box) and potential (broken-line box) NGFI-A DNA bindings sites are indicated.

site resulting in translation of the same GR protein (Figure 9.3) (Steiger *et al.* 2013, Turner and Muller 2005). The distal and proximal promoter regions locate 30 kb and 5 kb upstream of exon 2 that encodes the start codon (13 nucleotides downstream from its 5'-primed end). A CGI consisting of a high frequency of CpG residues straddles the proximal promoter including exon 1F whose homologue corresponds to differences in maternal care in rodents (Figure 9.3). The human 1F promoter (326 bp in size) and exon 1F (63 bp in size) harbor 47 CpG sites (1F promoter sites 1–43 and exon 1F sites 44–47) and are transcriptionally active in the hippocampus, B lymphocytes, and innate immune cells among others (Turner and Muller 2005).

Most alternative promoters and splice variants at the *NR3C1* locus are expressed in a tissue-specific, and partly species-specific, manner (Bockmühl *et al.* 2011, Turner and Muller 2005). The proximal *NR3C1* promoter region is, however, well-conserved between human, rat and mice (Bockmühl *et al.* 2011) and allows to compare epigenetic marking in response to different environmental exposures across species. Accordingly, preclinical animal studies can serve as signpost for refined analysis of valuable human materials (e.g. postmortem samples) and findings from there can be translated into appropriate animal studies to explore potential therapeutic interventions.

9.5.2 *Postnatal maternal care leaves a lasting footprint at the Nr3c1 promoter*

As early as 2004, a seminal study reported on epigenetic programming of *Nr3c1* in response to differences in the quality of postnatal maternal care in rodents (Weaver IC *et al.* 2004).

Specifically, rearing by high-care taking mothers (displaying a high frequency of pup licking and grooming) enhances serotonin (5-hydroxytryptamine [5-HT]) turnover in the hippocampus of newborn rats, where GR expression is site-specifically increased. 5-HT binds to G-protein coupled receptors stimulating via protein kinase A expression of the transcription factor EGR1 (early growth response 1). Following binding to its DNA recognition site at exon 1.7 (the homologue of 1F in human), EGR1 recruits general coactivators containing histone acetyltransferase activity. These enzymes promote an open chromatin structure, subsequently *Nr3c1* transcription and, importantly, demethylation of the EGR1 binding site. Such hypomethylation engraves a lasting memory trace of the quality of maternal care and underpins lasting increases in GR expression and distinct endocrine and behavioral changes in adulthood (Weaver IC *et al.* 2004). In contrast, pups reared by low-care taking mothers show unaltered *Nr3c1* promoter methylation and expression and a stress-vulnerable phenotype. Consistent with the dynamic nature of methylation marks, epigenetic programming of *Nr3c1* is reversible through timely pharmacological and/or behavioral interventions (Weaver *et al.* 2005).

9.6 Adverse Experiences During Prenatal Human Life

The impact of ELA on *NR3C1* 1F promoter methylation has been intensively studied over the last years and key findings from these, as well as their limitations, will be presented in this section. Since *NR3C1* methylation cannot be analyzed in utero or postnatally in offspring's brain tissues, the use of peripheral tissues (cord blood, peripheral blood mononuclear cells (PBMCs) and placenta) as a proxy to brain methylation has become common practice based on two assumptions. Firstly, biological processes of the condition under study can affect peripheral methylation independent from brain methylation offering valuable biomarkers (signature model). Secondly, some condition-related methylation patterns may be identical in the periphery and the brain (mirror model) possibly as a result of shared humoral factors (e.g. stress hormones) and/or of common developmental origins (e.g. *in utero*) (Aberg *et al.* 2013).

9.6.1 *Maternal depression and anxiety*

The effects of maternal mood and affective disorders (including comorbid substance abuse) on prenatal and/or postnatal child development may be mediated partly through epigenetic programming of *NR3C1*. In support of this hypothesis, increased neonatal methylation of CpG1–3, located at the 5'end in the potential EGR1 binding site of promoter 1F, is associated with prenatal exposure to third trimester maternal depression (Oberlander *et al.* 2008a). In this study, the methylation status of *NR3C1*, including exon 1F, was analyzed in genomic DNA isolated from cord blood mononuclear cells by bisulfite pyrosequencing in infants of depressed, antidepressant (serotonin uptake inhibitor, [SRI]) treated mothers (n = 33), infants of depressed untreated mothers (n = 13), and infants of non-depressed/untreated mothers (n = 36). Functionally, increased methylation at CpG3 was accompanied by heightened stress reactivity as evidenced by increased saliva cortisol at three months following exposure to a visual information processing task. In contrast, resting afternoon/early evening levels of cortisol were unaffected.

DNA methylation was additionally assessed at repetitive DNA sequences (LINE-1 elements), which did not reveal any detectable changes. This result disfavors global impairments in DNA methylation due to maternal depression and/or antidepressant treatment and points to a rather site-specific effect directed at promoter 1F of *NR3C1*.

Collectively, the association between CpG3 methylation at the 5'end of the potential EGR1 binding site and three month HPA-stress reactivity raises the possibility that epigenetic marking of this residue recalls exposure to late gestational maternal depression and predicts future risk for altered HPA-axis function in offspring. A weakness of the study is, however, the fact that methylation was assessed solely in cord blood cells but not at later ages in the offspring. Hence, it remained unresolved whether and how long epigenetic marks established in depressed mothers prenatally persist after birth in offspring.

9.6.2 *Maternal trauma and war-stress*

Although a sizeable fraction of ELA results from maternal depression, other causes such as maternal trauma exist additionally. In this respect, a recent study (Radtke *et al.* 2011) investigated the enduring effects of maternal aversive experiences on offspring promoter 1F methylation 10–19 years after birth by bisulfite sequencing of DNA isolated from PBMCs. A total of 25 women, aged between 29 and 51 years, all of them having children aged between 10 and 19 years, participated in this study. A retrospective evaluation of maternal exposure to intimate partner violence

(IPV), defined as 'any behavior with an intimate relationship that causes physical, psychological or sexual harm in the relationship' (Krug *et al.* 2002), was carried out by the composite abuse scale (Hegarty *et al.* 2005). Accordingly, three women were exposed to IPV before pregnancy, eight during pregnancy, and nine after pregnancy. None of them showed changes in *NR3C1* methylation status irrespective of the time during which IPV was experienced. In contrast, offspring DNA methylation at promoter 1F, particularly at CpG residues 1–3 at the known EGR1 DNA binding site, was significantly associated with maternal exposure to IPV during pregnancy ($n = 23$) but not before ($n = 21$) or after ($n = 22$) pregnancy.

Taken together, this study indicates that prenatal maternal trauma can lead to enduring, rather than short-lived, alterations in DNA methylation of an HPA-axis regulatory gene in the offspring. It remains arguable, however, whether *NR3C1* methylation in blood corresponds with *NR3C1* status in human hippocampus (or other relevant brain tissues) and HPA-axis function. Therefore, follow-up studies are desirable to evaluate concomitantly DNA methylation, stress reactivity and psychological functions in offspring exposed in utero to maternal trauma.

Although maternal depression and trauma are major sources of ELA in westernized societies, it is important to note that other conditions may prevail on a world-wide scale. These include exposure to severe socioeconomic disparities, natural disasters, civil war and war-related events among others. In this context, three forms of maternal stress including material deprivation (proposed to reflect availability of financial resources), mundane stressors (proposed to reflect daily psychosocial stress), and war stress (proposed to measure the influence of war-related events) have been investigated for newborn birth weight and *NR3C1* methylation (Mulligan *et al.* 2012). Maternal venous blood and umbilical cord blood was obtained from 25 mother–newborn dyads from Eastern Democratic Republic of Congo, a region plagued by ongoing war and extreme violence to women.

War stress had the largest effect on birth weight with personal experience of rape accounting for one-third of weight variance and dominating the effects of other war stressors (killed family members, refugee status, past kidnapping, etc.). The percentage of DNA methylation measured for 39 CpGs contained in promoter 1F was scrutinized by principal component analysis (PCA) to account for independent components that may relate to different environmental, genetic or biological impacts on DNA methylation. Firstly, methylated sequences from mothers and newborns were highly unrelated supporting that all maternal marks had been reset during embryogenesis. Secondly, no tissue-specific marks could be inferred from maternal venous blood versus umbilical cord blood excluding a major impact of tissue specific factors. Thirdly, the strongest correlations were detected between war stress and newborn site-specific 1F methylation, which in turn, strongly correlated with birth weight (Mulligan *et al.* 2012). In contrast, no correlations were detected for promoter 1F total methylation indicating that these factors associated with site-specific effects rather than a general shift in methylation.

In sum, this work shows that severe psychosocial stress can impose site-specific methylation changes at promoter 1F in newborn cord blood cells that correlate with birth weight as a proxy to overall fetal development. As noted above, follow-up studies in offspring are needed to assess the persistence of these findings jointly with functional HPA-axis and neurobehavioral tests to inform about infants' future development.

The dramatic impact of civil war on transgenerational transmission of maternal stress and concomitant epigenetic programming of *NR3C1* has been recently investigated in the context of the Rwandan genocide against ethnic Tutsi between April and June 1994, which resulted in the death of almost one million people (Perroud *et al.* 2014). In aftermath of this event, more than 20% of the general population met the diagnosis of depression and/or PTSD in 2012 (Munyandamutsa *et al.* 2012). Parental PTSD is a strong risk factor for the development of PTSD in offspring, which associates with HPA-axis deregulation manifest by low cortisol alongside increased cortisol suppression (following application of the synthetic glucocorticoid dexamethasone), and enhanced GR expression in PBMCs (Raabe and Spengler 2013). To investigate trauma-induced fetal programing, 25 widows, pregnant at the time of exposure to the Tutsi genocide, and 25 pregnant women from the same ethnicity but without trauma exposure and living abroad at the same time, were assessed for PTSD and depression severity, plasma cortisol and GR and MR protein levels (measured by ELISA) as well as *NR3C1* and *NR3C2* methylation (measured by pyro-sequencing) in mothers and offspring.

In accord with previous findings (Yehuda *et al.* 2007), severity of PTSD and depression correlated significantly between trauma exposed widows and their children. This correlation was even stronger in unexposed mother–child dyads probably due to a ceiling effect in the exposed group (Perroud *et al.* 2014). Lower cortisol was measured in trauma exposed, but not in unexposed, mother–child dyads and correlated strongly with PTSD and depression severity. Moreover, trauma exposed mothers and their children showed lower plasma GR and higher plasma MR expression when compared to unexposed dyads.

Interestingly, mothers exposed to genocide trauma showed higher exon 1F *NR3C1* promoter methylation (CpG1, CpG5 and CpG8) than unexposed mothers. Similarly, children born from traumatized mothers displayed higher 1F methylation (CpG3-9) compared to control children. However, these differences did not reach statistical significance in the trauma exposed groups, probably due to a ceiling effect, whereas the correlation between 1F methylation in mothers and children was highly significant in unexposed controls. Lastly, lower *NR3C2* methylation at CpG 1–8 (located within the overlap between a CpG island and the transcriptional start site) did not correlate significantly with higher MR protein expression in trauma exposed mother–child dyads.

All in all, this study supports the hypothesis that severe maternal trauma can be transmitted to offspring and associates with profound effects on neurobehavioral development possibly through HPA-axis dysfunction as evidenced by lower plasma cortisol (Yehuda and Bierer 2008). Mother–child dyads suffering from PTSD showed higher exon 1F *NR3C1* promoter methylation and reduced GR expression when compared to controls, whereas higher *NR3C2* methylation did not associated significantly with enhanced MR expression.

This study is to be recommended for assessing depression/PTSD in trauma exposed and unexposed mother–child dyads together with GR/MR protein expression, *NR3C1/NR3C2* methylation status and offspring cortisol levels. Limitations are confounding ceiling effects in questionnaires and methylation analysis and possible contributions from environmental factors unrelated to trauma such as higher socio-economic status for mother–child dyads living abroad. In any case, this study highlights the epidemic impact of the Tutsi genocide on maternal PTSD and depression and on mental well-being of generations to come.

9.6.3 *A role for cortisol in mediating maternal stress?*

Epigenetic programing of *NR3C1* in response to maternal depression (Oberlander *et al.* 2008a) or adversity (Mulligan *et al.* 2012, Perroud *et al.* 2014, Radtke *et al.* 2011) stimulates the question, which biological mechanisms may mediate these effects. A recent study investigated the association between maternal diurnal cortisol, pregnancy-related anxiety and *NR3C1* promoter methylation (Hompes *et al.* 2013). Healthy pregnant women (at about 6–12 weeks gestation) were asked to complete a number of self-report questionnaires on psychological well-being and to collect saliva samples for cortisol assessment each pregnancy trimester. Cord blood was taken at birth from 83 mother–child dyads for methylation analysis (exons 1B, 1D and 1F). Methylation status at the exon 1F *NR3C1* promoter, particularly CpG9, was most associated with prenatal maternal emotional state and cortisol levels. Additional, albeit weak, associations between psychological and methylation measurements, were detected for single CpG residues at exon 1B, 1D and 1F.

Mothers showed medium-levels of pregnancy-related anxiety and a progressive increase in salivary cortisol. However, neither depressive nor anxiety symptoms, nor attachment, were related to cortisol measurements indicating that cortisol is unlikely to directly mediate the effects of these conditions on exon 1F methylation. In this context it is important to note, that CRH, ACTH and cortisol increase physiologically during the course of pregnancy (Levine *et al.* 2007). Moreover, mothers of this study were healthy and free of exposure to severe adversity and/or stress.

One possible explanation for these findings is that physiological versus traumatic stress during pregnancy leads to disparate marking of exon 1F *NR3C1* promoter residues. Contrary to a strong surge of cortisol in response to acute trauma, physiological increases across pregnancy appear less likely to mediate epigenetic marking of *NR3C1,* which may occur through unrelated pathways.

9.6.4 *A role for the placenta in mediation of genes-environment interactions*

The placenta represents a multilayered interface between mother and fetus with a crucial role for growth and development. In addition to its endocrine and metabolic functions, the placenta transports nutrients, water, gas and waste products during fetal growth. Alterations in placental gene expression can affect intrauterine environment and ultimately child's health and performances throughout life (Barker and Clark 1997, Gluckman *et al.* 2008). Hence, deregulation of placental gene expression may record maternal stress exposures during pregnancy and thus serve as a forecast of future child development.

Low birth weight is a multifactorial outcome associated with increased infant morbidity and mortality as well as a predictor of later life health and disease, comprising metabolic and cardiovascular disease, but also mental disorders (Barker and Clark 1997, Gluckman *et al.* 2008). To investigate if epigenetic programming of *NR3C1* associates with infant growth, 480 human placentas from healthy mothers were analyzed for 13 CpG sites at the exon 1F *NR3C1* promoter for DNA methylation (Filiberto *et al.* 2011). Birth weights were classified as small for gestational age (SGA, n = 102), appropriate for gestational age (AGA, n = 343), and large for gestational age (LGA, n = 35) corresponding to $< 10^{th}$, $10 - 90^{th}$, and $> 90^{th}$ percentile of birth weight for gestational age, respectively. Whereas all CpG sites at the exon 1F *NR3C1* promoter were differentially methylated across these groups, only the mean across all CpGs and CpG7 reached significance dependent on birth weight classification. Specifically, infants considered LGA showed higher exon 1F and CpG7 methylation, whereas methylation at residues CpG3 and

CpG4, which map to the potential EGR1 binding site (previously reported to associate with maternal stress and mood (Oberlander *et al.* 2008a)), revealed non-significant increases in methylation when compared to SGA and AGA infants. In conclusion, these results suggest that environmental influences acting on birth weight affect preferentially CpG sites at the exon 1F *NR3C1* promoter distinct from the potential EGR1 DNA-binding site.

In sum, this study firstly demonstrates that infant growth — a proxy to intrauterine environment — couples to placental exon 1F *NR3C1* promoter methylation status and raises the possibility that DNA methylation is a critical determinant of placental function.

Since the effects of intrauterine development may well extend beyond embryonic growth to postnatal development, these findings stimulate the question whether exon 1F *NR3C1* promoter methylation status can also inform about infant neurobehavior. In a follow up study (Bromer *et al.* 2013), 186 healthy newborn infants were assessed for 1F methylation (13 CpGs analyzed by pyro-sequencing), a common single nucleotide polymorphism in the *NR3C1* promoter, and neurobehavioral performance (including neurological integrity, behavioral performance, and signs of stress/abstinence) as assessed by the Neonatal Intensive Care Unit (NICU) Network Neurobehavioral Scales (Liu *et al.* 2010). Results showed that an increased methylation of CpG7 and CpG10 associated with lower stress abstinence scores and increased infant habituation, respectively, whereas increased methylation of CpG sites 6, 7 and 8 associated with infant quality of movement scores. Controlling for confounders enhanced the significance of the association both overall and for specific CpG sites, indicating a complex interaction between genetics, epigenetics, environment and neurobehavioral outcome (Bromer *et al.* 2013). Functionally, protective methylation marks at single CpGs seem to add, causing poorer performance in other parts of the neurobehavioral examination. For example, infants who score well on habituation may lack some of the methylation marks promoting attention. In any case, this study is the first to assign to placental exon 1F *NR3C1* promoter methylation status a predictive role for infant neurobehavioral development. Together, these findings strengthen the crucial role of placental functions on infant development and identify *NR3C1* methylation as a critical variable in this process.

A limitation of this study is that biological mechanisms, which link intrauterine development to changes in placental DNA methylation, remain currently unknown. Similarly, translational studies are looked for to gain insight into the consequences from *NR3C1* methylation for placenta function at a cellular scale. Moreover, it would be interesting to know whether placental *NR3C1* methylation concurs with methylation in cord blood cells, PBMCs and postnatal stress responsivity.

Relatedly, the role of placental *NR3C1* methylation was studied in depressed mothers in order to explore the effect of maternal mood on neurobehavioral outcomes in offspring via epigenetic programming of placenta tissue (Conradt *et al.* 2013). This study included 27 mothers who reported depression only, 18 who reported anxiety only, 39 who reported comorbid anxiety and depression and 398 healthy mothers. These numbers correspond to 13.7% of pregnant mothers being diagnosed as depressive or anxious in accord with estimated prevalence rates in the general population (Gaynes *et al.* 2005). Interestingly, the group of infants whose mothers reported depression during pregnancy showed greater methylation of placental *NR3C1* at exon 1F CpG2 and poorer self-regulation than infants whose mothers were neither depressed nor anxious irrespective of exon 1F methylation status. Greater methylation at exon 1F CpG2 in infants associated also with more hypotonia and lethargy in both depressed and non-depressed mothers although this effect was more pronounced in depressed mothers (Conradt *et al.* 2013) (see also below).

Placental *NR3C1* exon 1F methylation thus corroborates previous findings from cord blood mononuclear cells during depression (Oberlander *et al.* 2008a) and supports that increased methylation negatively affects newborn neurobehavior. It is important to note, however, that increased methylation of exon 1F CpG2 associates with poorer newborn neurobehavior only in case infants have been exposed to maternal depression during pregnancy. Hence, increased exon 1F methylation rather represents a risk factor in combination with additional maternal stress or trauma than predetermining neurobehavioral impairments. If yes, postnatal parenting practice and newborn exposure to further stressors may superimpose risk for depression transmitted prenatally.

9.6.5 *Early life adversity affects placental glucocorticoid signaling*

Transcriptional activities of GR are tightly controlled at multiple levels including assembly with cytosolic chaperon complexes (regulating receptor maturation, nuclear translocation and recycling) and the availability of free steroid hormones (Storer *et al.* 2011).

In most mammalian species exposure of a pregnant female to stress (Cadet *et al.* 1986) or to glucocorticoids (Dean and Matthews 1999) elicits an enhanced maternal glucocorticoid secretion. Glucocorticoids can cross the placenta to reach the fetus (Figure 9.2(b)) and stimulate fetal HPA-axis activity. Additionally, they can influence the development of neuronal structures, circuitries, neuronal differentiation and maturation (reviewed in (Seckl 2008)). The placenta serves as a physiological interface between mother and fetus and stress-induced changes in placental phenotype may constitute a mechanisms for transmission of early experience to the fetus later in gestation and thus enable intrauterine programming of cerebral development long after the initial insult (Uno *et al.* 1989, 1990).

The placental enzyme 11-β hydroxysteroid dehydrogenase type 2 (HSD11B2) catalyzes the conversion of noxious maternal cortisol into the benign metabolite cortisone before passing it to the developing fetus and thus shields from excess glucocorticoid exposure (Figure 9.2(b)). Maternal stress leads to a significant decline in expression and activity of placental HSD11B2 and enhances glucocorticoid exposure of the placenta and the fetus (Mairesse *et al.* 2007, Welberg *et al.* 2005). Whilst a major proportion of maternal cortisol is converted to cortisone upon placental passage, minor increases in maternal cortisol following from maternal stress can still double fetal concentrations (Gitau *et al.* 1998). In fact, recent work has evidenced associations between prenatal mother's levels of anxiety and mood and reduced expression and activity of HSD11B2 (O'Donnell *et al.* 2012, 2014). Increased fetal glucocorticoids derived from elevated maternal glucocorticoids can lead to sustained deregulations of HPA-axis activity in the offspring and is thought to represent an important risk factor for later development of various psychiatric diseases (O'Connor *et al.* 2002a, 2002b, 2003). Hence, the placenta fulfills a key role in filtering out the harmful effects of maternal glucocorticoids that affect the fetus.

A recent study on 186 healthy mothers suggested that higher placental *HSD11B2* DNA methylation of 4 CpG residues at the CpG island straddling the *HSD11B2* promoter (determined by quantitative bisulfite pyrosequencing) associates inversely with infant quality of movement and low birth weight in healthy newborns (Marsit *et al.* 2012). Moreover, mean *HSD11B2* promoter methylation correlated moderately with reduced mRNA expression. Hence, reduced *HSD11B2* expression, and subsequent exposure of the fetus to maternal cortisol, may influence fetal growth and neurobehavioral outcomes. These findings agree with previous findings from the same cohort of healthy newborns (Bromer *et al.* 2013), which indicated that placental *NR3C1* 1F promoter methylation at specific CpG sites was associated with infant neurobehavior. Jointly, these results show for the first time that epigenetic changes in placental glucocorticoid signaling can impact on in utero fetal growth and postnatal neurodevelopment.

In addition to healthy mother–newborn dyads, placental *NR3C1* and *HSD11B2* DNA methylation has been analyzed among newborns whose mothers suffered from either depression or anxiety during pregnancy (Conradt *et al.* 2013) (see above). This work evidenced enhanced placental *NR3C1* and *HSD11B2* DNA methylation in diseased mothers when compared to controls and associated with impaired infant neurobehavioral performance.

A recent follow-up study on healthy newborns ($n = 373$) corroborated that placental *NR3C1* and *HSD11B2* methylation jointly influence infant neurobehavior (Appleton *et al.* 2015). Specifically, low *NR3C1* and high *HSD11B2* methylation — presumably corresponding to high GR activities — associated with lower excitability scores whilst the inverse methylation pattern — presumably corresponding to low GR activities — associated with more asymmetrical reflexes and high DNA methylation at both genes associated with higher habituation scores. Collectively, these results suggest that epigenetic marking of integrated determinants in glucocorticoid signaling may inform more accurately about fetal development and future neurobehavioral outcomes.

The HSP90 co-chaperone FK506 binding protein 5 (FKBP5) is part of the cytosolic GR holocomplex and serves as an ultra-short intracellular negative feedback loop with GR where FKBP5 reduces glucocorticoid binding affinity and nuclear translocation (Storer *et al.* 2011). Additionally, GR binds via glucocorticoid response elements (GRE) to *FKBP5* promoter and intronic regions 2 and 7 to confer transactivation (Figure 9.2(b)). This negative feedback mechanism curtails GR signaling, facilitates the development of GR resistance under conditions of protracted stress and ultimately predisposes to a number of psychiatric disorders.

A single nucleotide polymorphism (rs1360780) close to a functional GRE in intron 2 of *FKBP5* encodes a distinct transcriptional response to stress. Specifically, the T-allele of rs1360780 promotes an interaction between the intron 2 GRE and the upstream promoter and consequently facilitates transcription under heightened glucocorticoids. In contrast, this interaction is weakened in case of the C-allele of rs1360780 and results in dampened *FKBP5* transcription (Klengel and Binder 2015). The risk (T) and protective (C) alleles (encoding high and low *FKBP5* transcription, respectively) additionally differ in their response to ELA. DNA methylation in and around the intron 7 GRE is preserved for the protective C-allele, whereas early trauma-induced DNA demethylation of the risk T-allele enhances the interaction between intron 7 and the promoter, and subsequently, *FKBP5* transcription (Klengel and Binder 2015). Hence, ELA triggers derepression of *FKBP5*, GR resistance, and an increased risk to develop stress-related psychiatric disorders in risk T-allele carriers.

To investigate the role of placental *FKBP5* genetic and epigenetic variation for infant neurobehavioral performance, intron 7 methylation was measured by quantitative bisulfite pyrosequencing in 509 healthy infants born from healthy mothers (Paquette *et al.* 2014). As anticipated, infants homozygous for the risk T-allele showed higher FKBP5 expression compared to heterozygous and homozygous carriers of the protective C-allele. Hereby methylation of intron 7 correlated negatively with expression amongst newborns harboring the protective C-allele, but not in infants with the risk T-allele or the heterozygous genotype. A possible explanation for this finding is that the activating effect conferred by intron 2 overrides the inhibitory effect of methylation at intron 7 in placental tissue.

Interestingly, infants in the highest quartile of *FKBP5* intron 7 methylation showed an increased risk of high arousal as assessed by the Neonatal Intensive Care Unit (NICU) Network Neurobehavioral Scales when compared to infants in the lowest quartile. Moreover, infants harboring the risk T-allele had increased odds of high stress abstinence when investigated by this questionnaire.

In sum, these results show that *FKBP5* expression in the placenta depends on the presence of the risk T-allele (high expression) versus the protective C-allele (low expression). Moreover, DNA methylation inhibits *FKBP5* expression in a genotype-specific fashion with the transcriptional effects of intron 2 preventing the inhibitory effects of methylation from intron 7 in case of the risk T-allele. Hence, genetic and epigenetic differences jointly regulate *FKBP5* expression and GR signaling in the placenta as well as associated neurobehavioral performance. In view of previous findings that ELA triggers demethylation at intron 7 of the risk T-allele, but not of the protective C-allele (Klengel and Binder 2015), it will be important to assess placental *FKBP5* methylation status additionally in women suffering from depression or aversive exposures during pregnancy.

All in all, conditions of physiological or traumatic stress during pregnancy can lead to epigenetic programming of placental GR signaling through epigenetic marking of *HSD11B2*, *FKBP5* and *NR3C1*.

9.7 Conclusions and Outlook

Prenatal development comprises a period of heightened cellular and molecular plasticity and associates with great opportunities and great vulnerabilities. DNA methylation and histone marking underlie epigenetic plasticity in fetal neurodevelopment and expand as well to postnatal life to integrate various demands originating from an organism's interactions with the environment.

Maternal depression, anxiety or trauma are common sources of prenatal and postnatal ELA and represent important risk factors for psychiatric disease in offspring. All of these states typically associate with HPA-axis deregulation and can trigger epigenetic programming of HPA-axis regulatory genes in offspring. The GR plays an important role in terminating the stress response and numerous reports have investigated *NR3C1* methylation in relationship to early life experience and psychopathology (Table 9.1). Collectively, these studies support that physiological

Table 9.1. The impact of maternal stress on *NR3C1* methylation.

Study	Experimental group	Sample size	Tissue(s)	Age at sample collection	*NR3C1* Exon(s)	NGFI-A site	Additional genes
Oberlander *et al.* (2008a)	3rd trimester, maternal depression and anxiety	SRI exposed: 33 MDD, no SRI: 13 Control 36	Cord blood	At birth	1F ↑	+	
Radtke *et al.* (2011)	Maternal exposure to intimate partner violence	25	Blood	Mother: 25.2 (mean) Child: 14.1 (mean)	1F ↑	+	
Mulligan *et al.* (2012)	Different kinds of maternal stress versus war-stress	25 widow–child dyads	Cord blood	At birth	1F ↑	+	
Perroud *et al.* (2014)	Women exposed to the Tutsi genocide during pregnancy and their children	Exposed mother–child dyads: 25 Non-exposed mother–child dyads: 25	Blood	Mother: 39–52 Child: 17–18	1F ↑	+	*NR3C2* ↓
Hompes *et al.* (2013)	Pregnancy related anxiety over three trimesters	83 mother–child dyads	Cord blood	At birth	1D ↑, 1B ↓ 1F ↑↓[a]	+	
Filiberto *et al.* (2011)	Healthy pregnant women	480 mother–child dyads	Placenta	At birth	1F ↑[b]	–	
Bromer *et al.* (2013)	Healthy pregnant women	186 mother-child dyads	Placenta	At birth	1F ↑[c]	–	
Conradt *et al.* (2013)	Depressed pregnant women	27 depressed women, 18 anxious women, 39 depressed and anxious women, 389 healthy women	Placenta	At birth	1F ↑[d]	–	*HSD11B2* ↑[d]

Marsit *et al.* (2012)	Healthy pregnant women	186 mother–child dyads	Placenta	At birth			*HSD11B2* ↑[e]
Appleton *et al.* (2015)	Healthy pregnant women	373 mother–child dyads	Placenta	At birth	1F ↑[f]	–	*HSD11B2* ↑[f]
Paquette *et al.* (2014)	Healthy pregnant women	509 mother–child dyads	Placenta	At birth			*FKBP5* ↑[g]

NR3C1, glucocorticoid receptor; NR3C2, mineralocorticoid receptor, FKBP5, FK506 binding protein; HSD11B2, 11-β-hydroxysteroid dehydrogenase type 2, ↑↓, increase or decrease in CpG methylation, ss, site-specifically; +/− methylation changes at EGR1 DNA-binding site. References refer to those in the main text.

[a] Site-specific changes in1D, 1B, and 1F CpG methylation.

[b] Higher exon 1F and CpG7 methylation in newborns small for gestational age.

[c] Increased methylation of CpG7 associates with lower stress abstinence scores and increased infant habituation, whereas increased methylation of CpG sites 6, 7 and 8 associates with infant quality of movement scores.

[d] Depression and anxiety during pregnancy lead to higher placental *NR3C1* and *HSD11B2* methylation, respectively.

[e] *HSD11B2* methylation associates inversely with birth weight and neurobehavioral outcome.

[f] Low *NR3C1* and *HSD11B2* methylation associate with lower excitability scores, whereas high methylation with more asymmetrical reflexes.

[g] Increased *FKBP5* methylation associates with high arousal.

and traumatic stress during pregnancy can lead to epigenetic programming of *NR3C1*, particularly at the exon 1F promoter, in different tissues such as placenta, cord blood and newborn and adolescent PBMCs. Epigenetic marking of *NR3C1* can associate with lasting changes in stress reactivity and neurobehavior outcomes in offspring (Table 9.1). The role of heightened maternal cortisol in driving epigenetic programing of fetal *NR3C1* is still unresolved and may depend on the severity of stress exposure (mundane versus toxic stress).

Interestingly, fetal methylation changes at *NR3C1* seem to relate to maternal states and occur at selective CpG residues rather than a mean increase in exon 1F promoter methylation supporting specific responses in epigenetic programming (Table 9.1). A limitation for many of these studies remains the limited number of participants and the lack of comprehensive follow-up studies in offspring. Ideally, *NR3C1* methylation changes should be assessed in cord blood, PBMCs of mother-child dyads (postnatally and thereafter), correlated with respective GR expression changes, psychological and neurobehavioral tests, and HPA-axis reactivity.

In this context, it is important to note that integrated analysis of ELA in mice evidenced that epigenetic programming depends on the quality of the initial stressor, is confined to specific neuronal cell populations, and underpins context-dependent functional outcomes (Bockmühl *et al.* 2015, Murgatroyd *et al.* 2009, Wu *et al.* 2014). Therefore, integrated analysis of glucocorticoid signaling in various tissues from infant–mother dyads may advance our insight into the dynamics and long-term course of fetal programing (Table 9.1). GR transactivates *HSD11B2* and *FKBP5*, both of which curtail GR signaling, and all of these genes seem to be subject to epigenetic programing. All in all, integrated epigenetic profiles from offspring at risk may provide valuable epigenetic biomarkers to guide timely therapeutic interventions to ameliorate or even prevent, the development of future psychiatric diseases.

References

Aberg, K. A., Xie, L. Y., McClay, J. L., Nerella, S., Vunck, S., Snider, S., Beardsley, P. M. and van den Oord, E. J. (2013). Testing two models describing how methylome-wide studies in blood are informative for psychiatric conditions. *Epigenomics* 5, 367–377.

Allis, C. D., Caparros, M.-L., Jenuwein, T. and Reinberg, D. (2015). Epigenetics (Cold Spring Harbor NY: Cold Spring Harbor Laboratory Press).

Andrade, S. E., Raebel, M. A., Brown, J., Lane, K., Livingston, J., Boudreau, D., Rolnick, S. J., Roblin, D., Smith, D. H., Willy, M. E. *et al.* (2008). Use of antidepressant medications during pregnancy: a multisite study. *American Journal of Obstetrics and Gynecology* 198, 194.e1–e194.e5.

Appleton, A. A., Lester, B. M., Armstrong, D. A., Lesseur, C. and Marsit, C. J. (2015). Examining the joint contribution of placental NR3C1 and HSD11B2 methylation for infant neurobehavior. *Psychoneuroendocrinology* 52, 32–42.

Armstrong, L. (2013). *Epigenetics*. Garland Science Custom Publishing.

Auclair, G., Guibert, S., Bender, A. and Weber, M. (2014). Ontogeny of CpG island methylation and specificity of DNMT3 methyltransferases during embryonic development in the mouse. *Genome Biology* 15, 545.

Barker, D. J. and Clark, P. M. (1997). Fetal undernutrition and disease in later life. *Reviews of Reproduction* 2, 105.

Bateson, P., Gluckman, P. and Hanson, M. (2014). The biology of developmental plasticity and the predictive adaptive response hypothesis. *Journal of Physiology* 592, 2357–2368.

Bennett, H. A., Einarson, A., Taddio, A., Koren, G. and Einarson, T. R. (2004). Prevalence of depression during pregnancy: systematic review. *Obstetrics & Gynecology* 103, 698–709.

Beydoun, H. and Saftlas, A. F. (2008). Physical and mental health outcomes of prenatal maternal stress in human and animal studies: a review of recent evidence. *Paediatric and Perinatol Epidemiology* 22, 438–466.

Bockmühl, Y., Murgatroyd, C. A., Kuczynska, A., Adcock, I. M., Almeida, O. F. X. and Spengler, D. (2011). Differential regulation and function of 5′-untranslated GR-Exon 1 transcripts. *Journal of Molecular Endocrinology* 25, 1100–1110.

Bromer, C., Marsit, C. J., Armstrong, D. A., Padbury, J. F. and Lester, B. (2013). Genetic and epigenetic variation of the glucocorticoid receptor (NR3C1) in placenta and infant neurobehavior. *Developmental Psychobiology* 55, 673–683.

Cadet, R., Pradier, P., Dalle, M. and Delost, P. (1986). Effects of prenatal maternal stress on the pituitary adrenocortical reactivity in guinea-pig pups. *Journal of Developmental Physiology* 8, 467–475.

Cairns, B. R. (2007). Chromatin remodeling: insights and intrigue from single-molecule studies. *Nature Structural & Molecular Biology* 14, 989–996.

Charmandari, E., Achermann, J. C., Carel, J.-C., Soder, O. and Chrousos, G. P. (2012). Stress response and child health. *Science Signaling* 5, mr1.

Chrousos, G. P. (2009). Stress and disorders of the stress system. *Nature Reviews Endocrinology* 5, 374–381.

Conradt, E., Lester, B. M., Appleton, A. A., Armstrong, D. A. and Marsit, C. J. (2013). The roles of DNA methylation of NR3C1 and 11β-HSD2 and exposure to maternal mood disorder in utero on newborn neurobehavior. *Epigenetics* 8, 1321–1329.

Cougle, J. R., Timpano, K. R., Sachs-Ericsson, N., Keough, M. E. and Riccardi, C. J. (2010). Examining the unique relationships between anxiety disorders and childhood physical and sexual abuse in the National Comorbidity Survey-Replication. *Psychiatry Research*. 177, 150–155.

Dean, F. and Matthews, S. G. (1999). Maternal dexamethasone treatment in late gestation alters glucocorticoid and mineralocorticoid receptor mRNA in the fetal guinea pig brain. *Brain Research* 846, 253–259.

Eccleston, A., DeWitt, N., Gunter, C., Marte, B. and Nath, B. (2007). Nature insight: epigenetics. *Nature* 396–440.

Filiberto, A. C., Maccani, M. A., Koestler, D. C., Wilhelm-Benartzi, C., Avissar-Whiting, M., Banister, C. E., Gagne, L. A. and Marsit, C. J. (2011). Birthweight is associated with DNA promoter methylation of the glucocorticoid receptor in human placenta. *Epigenetics* 6, 566–572.

Fink, G. (2007). *Encyclopedia of Stress*. Amsterdam, Boston, Academic Press.

Fournier, A., Sasai, N., Nakao, M. and Defossez, P.-A. (2012). The role of methyl-binding proteins in chromatin organization and epigenome maintenance. *Briefings in Functional Genomics* 11, 251–264.

Gaynes, B. N., Gavin, N., Meltzer-Brody, S., Lohr, K. N., Swinson, T., Gartlehner, G., Brody, S. and Miller, W. C. (2005). Perinatal depression: prevalence, screening accuracy, and screening outcomes. *Evidence Report/Technology Assessment* 1–8.

Gerardin, P., Wendland, J., Bodeau, N., Galin, A., Bialobos, S., Tordjman, S., Mazet, P., Darbois, Y., Nizard, J., Dommergues, M. *et al.* (2011). Depression during pregnancy: is the developmental impact earlier in boys? A prospective case-control study. *Journal of Clinical Psychiatry* 72, 378–387.

Gibney, E. R. and Nolan, C. M. (2010). Epigenetics and gene expression. *Heredity* 105, 4–13.

Gitau, R., Cameron, A., Fisk, N. M. and Glover, V. (1998). Fetal exposure to maternal cortisol. *Lancet* 352, 707–708.

Gluckman, P. D., Hanson, M. A., Cooper, C. and Thornburg, K. L. (2008). Effect of in utero and early-life conditions on adult health and disease. *New England Journal of Medicine* 359, 61–73.

Gluckman, P. D., Hanson, M. A., Bateson, P., Beedle, A. S., Law, C. M., Bhutta, Z. A., Anokhin, K. V., Bougnères, P., Chandak, G. R., Dasgupta, P. *et al.* (2009). Towards a new developmental synthesis: adaptive developmental plasticity and human disease. *The Lancet* 373, 1654–1657.

Goldberg, A. D., Allis, C. D. and Bernstein, E. (2007). Epigenetics: a landscape takes shape. *Cell* 128, 635–638.

Green, J. G., McLaughlin, K. A., Berglund, P. A., Gruber, M. J., Sampson, N. A., Zaslavsky, A. M. and Kessler, R. C. (2010). Childhood adversities and adult psychiatric disorders in the national comorbidity survey replication I: associations with first onset of DSM-IV disorders. *Archives of General Psychiatry* 67, 113–123.

Hay, D. F., Pawlby, S., Waters, C. S., Perra, O. and Sharp, D. (2010). Mothers' antenatal depression and their children's antisocial outcomes. *Child Development*. 81, 149–165.

He, Y.-F., Li, B.-Z., Li, Z., Liu, P., Wang, Y., Tang, Q., Ding, J., Jia, Y., Chen, Z., Li, L. *et al.* (2011). Tet-mediated formation of 5-carboxylcytosine and its excision by TDG in mammalian DNA. *Science* 333, 1303–1307.

Hegarty, K., Fracgp, Bush, R. and Sheehan, M. (2005). The composite abuse scale: further development and assessment of reliability and validity of a multidimensional partner abuse measure in clinical settings. *Violence and Victims* 20, 529–547.

Heim, C. and Binder, E. B. (2012). Current research trends in early life stress and depression: review of human studies on sensitive periods, gene-environment interactions, and epigenetics. *Experimental Neurology* 233, 102–111.

Hoffmann, A. and Spengler, D. (2014). DNA memories of early social life. *Neuroscience* 264, 64–75.

Holliday, R. (1979). A new theory of carcinogenesis. *British Journal of Cancer* 40, 513–522.

Hompes, T., Izzi, B., Gellens, E., Morreels, M., Fieuws, S., Pexsters, A., Schops, G., Dom, M., Van Bree, R., Freson, K. *et al.* (2013). Investigating the influence of maternal cortisol and emotional state during pregnancy on the DNA methylation status of the glucocorticoid receptor gene (NR3C1) promoter region in cord blood. *Journal of Psychiatric Research* 47, 880–891.

Hunter, R. G. and McEwen, B. S. (2013). Stress and anxiety across the lifespan: structural plasticity and epigenetic regulation. *Epigenomics* 5, 177–194.

Innocenti, G. M. and Price, D. J. (2005). Exuberance in the development of cortical networks. *Nature Reviews Neuroscience* 6, 955–965.

Ito, S., D'Alessio, A. C., Taranova, O. V., Hong, K., Sowers, L. C. and Zhang, Y. (2010). Role of Tet proteins in 5mC to 5hmC conversion, ES-cell self-renewal and inner cell mass specification. *Nature* 466, 1129–1133.

Ito, S., Shen, L., Dai, Q., Wu, S. C., Collins, L. B., Swenberg, J. A., He, C. and Zhang, Y. (2011). Tet proteins can convert 5-methylcytosine to 5-formylcytosine and 5-carboxylcytosine. *Science* 333, 1300–1303.

Jeltsch, A. and Jurkowska, R. Z. (2014). New concepts in DNA methylation. *Trends in Biochemical Sciences* 39, 310–318.

Johnson, T. B. and Coghill, R. D. (1925). Researches on pyrimidines. C111. The discovery of 5-methyl-cytosine in tuberculinic acid, the nucleic acid of the tubercle bacillus1. *Journal of American Chemical Society* 47, 2838–2844.

Kendler, K. S., Karkowski, L. M. and Prescott, C. A. (1999). Causal relationship between stressful life events and the onset of major depression. *The American Journal of Psychiatry* 156, 837–841.

Kessler, R. C., Davis, C. G. and Kendler, K. S. (1997). Childhood adversity and adult psychiatric disorder in the US National Comorbidity Survey. *Psychological Medicine* 27, 1101–1119.

Klengel, T. and Binder, E. B. (2015). FKBP5 allele-specific epigenetic modification in gene by environment interaction. *Neuropsychopharmacology: Official Publication of the American College of Neuropsychopharmacology* 40, 244–246.

De Kloet, E. R., Joëls, M. and Holsboer, F. (2005). Stress and the brain: from adaptation to disease. *Nature Reviews Neuroscience* 6, 463–475.

Kouzarides, T. (2007). Chromatin modifications and their function. *Cell* 128, 693–705.

Krug, E. G. Mercy, J. A., Dahlberg, L. L. and Zwi, A. B. (2002). The world report on violence and health. *Lancet* 360, 1083–1088.

Lang, A. J., Aarons, G. A., Gearity, J., Laffaye, C., Satz, L., Dresselhaus, T. R. and Stein, M. B. (2008). Direct and indirect links between childhood maltreatment, posttraumatic stress disorder, and women's health. *Behavioral Medicine (Washington, D. C.)* 33, 125–135.

Levine, A., Zagoory-Sharon, O., Feldman, R., Lewis, J. G. and Weller, A. (2007). Measuring cortisol in human psychobiological studies. *Physiology Behavior* 90, 43–53.

Li, B., Carey, M. and Workman, J. L. (2007). The role of chromatin during transcription. *Cell* 128, 707–719.

Lister, R., Pelizzola, M., Dowen, R. H., Hawkins, R. D., Hon, G., Tonti-Filippini, J., Nery, J. R., Lee, L., Ye, Z., Ngo, Q.-M. *et al.* (2009). Human DNA methylomes at base resolution show widespread epigenomic differences. *Nature* 462, 315–322.

Lister, R., Mukamel, E. A., Nery, J. R., Urich, M., Puddifoot, C. A., Johnson, N. D., Lucero, J., Huang, Y., Dwork, A. J., Schultz, M. D. *et al.* (2013). Global epigenomic reconfiguration during mammalian brain development. *Science* 341, 1237905.

Liu, J., Bann, C., Lester, B., Tronick, E., Das, A., Lagasse, L., Bauer, C., Shankaran, S. and Bada, H. (2010). Neonatal neurobehavior predicts medical and behavioral outcome. *Pediatrics* 125, e90–e98.

Lupien, S. J., McEwen, B. S., Gunnar, M. R. and Heim, C. (2009). Effects of stress throughout the lifespan on the brain, behaviour and cognition. *Nature Reviews Neuroscience* 10, 434–445.

Mairesse, J., Lesage, J., Breton, C., Bréant, B., Hahn, T., Darnaudéry, M., Dickson, S. L., Seckl, J., Blondeau, B., Vieau, D. *et al.* (2007). Maternal stress alters endocrine function of the feto-placental unit in rats. *American Journal of Physiology-Endocrinology and Metabalism* 292, E1526–E1533.

Maiti, A. and Drohat, A. C. (2011). Thymine DNA glycosylase can rapidly excise 5-formylcytosine and 5-carboxylcytosine: potential implications for active demethylation of CpG sites. *Journal of Biological Chemistry* 286, 35334–35338.

Marsit, C. J., Maccani, M. A., Padbury, J. F. and Lester, B. M. (2012). Placental 11-beta hydroxysteroid dehydrogenase methylation is associated with newborn growth and a measure of neurobehavioral outcome. *PloS One* 7, e33794.

Maunakea, A. K., Nagarajan, R. P., Bilenky, M., Ballinger, T. J., D'Souza, C., Fouse, S. D., Johnson, B. E., Hong, C., Nielsen, C., Zhao, Y. *et al.* (2010). Conserved role of intragenic DNA methylation in regulating alternative promoters. *Nature* 466, 253–257.

McEwen, B. S. (2007). Physiology and neurobiology of stress and adaptation: central role of the *Brain Physiological Review* 87, 873–904.

McLaughlin, K. A., Green, J. G., Gruber, M. J., Sampson, N. A., Zaslavsky, A. M. and Kessler, R. C. (2010). Childhood adversities and adult psychiatric disorders in the national comorbidity survey replication II: associations with persistence of DSM-IV disorders. *Archives of General Psychiatry* 67, 124–132.

Misteli, T. (2007). Beyond the sequence: cellular organization of genome function. *Cell* 128, 787–800.

Moses-Kolko, E. L., Bogen, D., Perel, J., Bregar, A., Uhl, K., Levin, B. and Wisner, K. L. (2005). Neonatal signs after late in utero exposure to serotonin reuptake inhibitors: literature review and implications for clinical applications. *Journal of American Medical Association* 293, 2372–2383.

Mulligan, C., D'Errico, N., Stees, J. and Hughes, D. (2012). Methylation changes at NR3C1 in newborns associate with maternal prenatal stress exposure and newborn birth weight. *Epigenetics* 7, 853–857.

Munyandamutsa, N., Nkubamugisha, P. M., Gex-Fabry, M. and Eytan, A. (2012). Mental and physical health in Rwanda 14 years after the genocide. *Social Psychiatry and Psychiatric Epidemiology* 47, 1753–1761.

Murgatroyd, C., Wu, Y., Bockmühl, Y. and Spengler, D. (2010). Genes learn from stress: How infantile trauma programs us for depression. *Epigenetics: Official Journal of the DNA Methylation Society* 5, 194–199.

Murray, C. J. and Lopez, A. D. (1996). Evidence-based health policy — lessons from the Global Burden of Disease Study. *Science* 274, 740–743.

Jones, N., Field, T., Fox, N., Davalos, M., Lundy, B. L. and Hart, S. (1998). Newborns of mothers with depressive symptoms are physiologically less developed. *Infant Behavior and Development* 21, 537–541.

NCANDS (2013). Child Maltreatment 2012. 370 L'Enfant Promenade, S.W. Washington, D.C. 20447, US Department of Health and Human Services.

Oberlander, T. F., Warburton, W., Misri, S., Aghajanian, J. and Hertzman, C. (2006). Neonatal outcomes after prenatal exposure to selective serotonin reuptake inhibitor antidepressants and maternal depression using population-based linked health data. *Archives of General Psychiatry* 63, 898–906.

Oberlander, T. F., Weinberg, J., Papsdorf, M., Grunau, R., Misri, S. and Devlin, A. M. (2008a). Prenatal exposure to maternal depression, neonatal methylation of human glucocorticoid receptor gene (NR3C1) and infant cortisol stress responses. *Epigenetics* 3, 97–106.

Oberlander, T. F., Warburton, W., Misri, S., Aghajanian, J. and Hertzman, C. (2008b). Effects of timing and duration of gestational exposure to serotonin reuptake inhibitor antidepressants: population-based study. *The British Journal of Psychiatry* 192, 338–343.

O'Connor, T. G., Heron, J., Glover, V. and Alspac Study Team (2002a). Antenatal anxiety predicts child behavioral/emotional problems independently of postnatal depression. *Journal of the American Academy of Child & Adolescent Psychiatry* 41, 1470–1477.

O'Connor, T. G., Heron, J., Golding, J., Beveridge, M. and Glover, V. (2002b). Maternal antenatal anxiety and children's behavioural/emotional problems at 4 years. Report from the Avon Longitudinal Study of Parents and Children. *British Journal of Psychiatry: Journal of Mentol Science* 180, 502–508.

O'Connor, T. G., Heron, J., Golding, J., Glover, V. and ALSPAC Study Team (2003). Maternal antenatal anxiety and behavioural/emotional problems in children: a test of a programming hypothesis. *Journal of Child Psychology Psychiatry* 44, 1025–1036.

O'Donnell, K. J., Bugge Jensen, A., Freeman, L., Khalife, N., O'Connor, T. G. and Glover, V. (2012). Maternal prenatal anxiety and downregulation of placental 11β-HSD2. *Psychoneuroendocrinology* 37, 818–826.

O'Donnell, K. J., Glover, V., Barker, E. D. and O'Connor, T. G. (2014). The persisting effect of maternal mood in pregnancy on childhood psychopathology. *Development and Psychopathology* 26, 393–403.

Paquette, A. G., Lester, B. M., Koestler, D. C., Lesseur, C., Armstrong, D. A. and Marsit, C. J. (2014). Placental FKBP5 genetic and epigenetic variation is associated with infant neurobehavioral outcomes in the RICHS cohort. *PloS One* 9, e104913.

Patchev, A. V., Rodrigues, A. J., Sousa, N., Spengler, D. and Almeida, O. F. X. (2014). The future is now: early life events preset adult behaviour. *Acta Physiologica* 210, 46–57.

Perroud, N., Rutembesa, E., Paoloni-Giacobino, A., Mutabaruka, J., Mutesa, L., Stenz, L., Malafosse, A. and Karege, F. (2014). The Tutsi genocide and transgenerational transmission of maternal stress: epigenetics and biology of the HPA axis. *World Journal of Biological, Psychiatry* 15, 334–345.

Raabe, F. J. and Spengler, D. (2013). Epigenetic risk factors in PTSD and depression. *Frontiers in Psychiatry* 4, 80.

Radtke, K. M., Ruf, M., Gunter, H. M., Dohrmann, K., Schauer, M., Meyer, A. and Elbert, T. (2011). Transgenerational impact of intimate partner violence on methylation in the promoter of the glucocorticoid receptor. *Translational Psychiatry* 1, e21.

Reiss, A. L., Abrams, M. T., Singer, H. S., Ross, J. L. and Denckla, M. B. (1996). Brain development, gender and IQ in children. A volumetric imaging study. *Brain A Journal of Neurology* 119(5), 1763–1774.

Riggs, A. D. (1975). X inactivation, differentiation, and DNA methylation. *Cytogenetics and Cell Genetics* 14, 9–25.

Ross, L. E. and McLean, L. M. (2006). Anxiety disorders during pregnancy and the postpartum period: A systematic review. [CME]. *Journal of Clinical Psychiatry* 67, 1285–1298.

Ross, L. E., Grigoriadis, S., Mamisashvili, L., Vonderporten, E. H., Roerecke, M., Rehm, J., Dennis, C.-L., Koren, G., Steiner, M., Mousmanis, P. *et al.* (2013). Selected pregnancy and delivery outcomes after exposure to antidepressant medication: a systematic review and meta-analysis. *Journal of American Medical Assocation Psychiatry* 70, 436–443.

Russo, V. E. A., Martienssen, R. A. and Riggs, A. D. (1996). *Epigenetic Mechanisms of Gene Regulation* Plainview NY., Cold Spring Harbor Laboratory Press.

Sameroff, A. (2010). A unified theory of development: a dialectic integration of nature and nurture. *Child Development* 81, 6–22.

Saxonov, S., Berg, P. and Brutlag, D. L. (2006). A genome-wide analysis of CpG dinucleotides in the human genome distinguishes two distinct classes of promoters. *Proceedings of the National Academy of Sciences USA* 103, 1412–1417.

Schübeler, D. (2015). Function and information content of DNA methylation. *Nature* 517, 321–326.

Seckl, J. R. (2008). Glucocorticoids, developmental "programming" and the risk of affective dysfunction. *Progress in Brain Research* 167, 17–34.

Shonkoff, J. P. (2003). From neurons to neighborhoods: old and new challenges for developmental and behavioral pediatrics. *Journal of Developmental and Behavioral Pediatrics* 24, 70–76.

Silberg, J. L., Maes, H. and Eaves, L. J. (2010). Genetic and environmental influences on the transmission of parental depression to children's depression and conduct disturbance: an extended children of twins study. *Journal of Child Psychology and Psychiatry* 51, 734–744.

Spruijt, C. G., Gnerlich, F., Smits, A. H., Pfaffeneder, T., Jansen, P. W. T. C., Bauer, C., Münzel, M., Wagner, M., Müller, M., Khan, F. *et al.* (2013). Dynamic readers for 5-(hydroxy)methylcytosine and its oxidized derivatives. *Cell* 152, 1146–1159.

Steiger, H., Labonté, B., Groleau, P., Turecki, G. and Israel, M. (2013). Methylation of the glucocorticoid receptor gene promoter in bulimic women: Associations with borderline personality disorder, suicidality, and exposure to childhood abuse. *International Journal of Eating Disorders* 46, 246–255.

Stiles, J. and Jernigan, T. L. (2010). The basics of brain development. *Neuropsychology Reviews* 20, 327–348.

Storer, C. L., Dickey, C. A., Galigniana, M. D., Rein, T. and Cox, M. B. (2011). FKBP51 and FKBP52 in signaling and disease. *Trends in Endocrinology & Metabolism TEM* 22, 481–490.

Szyf, M. and Bick, J. (2013). DNA methylation: A mechanism for embedding early life experiences in the genome. *Child Development* 84, 49–57.

Tahiliani, M., Koh, K. P., Shen, Y., Pastor, W. A., Bandukwala, H., Brudno, Y., Agarwal, S., Iyer, L. M., Liu, D. R., Aravind, L. *et al.* (2009). Conversion of 5-methylcytosine to 5-hydroxymethylcytosine in mammalian DNA by MLL partner TET1. *Science* 324, 930–935.

Teicher, M. H., Andersen, S. L., Polcari, A., Anderson, C. M. and Navalta, C. P. (2002). Developmental neurobiology of childhood stress and trauma. *Psychiatric Clinics of North America* 25, 397–426, vii–viii.

Tollefsbol, T. (2011). *Handbook of Epigenetics*: Amsterdam, Elsevier.

Turner, J. D. and Muller, C. P. (2005). Structure of the glucocorticoid receptor (NR3C1) gene 5′ untranslated region: identification, and tissue distribution of multiple new human exon 1. *Journal of Molecular Endocrinology* 35, 283–292.

Uno, H., Tarara, R., Else, J. G., Suleman, M. A. and Sapolsky, R. M. (1989). Hippocampal damage associated with prolonged and fatal stress in primates. *Journal of Neuroscience Official Journal Society Neuroscience* 9, 1705–1711.

Uno, H., Lohmiller, L., Thieme, C., Kemnitz, J. W., Engle, M. J., Roecker, E. B. and Farrell, P. M. (1990). Brain damage induced by prenatal exposure to dexamethasone in fetal rhesus macaques. I. Hippocampus. *Brain Research Developmental Brain Research* 53, 157–167.

Waddington C. H. (2014). *The Strategy of the Genes*. 7, New York, NY, Routledge.

Wadhwa, P. D., Culhane, J. F., Rauh, V., Barve, S. S., Hogan, V., Sandman, C. A., Hobel, C. J., Chicz-DeMet, A., Dunkel-Schetter, C., Garite, T. J. *et al.* (2001). Stress, infection and preterm birth: a biobehavioural perspective. *Paediatric and Perinatal Epidemiology* 15, 17–29.

Weaver, I. C. G., Champagne, F. A., Brown, S. E., Dymov, S., Sharma, S., Meaney, M. J. and Szyf, M. (2005). Reversal of maternal programming of stress responses in adult offspring through methyl supplementation: altering epigenetic marking later in life. *Journal of Neuroscience* 25, 11045–11054.

Weaver, I. C., Cervoni N., Champagne, F. A., D'Alessio, A. C., Sharma S., Seckl, J. R., Dymov, S., Szyf, M. and Meaney, M. J (2004). Epigenetic programming by maternal behavior. *Nature Neuroscience* 7, 847–854.

Weber, M., Hellmann, I., Stadler, M. B., Ramos, L., Pääbo, S., Rebhan, M. and Schübeler, D. (2007). Distribution, silencing potential and evolutionary impact of promoter DNA methylation in the human genome. *Nature Genetics* 39, 457–466.

Welberg, L. A. M., Thrivikraman, K. V. and Plotsky, P. M. (2005). Chronic maternal stress inhibits the capacity to up-regulate placental 11beta-hydroxysteroid dehydrogenase type 2 activity. *Journal of Endocrinology* 186, R7–R12.

Wittchen, H. U., Jacobi, F., Rehm, J., Gustavsson, A., Svensson, M., Jönsson, B., Olesen, J., Allgulander, C., Alonso, J., Faravelli, C. *et al.* (2011). The size and burden of mental disorders and other disorders of the brain in Europe 2010. *European Neuropsychopharmacology* 21, 655–679.

Wu, H. and Zhang, Y. (2014). Reversing DNA methylation: mechanisms, genomics, and biological functions. *Cell* 156, 45–68.

Wyatt, G. R. (1950). Occurrence of 5-methyl-cytosine in nucleic acids. *Nature* 166, 237–238.

Xie, W., Barr, C. L., Kim, A., Yue, F., Lee, A. Y., Eubanks, J., Dempster, E. L. and Ren, B. (2012). Base-resolution analyses of sequence and parent-of-origin dependent DNA methylation in the mouse genome. *Cell* 148, 816–831.

Yehuda, R. and Bierer, L. M. (2008). Transgenerational transmission of cortisol and PTSD risk. *Progress in Brain Research* 167, 121–135.

Yehuda, R., Teicher, M. H., Seckl, J. R., Grossman, R. A., Morris, A. and Bierer, L. M. (2007). Parental posttraumatic stress disorder as a vulnerability factor for low cortisol trait in offspring of holocaust survivors. *Archives of General Psychiatry* 64, 1040–1048.

Epigenetic Consequences of Human Assisted Reproductive Technologies

10

Debra L. Fulton[*], *Seang Lin Tan*[†] *and Teruko Taketo*[*,†,‡]

[*]*Department of Surgery, McGill University Montreal, Quebec, Canada*
[†]*Department of Obstetrics and Gynecology, McGill University Montreal, Quebec, Canada*
[‡]*Department of Biology, McGill University, Montreal, Quebec, Canada*

One in six couples is diagnosed with infertility. Many infertile couples are turning to assisted reproductive technologies (ART) to facilitate conception and child bearing. While the majority of ART offspring are healthy, there is a concern that ART interventions during major epigenomic reprogramming in oocytes and embryos can have harmful effects on the health of ART-conceived children. This concern has spawned a multitude of ART-related epigenetic-based studies in animals and human. While studies of animal models suggest that there is an association between ART interventions and epigenetic aberrations, these findings may not be fully applicable to humans. This chapter reviews current findings for ART-affected epigenomic perturbations based on human-centric studies.

10.1 Introduction

Infertility is generally defined as the persistent inability to conceive, using natural methods, for over a year and it afflicts over 16% of couples (Boivin *et al.* 2007). Many infertile couples are now turning to assisted reproductive technologies (ART) to enable conception and child bearing. The term ART refers to the procedures used to achieve pregnancy through clinical interventions. It is estimated that over five million babies have been born to date using ART although the use of ART approaches are adopted in only a portion of infertility cases (Adamson *et al.* 2013).

ART manipulations can be employed at several stages leading up to conception. Such procedures involve one or more of the following methods: controlled ovarian hyperstimulation (COH) to allow the development of multiple ovarian follicles (sometimes referred to as superovulation), intrauterine insemination (IUI), collection of oocytes from follicles, *in vitro* maturation (IVM) of oocytes, *in vitro* fertilization (IVF) of oocytes with sperm or intracytoplasmic sperm injection (ICSI), culturing and monitoring of embryos, selection of pre-implantation embryos for live transfer and/or cryopreservation, and transfer of one or more embryos to a recipient (Figure 10.1).

Most ART offspring are healthy and their cognitive abilities and behavioral characteristics are no different than that of naturally conceived children. However, in lieu of ART-related invasive manipulations and, in particular, the temporal overlap of ART procedures during essential genomic reprogramming events, there is a concern that the health of ART-conceived children can be affected through environmentally-induced epigenetic alterations. The term *epigenetics* encompasses heritable genomic modifications that do not alter DNA sequence. Epigenetic modifications include: (i) the addition of methyl groups to DNA, (ii) post-translational modifications of histone tails such as methylation, acetylation, phosphorylation, ubiquitination, and sumoylation and (iii) changes in chromatin and nuclear architecture (Bernstein *et al.* 2007, Daxinger and Whitelaw 2012, Schneider and Grosschedl 2007). Epigenetic mechanisms direct temporally- and spatially-specific gene expression by modulating physical access of regulatory molecules (e.g. transcription factors, microRNAs) to their genetic targets.

DNA methylation and histone posttranslational modifications are well-studied in the context of reproductive biology. For a zygote to achieve totipotency, germ cells and embryos must undergo extensive epigenetic reprogramming (Morgan *et al.* 2005). Specifically, epigenetic reprogramming occurs at two stages of development: during germ cell differentiation and during the early stages of embryonic development (reviewed in (Sasaki and Matsui 2008)). At each stage, DNA methylation marks are largely removed and histone modifications are altered and then a new methylation pattern is established. Primordial germ cells (PGCs) originate in the epiblast and migrate to the gonads. In lieu of their somatic embryonic cell origin, PGCs must undergo significant remodeling to facilitate their germ cell fate. DNA methylation patterns, including parental imprints, are removed along with the reactivation of the inactive X chromosome to facilitate *de novo* DNA methylation during germ cell development. After gonadal and then germ cell sex determination, *de novo* methylation and gene imprinting take place differentially in

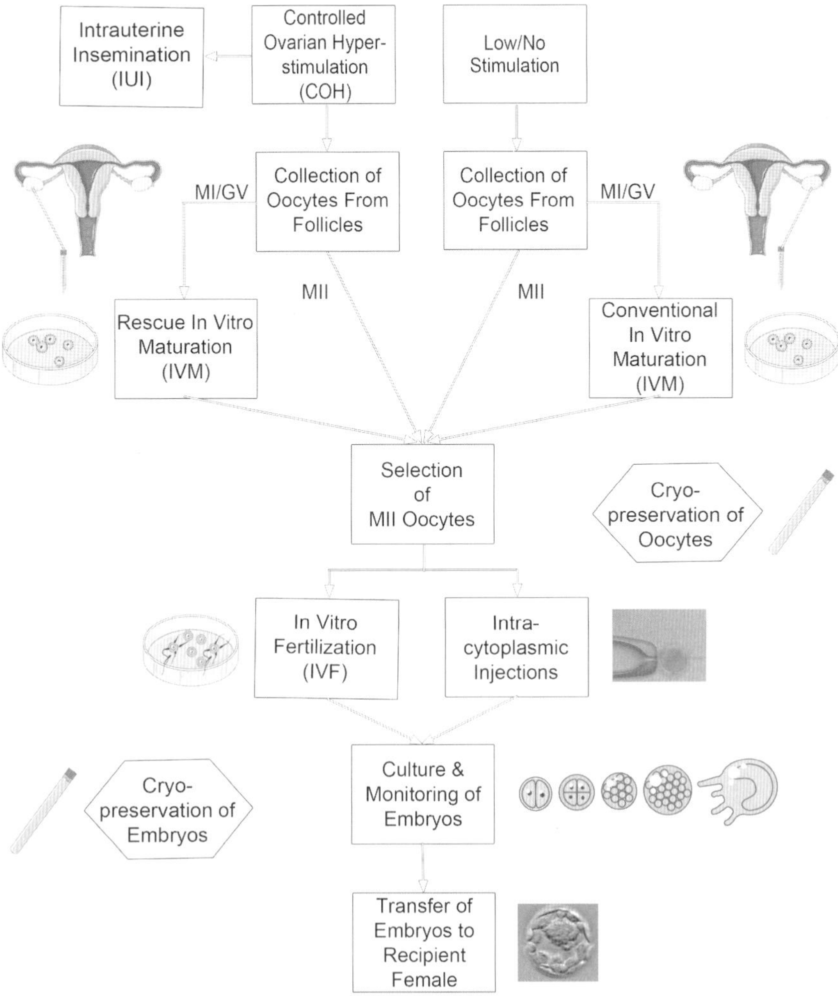

Figure 10.1. Schematic of ART steps and technologies that may be employed during a fertility cycle.[1]

male and female germ cells: starting from the prenatal period in mitotically arrested prospermatogonia and during the postnatal growth phase of oocytes

[1]This figure was produced, in part, by using Servier Medical Art. The ICSI image is courtesy of H.-Y Chen, Originelle Fertility Clinic & Women's Health Centre. The image of the day 5 human blastocyst was kindly provided by Dr. C. Racowsky, ART Laboratory — Brigham & Women's Hospital.

arrested in meiotic prophase I (Davis *et al.* 2000, La Salle *et al.* 2004, Sasaki and Matsui 2008, Smith and Meissner 2013). Another phase of epigenetic reprogramming occurs during early embryo development following fertilization. An overall DNA demethylation occurs during preimplantation development, with the exception of imprinted genes and certain repetitive sequences that persists until the blastocyst stage. At implantation, genome-wide *de novo* methylation begins in a lineage specific manner to regulate the expression of early embryonic genes, cell cleavage, and cell differentiation (Howlett and Reik 1991, Rossant *et al.* 1986, Smallwood *et al.* 2011, Smith and Meissner 2013, Tomizawa *et al.* 2011).

With the number of ART-conceived children on the rise, focused efforts are being applied to understand what effects ART may have on both the short- and long-term health status of offspring. Results from several animal model and human-based studies both support and refute the hypothesis that ART manipulations cause epigenomic perturbations in children conceived using *in vitro* methods. The difficulty in obtaining human embryos for research and ethical issues associated with applying interventions during embryo development and gestation significantly limits the scope of human-based studies. Consequently, the majority of human-based studies involve retrospective analyses of clinical data, which often limits analysis sample sizes. While studies of animal models suggest a link between epigenetic perturbations and ART manipulations, their findings may not be directly applicable to and/or are difficult to interpret in a human context. Importantly, the extrapolation of outcomes from animal-based studies must be considered carefully as we await the collection and analysis of much needed human-based longitudinal studies in this area. Here we will focus on the current findings regarding ART-affected epigenomic perturbations, based on human-centric studies. The first section will focus on the health risks and epigenetic consequences in children following ART treatment and the second section will outline the findings of epigenetic studies in relation to specific ART interventions.

10.2 Adverse Health Consequences and Epigenetic Disturbances Following ART

10.2.1 *Health outcomes*

Retrospective analyses of children born using ART suggest that there is an increased risk for birth defects (Hansen *et al.* 2013) and low birth weight (Schieve *et al.* 2002), preterm delivery and complications after birth (Wisborg *et al.* 2010). Notably, there is both evidence for and against a more general association between epigenetic modifications and birth weight outcomes in the human population (Engel *et al.*

2014, Ferreira *et al.* 2011, Filiberto *et al.* 2011, Haworth *et al.* 2014, Koukoura *et al.* 2011, Murphy *et al.* 2014, Rangel *et al.* 2014, Tan *et al.* 2014, Tobi *et al.* 2011) and epigenetic perturbations in preterm births (reviewed in (Parets *et al.* 2014)). One potential factor that may influence unfavorable outcomes for ART-conceived children is whether one or multiple embryos were transferred. Indeed, the rate of premature births and low birth weights, along with other maternal and pediatric complications, is substantially reduced in single embryo transfer cases (Kjellberg *et al.* 2006, Tan *et al.* 1992). However, a large Danish cohort study comprising of 13,692 pairs of singleton siblings with one conceived by IVF, ICSI or frozen embryo transfer and the other by natural conception found that ART-conceived children have a lower birth weight in general and IVF/ICSI-conceived children have a higher rate of preterm birth (Henningsen *et al.* 2011). Interestingly, a recent study found that, while hospital admission rates of ART-conceived children before nine months of age are higher than naturally conceived babies, the frequency of visits is comparable after nine months of age (Sutcliffe *et al.* 2014). The authors speculate that higher rates of hospital admission during the first nine months for ART-conceived babies could in part be due to parents who, after having undergone ART/sub-fertility interventions, may feel overly anxious about the health status of their babies such that their visits to the hospital are excessively increased.

10.2.2 *ART and genomic imprinting*

There is elevated albeit low risk of the occurrence of imprinting defects (absolute incidence <1:12000 births) in ART-conceived children (Manipalviratn *et al.* 2009), including Angelman (Ludwig *et al.* 2005), Beckwith-Wiedemann (Maher *et al.* 2003) and Silver–Russell (Hiura *et al.* 2012) syndromes (reviewed in (Laprise 2009)). While analyses of DNA methylation patterns in selected differentially methylated regions (DMRs) of aborted fetuses and in newborns after ART procedures did not detect higher levels of epigenetic perturbations (Feng *et al.* 2011, Li *et al.* 2011, Tierling *et al.* 2010), other studies have found epigenetic differences at specific genomic loci (Gomes *et al.* 2009, Shi *et al.* 2011, Turan *et al.* 2010). For example, dysregulated methylation patterns in the gene *H19*, Imprinted Maternally Expressed Transcript (*H19*), were detected in human preimplantation embryos derived using ART (Ibala-Romdhane *et al.* 2011). That said, recent studies question the causal association between the use of ART and imprinting diseases, independent of subfertility (reviewed in (Uyar and Seli 2014)). The disagreement between studies could stem from small sample sizes and temporal assessment differences used to detect methylation. Significantly, methylation aberrations that co-occur with ART can be associated with infertility-related parental factors. For example, aberrant loss of DNA methylation at imprinted loci detected in 6–9 weeks embryos was also identified in the parental sperm (Kobayashi *et al.* 2009). However, parental fertility problems need not affect ART outcomes. For instance, a recent study evaluating the sperm DNA methylation profiles of over 100 infertile men found that, while men with fertility problems have a higher proportion of sperm with aberrant DNA methylation patterns in imprinted regions, these methylation changes do not alter ART outcomes (Camprubi *et al.* 2012).

10.2.3 *ART-affected placental epigenetics*

The placenta couples the mother and fetus to facilitate the exchange of nutrients, gas, and waste by-products and, as such, its proper function is important for healthy fetal development and growth. Placenta dysfunction can lead to harmful consequences such as pre-clampsia, gestational trophoblastic disease and fetal growth retardation. Epigenetic regulation is responsible for controlling the expression of a number of imprinted genes in the placenta, and accordingly, plays an important role in placental development and function (Cotton *et al.* 2009, Kaneko-Ishino and Ishino 2010, Rossant and Cross 2001). Katari *et al.* (2009) assessed the methylation status of greater than 700 genes in a small set of placenta and cord blood samples and identified changes in CpG site methylation levels in ART-conceived children (Katari *et al.* 2009). Further, a recent analysis by Song *et al.* (2015) found that ART procedures were associated with methylation status changes in 16 genes in the placenta of ART-conceived children corroborating their own previous DNA methylation and transcript profiling results (Song *et al.* 2015, Turan *et al.* 2012). In contrast, a large ART cohort study, which included cases of intrauterine growth restriction (IUGR) and premature birth, evaluated the methylation status within the loci of a set of imprinting genes expressed in placenta and cord blood. This study found no increase in epigenetic variability at imprinted DMRs and concluded that IUGR and prematurity were not associated with imprinting defects in ART- and naturally- conceived populations (Camprubi *et al.* 2013). However, in another recent study, a broader assessment of CpG sites using a smaller number of individuals found significant methylation differences between ART-conceived children and controls (Melamed *et al.* 2015). Moreover, the group also detected a tendency towards hypomethylation in ART-conceived children and generally higher variability in genome-wide methylation marks. The disagreement between these studies motivates additional broad scope CpG site analyses using large sample cohorts.

10.3 Evidence for Epigenetic Alterations Associated with ART Manipulations

10.3.1 *Controlled ovarian hyperstimulation*

Controlled ovarian hyperstimulation, also termed simply ovarian hyperstimulation or superovulation, is a technique that is used to induce ovulation from multiple ovarian follicles using gonadotropins. Such treatment could allow oocytes to develop and mature too rapidly or rescue oocytes that would otherwise have been eliminated by atresia and so may be unable to properly establish or maintain genomic imprints during oocyte maturation and embryonic development (Fortier *et al.* 2008). Several birth phenotypes have been linked to superovulation. A recent analysis found that IVF after ovarian stimulation, compared to both no stimulation and natural conception, results in lower birth weight (Pelinck *et al.* 2010). Epigenetic disorders observed in children born after IVF, ICSI and IUI suggest that superovulation may be a cause (Fauque 2013).

DMR DNA methylation begins in growing oocytes found in primary to antral follicles after puberty. As such, hormones administered for ovarian hyperstimulation may perturb the imprinting processes. Accordingly, studies have reported differentially methylated sites in oocytes retrieved from hormonally-stimulated fertility cycles. However, these findings are not consistent. For example, the KCNQ1 Opposite Strand/Antisense Transcript 1 (*KCNQ1OT1*) DMR was more methylated in human GV and M1 oocytes from natural cycles than oocytes from stimulated cycles (Khoueiry *et al.* 2008), although the effects of oocyte culture *per se* could not be excluded (see Culture Parameters section). In contrast, methylation marks in this same DMR were found to be normal in the majority of GV/M1 and MII oocytes after ovarian stimulation (Geuns *et al.* 2007). Similarly, in both mouse and human oocytes, no differences were identified in Maternal Imprinting of Mesoderm Specific Transcript (*MEST*), *KCNQ1OT1* and Pleiomorphic Adenoma Gene-like 1 (*PLAGL1*) DMRs after ovarian hyperstimulation (Sato *et al.* 2007a). However, this study did find DNA methylation on *H19* DMR of superovulated oocytes, which is normally unmethylated in oocytes. In agreement with this result, another study by the same group reported that human oocytes collected after ovarian stimulation had imprinting errors in *MEST* and *H19* genes (Sato *et al.* 2007b). Notably, differing methylation status may be due in part to different ovarian stimulation protocols being employed and/or the variable susceptibilities of the woman that they are administered to. Large systematic studies will be necessary to comprehensively elucidate the effect of ovarian hyperstimulation on epigenetic outcomes in oocytes.

10.3.2 *In vitro maturation*

Since its establishment in 1991 (Cha *et al.* 1991), over 1,400 children have been born after IVM (Chian *et al.* 2014). In general, no major medical issues have been reported in IVM offspring (Basatemur and Sutcliffe 2011, Buckett *et al.* 2007). IVM differs from conventional IVF in that immature oocytes are collected at the GV/metaphase 1 (M1) stage and then matured in culture. Specifically, IVM involves the collection of cumulus cell-oocyte complexes (COCs) from small diameter antral follicles (not greater than 10 mm) in unstimulated (i.e. natural cycle IVF) or low stimulated fertility cycles (i.e. minimal IVF). Natural cycle IVF does not use hormones to increase the production of oocytes and minimal IVF uses low doses of hormones to encourage the growth of a small number of oocytes. With either approach, administration of human chorionic gonadotropin (hCG) may be used to help oocytes complete the maturation process. Since IVM requires less ovary-stimulating hormones before oocyte collection, it is considered a safer approach than conventional IVF in general for women undergoing ART treatment (Chian *et al.* 2004). Although IVM treatments reduce pregnancy success rates in woman with normo-ovulatory cycles (Fadini *et al.* 2013), this milder approach is recommended for woman with polycystic ovarian syndrome (PCOS) given a higher associated risk of developing ovarian hyper stimulation syndrome (OHSS) after conventional IVF, which can result in fatality (Child *et al.* 2001, Fadini *et al.* 2009, Gremeau *et al.* 2012, Junk and Yeap 2012). Notably, one retrospective study found that there are no differences in pregnancy rates for fresh or frozen embryo transfer (FET) in PCOS-afflicted women and, specifically, in live birth rates of PCOS patients using FET-IVF and FET-IVM (Walls *et al.* 2015).

IVM requires lengthy culture of oocytes, exposing them to a substantially different environment than that found endogenously. Gene expression profiling of human oocytes has shown that IVM oocytes, in part, retain an immature gene expression profile despite their MII morphology (Wells and Patrizio 2008). Both human and murine studies have identified down-regulation of HDAC1 histone acetylation epigenetic modifiers in oocytes that have undergone IVM (Huang *et al.* 2012, Wang *et al.* 2010) and deficient histone-3 lysine-9 (H3K9) deacetylation in human IVM oocytes (Wang *et al.* 2010). Moreover, altered methylation patterns have been found in *H19* DMR, which is normally maternally unmethylated, and *KCNQ1OT1* DMR, which is normally maternally methylated (Borghol *et al.* 2006, Khoueiry *et al.* 2008). However, it should be noted that these studies were performed on rescued, low quality oocytes that had not yet matured following ovarian hyperstimulation. A subsequent study from the same group demonstrated a reduction in methylation differences in the same imprinted genes when IVM of a similar pool of rescued oocytes was extended up to 12 hours (Al-Khtib *et al.* 2011). Another study that looked at the methylation status of maternally methylated: *KCNQ1OT1*, small nuclear ribonucleoprotein polypeptide N (*SNRPN*), paternally expressed 3 (*PEG3*) and paternally methylated: Maternally Expressed 3 (*MEG3*) imprinting genes in oocytes retrieved from PCOS patients using conventional IVM (without an hCG trigger) found no increase in DNA methylation in imprinted genes after IVM compared to oocytes retrieved after ovarian hyperstimulation (Kuhtz *et al.* 2014). Following this novel result, a more recent study evaluating the methylation status of numerous genes (see following list) in chorionic villi and cord-blood of children born after IVM also found no statistically significant differences compared to *in vivo* matured oocyte controls collected after conventional ovarian hyperstimulation; gene list: 6 imprinted genes (*KCNQ1OT1*, Protein Tyrosine Phosphatase, Non-receptor Type 4 — *PTPN4*, *MEST*, GNAS Antisense RNA 1 — *GNAS-AS1*), *PEG3* and *SNRPN*), 5 tumor-suppressor genes (Adenomatous Polyposis Coli — *APC*, Adenomatous Polyposis Coli — *ATM*, Breast cancer 1, Early Onset — *BRCA1*, RAD51 Paralog C — *RAD51C* and Tumor Protein p53 — *TP53*), 2 genes required for pluripotency (Nanog Homeobox — *NANOG* and POU Class 5 Homeobox 1 — *POU5F1*) and two metabolic genes (Leptin — *LEP* and Nuclear Receptor Subfamily 3, Group C, Member 1 — *NR3C1*), in addition to two repetitive elements (ALU and Long Interspersed Element 1 — LINE1) (Pliushch *et al.* 2015). Broader scoped genome-wide methylation analyses may help clarify the association between conventional and/or rescued-oocyte IVM protocols and epigenetic mutations.

10.3.3 *In vitro fertilization and intracytoplasmic sperm injection*

Oocytes may be fertilized by combining both eggs and sperm in a dish i.e. IVF or injecting sperm head into an egg i.e. ICSI. ICSI is often the method of choice for cases of low sperm concentration, and motility or morphology issues, as well as frozen donor sperm. In addition, an ICSI approach is more suitable for oocytes that have been subjected to IVM and/or cryopreservation mainly due to hardening of the zona pellucida. An evaluation of height and weight for IVF- and ICSI-conceived children at different ages (birth, 5 years, 7–9 years, and 10–12 years) determined that there were no significant differences compared to naturally conceived controls (Basatemur *et al.* 2010). In concordance with this observation, another recent study comparing human DNA methylation and chromatin architecture of both arrested/delayed embryos and blastocyst embryos after IVF or ICSI reported no significant epigenetic differences as a result of these insemination techniques (Santos *et al.* 2010). The *H19*/Insulin-Like Growth Factor 2 (*IGF2*) and *KCNQ1OT1* DMRs were compared between twins derived using IVF and those conceived naturally, and while differences were noted, they were not considered significant (Li *et al.* 2011). However, another study that quantitatively analyzed the mRNA expression levels of several growth-related imprinted genes: *H19*, *IGF2*, pleckstrin nomology-like domain, Family A, Member 2 (*PHLDA2*), cyclin-dependent kinase inhibitor 1C (*CDKN1C*) and *MEST* in placental samples noted that both *H19* and *PHLDA2* showed significant increases in mRNA expression levels in the IVF/ICSI versus control groups despite no evidence for loss of imprinting (Nelissen *et al.* 2014). The authors acknowledged that the reported differences may not be entirely attributable to the ART technique and may in part be caused by the subfertility of the parents. Similarly, epigenetic assessment of placental tissue from children born from ICSI and IVF found no altered methylation patterns in the genomic sequence for the Prader–Willi/Angelman Syndrome Critical Region (locus q11–13 on chromosome 15 controlled by *SNRPN* ICR) compared to naturally conceived controls for the maternal region (Angelman syndrome) and the paternal region (Prader–Willi syndrome) (Palermo *et al.* 2008). However, gene expression analyses by this group identified down-regulated *H19* and *CDKN1C* genes in the ART-conceived group. This group's expanded genome-wide microarray analyses of the IVF/ICSI conceptuses revealed 54 differentially up-regulated genes by 10-fold or more compared to controls. A study evaluating DNA methylation in buccal cell DNA from children born following IVF and ICSI found higher levels of DNA methylation in the imprinted gene *SNRPN* of ICSI-conceived babies and comparable levels were found in those children born to women with the longest time period of infertility regardless of the conception method (Whitelaw *et al.* 2014). A retrospective meta-analysis of published research concluded that there is no causal relationship between IVF and ICSI and imprinted diseases, and any increased risk for imprinting diseases is likely due to parental fertility problems (Vermeiden and Bernardus 2013). Furthermore, methylation reprogramming of male germ cells is largely established before IVF and ICSI manipulations suggesting that associated dysregulation of methylation detected in IVF/ICSI sperm is more likely due to male infertility factors. Studies including oocytes and sperm from individuals without fertility issues, along with wider epigenetic analyses of CpGs, would help distinguish parental factor contributions from ART-application associated methylation outcomes.

10.3.4 *Culture parameters*

In vitro culture is an integral step of ART protocols, exposing both oocytes and embryos to an environment that is significantly different than the endogenous milieu. A variety of commercial media are available for oocyte collection, IVF or ICSI and embryo culture. Embryos are often cultured together in a droplet under oil for up to six days or until they have reached the blastocyst stage. In addition to culture media, there are several culture parameters that can influence the state of oocytes and embryos in culture, including the type of incubator used (small versus large chamber), gas composition, environmental temperature fluctuations and light-induced perturbations from visualization equipment. A recent study found that the *H19* DMR in preimplantation embryos is resistant to culture-induced DNA methylation changes (Ibala-Romdhane *et al.* 2011). There is both evidence for and against whether the embryo culture system can influence birth weight (Eaton *et al.* 2012, Lin *et al.* 2013, Nelissen *et al.* 2012, Vergouw *et al.* 2012, Zhu *et al.* 2014). Recent studies suggest that culture medium has an effect on intrauterine fetal growth (Nelissen *et al.* 2013) and length of culture can affect birth weight (Makinen *et al.* 2013). Given that a set of culture conditions will likely vary across fertility centres, accompanied by inherent technical variation, unraveling each of their effects on the epigenetic status of human oocytes and embryos will be challenging.

10.3.5 *Oocyte and embryo cryopreservation*

Cryopreservation is commonly used in ART to store gametes and embryos for future use. Two freezing techniques can be employed: slow-rate freezing and vitrification. With the slow freezing method, successful cryopreservation is contingent upon the removal of water from individual cells. If water remains in the cell, the freezing process produces crystals that damage the cell. To mitigate formation of water crystals, a cryoprotectant replaces most of the water and under optimal conditions the cryoprotectant will resist crystal formation. The reverse process must occur after removal from cryogenic storage; the cryoprotectant must be replaced by water. If either directional exchange does not occur optimally across the cell membrane, gametes and embryos do not survive the process. With vitrification, gametes and embryos are immersed in liquid nitrogen, a process termed "flash freezing", upon which water takes on a glass-like form circumventing water crystal formation. Vitrification is currently the most commonly used method because it generally results in less damage to a cell (Edgar and Gook 2012).

Embryos can be frozen at Day 1, 2, 3 and blastocyst stages (Day 5 or 6). Blastocyst cryopreservation is more commonly employed because this stage offers a sufficient window of time and access to multiple cells (e.g. trophectoderm) to monitor and select euploid embryos. Oocytes that are to be stored for future use may be cryopreserved at Metaphase II (MII) or GV/M1 stages. It is challenging to cryopreserve MII-stage oocytes due to their large size, advanced complexity (spindle and other cell organelles), higher water volume, temperature sensitivity and suboptimal plasma membrane permeability to cryoprotectant and water, and, as such, MII oocytes are susceptible to freezing-induced damage (Mandelbaum *et al.* 2004). In contrast, GV-stage oocytes have a prominent nucleus and lack spindles, thus it was previously believed that cryopreservation at this stage could mitigate chromosomal damage. However, studies suggests that other GV-stage oocyte structures may be at risk for injury during cryopreservation, such as transzonal projections (TZPs) and cumulus cell compartments, which may explain issues related to their maturation, fertilization and developmental capacity after cryopreservation (Son *et al.* 1996, Toth *et al.* 1994a, 1994b). Moreover, the requirement to subject GV-stage oocytes to an IVM culture protocol upon their removal from cryostorage diminishes the success of its application in IVM-assisted human ART. Therefore, the current preferred approach is to mature GV oocytes to MII stage before vitrifying them (Chian *et al.* 2009).

Few studies have been reported as to the effect of cryopreservation on the epigenetic status of oocytes and embryos. An oocyte vitrification study demonstrated that in a sample of vitrified GV-stage oocytes, where a little over half of oocytes survived the cryopreservation process and three-quarters matured to MII-stage after IVM, there were generally normal methylation signatures in *H19* and fully maintained methylation patterns in *KCNQ1OT1* imprinting control regions (Al-Khtib *et al.* 2011). Cryopreservation of pronuclear stage embryos has been shown to have some impact on gene expression (Shaw *et al.* 2012). A study evaluating the expression of DMNTs in cryopreserved IVF/ICSI-derived preimplantation embryos found that spatio-temporal expression patterns of DNMTs is perturbed in embryos that have been vitrified (Petrussa *et al.* 2014). There is a need for further comprehensive analyses regarding the impact of cryopreservation and the thawing process on the epigenetic status of oocytes and embryos. While it appears that long-term cryopreservation does not significantly decrease implantation and pregnancy rates (Aflatoonian *et al.* 2013), which is highlighted by a recent precedent setting case of a child born from an embryo that was frozen for 20 years (Dowling-Lacey *et al.* 2011), longitudinal studies to evaluate the effects of long term storage on the health outcomes of babies should also be pursued.

10.4 Concluding Remarks

Since the birth of the first IVF child in 1978, the use of ART to overcome infertility problems has been on the rise. Coupled with the growth of ART applications is an increasing concern and interest regarding both the short- and long-term impact on ART-conceived children's health. These interests have spawned a multitude of both animal- and human-based studies. What is striking is the discordant outcomes of studies with similar research objectives. It is plausible that ART protocols and associated parameters are more congruent within a geographic cohort than across geographic cohorts, which could account for different epigenetic alterations. For instance, the use of a specific media culture could influence epigenomic alterations, which is just one parameter of several involved in the step of culturing oocytes and embryos (Pool *et al.* 2012). Further, it has been reported that different racial/ethnic groups may respond differently to ART interventions (Wellons *et al.* 2012). The other aspect that differs between studies is sample size, whether it be the number of samples from individuals or the range of potentially methylated CpG sites assayed. Several studies have used small biological sample sizes of less than 15, which may not be representative of the population being studied. Further, epigenomic analyses may be skewed by application of the assays themselves. The majority of methylation analysis protocols incorporate bisulfite-conversion of DNA from a small cell population, followed by amplification, cloning and pyroseqencing. This process can be skewed by the preferential amplification of only a few molecules and/or a bias for methylated or unmethylated DNA. There are many variables that must be controlled to perform robust comparisons across cohorts and registries. Well-designed longitudinal human studies will be necessary to clarify the potential impact that ART interventions have on the health outcomes in an ever-growing population of ART-conceived children.

References

Adamson, G. D., Tabangin, M., Macaluso, M. and de Mouzon, J. (2013). The number of babies born globally after treatment with the assisted reproductive technologies (art). *Fertility and Sterility* 100, S42–S42.

Aflatoonian, N., Pourmasumi, S., Aflatoonian, A. and Efteskhar, M. (2013). Duration of storage does not influence pregnancy outcome in cryopreserved human embryos. *Iranian Journal of Reproductive Medicine* 11, 843–846.

Al-Khtib, M., Perret, A., Khoueiry, R., Ibala-Romdhane, S., Blachere, T., Greze, C., Lornage, J. and Lefevre, A. (2011). Vitrification at the germinal vesicle stage does not affect the methylation profile of H19 and KCNQ1OT1 imprinting centers in human oocytes subsequently matured in vitro. *Fertility and Sterility* 95, 1955–1960.

Basatemur, E., Shevlin, M. and Sutcliffe, A. (2010). Growth of children conceived by IVF and ICSI up to 12years of age. *Reproductive Biomedicine Online* 20, 144–149.

Basatemur, E. and Sutcliffe, A. (2011). Health of IVM children. *Journal of Assisted Reproduction and Genetics* 28, 489–493.

Bernstein, B. E., Meissner, A. and Lander, E. S. (2007). The mammalian epigenome. *Cell* 128, 669–681.

Boivin, J., Bunting, L., Collins, J. A. and Nygren, K. G. (2007). International estimates of infertility prevalence and treatment-seeking: Potential need and demand for infertility medical care. *Human Reproduction* 22, 1506–1512.

Borghol, N., Lornage, J., Blachere, T., Sophie Garret, A. and Lefevre, A. (2006). Epigenetic status of the H19 locus in human oocytes following in vitro maturation. *Genomics* 87, 417–426.

Buckett, W. M., Chian, R. C., Holzer, H., Dean, N., Usher, R. and Tan, S. L. (2007). Obstetric outcomes and congenital abnormalities after in vitro maturation, in vitro fertilization, and intracytoplasmic sperm injection. *Obstetrics & Gynecology* 110, 885–891.

Camprubi, C., Iglesias-Platas, I., Martin-Trujillo, A., Salvador-Alarcon, C., Rodriguez, M. A., Barredo, D. R., Court, F. and Monk, D. (2013). Stability of genomic imprinting and gestational-age dynamic methylation in complicated pregnancies conceived following assisted reproductive technologies. *Biology of Reproduction* 89, 50.

Camprubi, C., Pladevall, M., Grossmann, M., Garrido, N., Pons, M. C. and Blanco, J. (2012). Semen samples showing an increased rate of spermatozoa with imprinting errors have a negligible effect in the outcome of assisted reproduction techniques. *Epigenetics* 7, 1115–1124.

Cha, K. Y., Koo, J. J., Ko, J. J., Choi, D. H., Han, S. Y. and Yoon, T. K. (1991). Pregnancy after in vitro fertilization of human follicular oocytes collected from nonstimulated cycles, their culture in vitro and their transfer in a donor oocyte program. *Fertility and Sterility* 55, 109–113.

Chian, R. C., Buckett, W. M., Abdul Jalil, A. K., Son, W. Y., Sylvestre, C., Rao, D. and Tan, S. L. (2004). Natural-cycle in vitro fertilization combined with in vitro maturation of immature oocytes is a potential approach in infertility treatment. *Fertility and Sterility* 82, 1675–1678.

Chian, R. C., Huang, J. Y., Gilbert, L., Son, W. Y., Holzer, H., Cui, S. J., Buckett, W. M., Tulandi, T. and Tan, S. L. (2009). Obstetric outcomes following vitrification of in vitro and in vivo matured oocytes. *Fertility and Sterility* 91, 2391–2398.

Chian, R. C., Xu, C. L., Huang, J. Y. and Ata, B. (2014). Obstetric outcomes and congenital abnormalities in infants conceived with oocytes matured in vitro. *Facts, Views and Vision in ObGyn* 6, 15–18.

Child, T. J., Abdul-Jalil, A. K., Gulekli, B. and Tan, S. L. (2001). In vitro maturation and fertilization of oocytes from unstimulated normal ovaries, polycystic ovaries, and women with polycystic ovary syndrome. *Fertility and Sterility* 76, 936–942.

Cotton, A. M., Avila, L., Penaherrera, M. S., Affleck, J. G., Robinson, W. P. and Brown, C. J. (2009). Inactive X chromosome-specific reduction in placental DNA methylation. *Human Molecular Genetics* 18, 3544–3552.

Davis, T. L., Yang, G. J., McCarrey, J. R. and Bartolomei, M. S. (2000). The H19 methylation imprint is erased and re-established differentially on the parental alleles during male germ cell development. *Human Molecular Genetics* 9, 2885–2894.

Daxinger, L. and Whitelaw, E. (2012). Understanding transgenerational epigenetic inheritance via the gametes in mammals. *Nature Reviews Genetics* 13, 153–162.

Dowling-Lacey, D., Mayer, J. F., Jones, E., Bocca, S., Stadtmauer, L. and Oehninger, S. (2011). Live birth from a frozen-thawed pronuclear stage embryo almost 20 years after its cryopreservation. *Fertility and Sterility* 95, 1120 e1121–1123.

Eaton, J. L., Lieberman, E. S., Stearns, C., Chinchilla, M. and Racowsky, C. (2012). Embryo culture media and neonatal birthweight following IVF. *Human Reproduction* 27, 375–379.

Edgar, D. H. and Gook, D. A. (2012). A critical appraisal of cryopreservation (slow cooling versus vitrification) of human oocytes and embryos. *Human Reproduction Update* 18, 536–554.

Engel, S. M., Joubert, B. R., Wu, M. C., Olshan, A. F., Haberg, S. E., Ueland, P. M., Nystad, W., Nilsen, R. M., Vollset, S. E., Peddada, S. D. *et al.* (2014). Neonatal genome-wide methylation patterns in relation to birth weight in the Norwegian Mother and Child Cohort. *American Journal of Epidemiology* 179, 834–842.

Fadini, R., Dal Canto, M. B., Mignini Renzini, M., Brambillasca, F., Comi, R., Fumagalli, D., Lain, M., Merola, M., Milani, R. and De Ponti, E. (2009). Effect of different gonadotrophin priming on IVM of oocytes from women with normal ovaries: A prospective randomized study. *Reproductive Biomedicine Online* 19, 343–351.

Fadini, R., Mignini Renzini, M., Dal Canto, M., Epis, A., Crippa, M., Caliari, I., Brigante, C. and Coticchio, G. (2013). Oocyte in vitro maturation in normo-ovulatory women. *Fertility and Sterility* 99, 1162–1169.

Fauque, P. (2013). Ovulation induction and epigenetic anomalies. *Fertility and Sterility* 99, 616–623.

Feng, C., Tian, S., Zhang, Y., He, J., Zhu, X. M., Zhang, D., Sheng, J. Z. and Huang, H. F. (2011). General imprinting status is stable in assisted reproduction-conceived offspring. *Fertility and Sterility* 96, 1417–1423.

Ferreira, J. C., Choufani, S., Grafodatskaya, D., Butcher, D. T., Zhao, C., Chitayat, D., Shuman, C., Kingdom, J., Keating, S. and Weksberg, R. (2011). WNT2 promoter methylation in human placenta is associated with low birthweight percentile in the neonate. *Epigenetics* 6, 440–449.

Filiberto, A. C., Maccani, M. A., Koestler, D., Wilhelm-Benartzi, C., Avissar-Whiting, M., Banister, C. E., Gagne, L. A. and Marsit, C. J. (2011). Birthweight is associated with DNA promoter methylation of the glucocorticoid receptor in human placenta. *Epigenetics* 6, 566–572.

Fortier, A. L., Lopes, F. L., Darricarrere, N., Martel, J. and Trasler, J. M. (2008). Superovulation alters the expression of imprinted genes in the midgestation mouse placenta. *Human Molecular Genetics* 17, 1653–1665.

Geuns, E., Hilven, P., Van Steirteghem, A., Liebaers, I. and De Rycke, M. (2007). Methylation analysis of KvDMR1 in human oocytes. *Journal of Medical Genetics* 44, 144–147.

Gomes, M. V., Huber, J., Ferriani, R. A., Amaral Neto, A. M. and Ramos, E. S. (2009). Abnormal methylation at the KvDMR1 imprinting control region in clinically normal children conceived by assisted reproductive technologies. *Molecular Human Reproduction* 15, 471–477.

Gremeau, A. S., Andreadis, N., Fatum, M., Craig, J., Turner, K., McVeigh, E. and Child, T. (2012). In vitro maturation or in vitro fertilization for women with polycystic ovaries? A case-control study of 194 treatment cycles. *Fertility and Sterility* 98, 355–360.

Hansen, M., Kurinczuk, J. J., Milne, E., de Klerk, N. and Bower, C. (2013). Assisted reproductive technology and birth defects: A systematic review and meta-analysis. *Human Reproduction Update* 19, 330–353.

Haworth, K. E., Farrell, W. E., Emes, R. D., Ismail, K. M., Carroll, W. D., Hubball, E., Rooney, A., Yates, A. M., Mein, C. and Fryer, A. A. (2014). Methylation of the FGFR2 gene is associated with high birth weight centile in humans. *Epigenomics* 6, 477–491.

Henningsen, A. K., Pinborg, A., Lidegaard, O., Vestergaard, C., Forman, J. L. and Andersen, A. N. (2011). Perinatal outcome of singleton siblings born after assisted reproductive technology and spontaneous conception: Danish national sibling-cohort study. *Fertility and Sterility* 95, 959–963.

Hiura, H., Okae, H., Miyauchi, N., Sato, F., Sato, A., Van De Pette, M., John, R. M., Kagami, M., Nakai, K., Soejima, H. *et al.* (2012). Characterization of DNA methylation errors in patients with imprinting disorders conceived by assisted reproduction technologies. *Human Reproduction* 27, 2541–2548.

Howlett, S. K. and Reik, W. (1991). Methylation levels of maternal and paternal genomes during preimplantation development. *Development* 113, 119–127.

Huang, J., Li, T., Ding, C. H., Brosens, J., Zhou, C. Q., Wang, H. H. and Xu, Y. W. (2012). Insufficient histone-3 lysine-9 deacetylation in human oocytes matured in vitro is associated with aberrant meiosis. *Fertility and Sterility* 97, 178–184 e173.

Ibala-Romdhane, S., Al-Khtib, M., Khoueiry, R., Blachere, T., Guerin, J. F. and Lefevre, A. (2011). Analysis of H19 methylation in control and abnormal human embryos, sperm and oocytes. *European Journal of Human Genetics* 19, 1138–1143.

Junk, S. M. and Yeap, D. (2012). Improved implantation and ongoing pregnancy rates after single-embryo transfer with an optimized protocol for in vitro oocyte maturation in women with polycystic ovaries and polycystic ovary syndrome. *Fertility and Sterility* 98, 888–892.

Kaneko-Ishino, T. and Ishino, F. (2010). Retrotransposon silencing by DNA methylation contributed to the evolution of placentation and genomic imprinting in mammals. *Development Growth & Differentiation* 52, 533–543.

Katari, S., Turan, N., Bibikova, M., Erinle, O., Chalian, R., Foster, M., Gaughan, J. P., Coutifaris, C. and Sapienza, C. (2009). DNA methylation and gene expression differences in children conceived in vitro or in vivo. *Human Molecular Genetics* 18, 3769–3778.

Khoueiry, R., Ibala-Rhomdane, S., Mery, L., Blachere, T., Guerin, J. F., Lornage, J. and Lefevre, A. (2008). Dynamic CpG methylation of the KCNQ1OT1 gene during maturation of human oocytes. *Journal of Medical Genetics* 45, 583–588.

Kjellberg, A. T., Carlsson, P. and Bergh, C. (2006). Randomized single versus double embryo transfer: Obstetric and paediatric outcome and a cost-effectiveness analysis. *Human Reproduction* 21, 210–216.

Kobayashi, H., Hiura, H., John, R. M., Sato, A., Otsu, E., Kobayashi, N., Suzuki, R., Suzuki, F., Hayashi, C., Utsunomiya, T. *et al.* (2009). DNA methylation errors at imprinted loci after assisted conception originate in the parental sperm. *European Journal of Human Genetics* 17, 1582–1591.

Koukoura, O., Sifakis, S., Soufla, G., Zaravinos, A., Apostolidou, S., Jones, A., Widschwendter, M. and Spandidos, D. A. (2011). Loss of imprinting and aberrant methylation of IGF2 in placentas from pregnancies complicated with fetal growth restriction. *International Journal of Molecular Medicine* 28, 481–487.

Kuhtz, J., Romero, S., De Vos, M., Smitz, J., Haaf, T. and Anckaert, E. (2014). Human in vitro oocyte maturation is not associated with increased imprinting error rates at LIT1, SNRPN, PEG3 and GTL2. *Human Reproduction* 29, 1995–2005.

La Salle, S., Mertineit, C., Taketo, T., Moens, P. B., Bestor, T. H. and Trasler, J. M. (2004). Windows for sex-specific methylation marked by DNA methyltransferase expression profiles in mouse germ cells. *Developmental Biology* 268, 403–415.

Laprise, S. L. (2009). Implications of epigenetics and genomic imprinting in assisted reproductive technologies. *Molecular Reproduction and Development* 76, 1006–1018.

Li, L., Wang, L., Le, F., Liu, X., Yu, P., Sheng, J., Huang, H. and Jin, F. (2011). Evaluation of DNA methylation status at differentially methylated regions in IVF-conceived newborn twins. *Fertility and Sterility* 95, 1975–1979.

Lin, S., Li, M., Lian, Y., Chen, L. and Liu, P. (2013). No effect of embryo culture media on birthweight and length of newborns. *Human Reproduction* 28, 1762–1767.

Ludwig, M., Katalinic, A., Gross, S., Sutcliffe, A., Varon, R. and Horsthemke, B. (2005). Increased prevalence of imprinting defects in patients with Angelman syndrome born to subfertile couples. *Journal of Medical Genetics* 42, 289–291.

Maher, E. R., Brueton, L. A., Bowdin, S. C., Luharia, A., Cooper, W., Cole, T. R., Macdonald, F., Sampson, J. R., Barratt, C. L., Reik, W. *et al.* (2003). Beckwith-Wiedemann syndrome and assisted reproduction technology (ART). *Journal of Medical Genetics* 40, 62–64.

Makinen, S., Soderstrom-Anttila, V., Vainio, J., Suikkari, A. M. and Tuuri, T. (2013). Does long in vitro culture promote large for gestational age babies? *Human Reproduction* 28, 828–834.

Mandelbaum, J., Anastasiou, O., Levy, R., Guerin, J. F., de Larouziere, V. and Antoine, J. M. (2004). Effects of cryopreservation on the meiotic spindle of human oocytes. *European Journal of Obstetrics Gynecology and Reproductive Biology* 113 Suppl 1, S17–23.

Manipalviratn, S., DeCherney, A. and Segars, J. (2009). Imprinting disorders and assisted reproductive technology. *Fertility and Sterility* 91, 305–315.

Melamed, N., Choufani, S., Wilkins-Haug, L. E., Koren, G. and Weksberg, R. (2015). Comparison of genome-wide and gene-specific DNA methylation between ART and naturally conceived pregnancies. *Epigenetics* 10, 474–483.

Morgan, H. D., Santos, F., Green, K., Dean, W. and Reik, W. (2005). Epigenetic reprogramming in mammals. *Human Molecular Genetics* 14 Spec No 1, R47–58.

Murphy, R., Thompson, J. M., Tost, J., Mitchell, E. A. and Auckland Birthweight Collaborative Study, G. (2014). No evidence for copy number and methylation variation in H19 and KCNQ10T1 imprinting control regions in children born small for gestational age. *BMC Medical Genetics* 15, 67.

Nelissen, E. C., Dumoulin, J. C., Busato, F., Ponger, L., Eijssen, L. M., Evers, J. L., Tost, J. and van Montfoort, A. P. (2014). Altered gene expression in human placentas after IVF/ICSI. *Human Reproduction* 29, 2821–2831.

Nelissen, E. C., Van Montfoort, A. P., Coonen, E., Derhaag, J. G., Geraedts, J. P., Smits, L. J., Land, J. A., Evers, J. L. and Dumoulin, J. C. (2012). Further evidence that culture media affect perinatal outcome: Findings after transfer of fresh and cryopreserved embryos. *Human Reproduction* 27, 1966–1976.

Nelissen, E. C., Van Montfoort, A. P., Smits, L. J., Menheere, P. P., Evers, J. L., Coonen, E., Derhaag, J. G., Peeters, L. L., Coumans, A. B. and Dumoulin, J. C. (2013). IVF culture medium affects human intrauterine growth as early as the second trimester of pregnancy. *Human Reproduction* 28, 2067–2074.

Palermo, G. D., Neri, Q. V., Takeuchi, T., Squires, J., Moy, F. and Rosenwaks, Z. (2008). Genetic and epigenetic characteristics of ICSI children. *Reproductive Biomedicine Online* 17, 820–833.

Parets, S. E., Bedient, C. E., Menon, R. and Smith, A. K. (2014). Preterm birth and its long-term effects: Methylation to mechanisms. *Biology* (*Basel*) 3, 498–513.

Pelinck, M. J., Hadders-Algra, M., Haadsma, M. L., Nijhuis, W. L., Kiewiet, S. M., Hoek, A., Heineman, M. J. and Middelburg, K. J. (2010). Is the birthweight of singletons born after IVF reduced by ovarian stimulation or by IVF laboratory procedures? *Reproductive Biomedicine Online* 21, 245–251.

Petrussa, L., Van de Velde, H. and De Rycke, M. (2014). Dynamic regulation of DNA methyltransferases in human oocytes and preimplantation embryos after assisted reproductive technologies. *Molecular Human Reproduction* 20, 861–874.

Pliushch, G., Schneider, E., Schneider, T., El Hajj, N., Rosner, S., Strowitzki, T. and Haaf, T. (2015). In vitro maturation of oocytes is not associated with altered deoxyribonucleic acid methylation patterns in children from in vitro fertilization or intracytoplasmic sperm injection. *Fertility and Sterility* 103, 720–727.

Pool, T. B., Schoolfield, J. and Han, D. (2012). Human embryo culture media comparisons. *Methods in Molecular Biology* 912, 367–386.

Rangel, M., dos Santos, J. C., Ortiz, P. H., Hirata, M., Jasiulionis, M. G., Araujo, R. C., Ierardi, D. F. and Franco Mdo, C. (2014). Modification of epigenetic patterns in low birth weight children: Importance of hypomethylation of the ACE gene promoter. *PLoS One* 9, e106138.

Rossant, J. and Cross, J. C. (2001). Placental development: Lessons from mouse mutants. *Nature Review Genetics* 2, 538–548.

Rossant, J., Sanford, J. P., Chapman, V. M. and Andrews, G. K. (1986). Undermethylation of structural gene sequences in extraembryonic lineages of the mouse. *Developmental Biology* 117, 567–573.

Santos, F., Hyslop, L., Stojkovic, P., Leary, C., Murdoch, A., Reik, W., Stojkovic, M., Herbert, M. and Dean, W. (2010). Evaluation of epigenetic marks in human embryos derived from IVF and ICSI. *Human Reproduction* 25, 2387–2395.

Sasaki, H. and Matsui, Y. (2008). Epigenetic events in mammalian germ-cell development: reprogramming and beyond. *Nature Review Genetics* 9, 129–140.

Sato, A., Otsu, E., Negishi, H., Utsunomiya, T. and Arima, T. (2007a). Aberrant DNA methylation of imprinted loci in superovulated oocytes. *Human Reproduction* 22, 26–35.

Sato, C., Shimada, M., Mori, T., Kumasako, Y., Otsu, E., Watanabe, H. and Utsunomiya, T. (2007b). Assessment of human oocyte developmental competence by cumulus cell morphology and circulating hormone profile. *Reproductive Biomedicine Online* 14, 49–56.

Schieve, L. A., Meikle, S. F., Ferre, C., Peterson, H. B., Jeng, G. and Wilcox, L. S. (2002). Low and very low birth weight in infants conceived with use of assisted reproductive technology. *New England Journal of Medicine* 346, 731–737.

Schneider, R. and Grosschedl, R. (2007). Dynamics and interplay of nuclear architecture, genome organization, and gene expression. *Genes Development* 21, 3027–3043.

Shaw, L., Sneddon, S. F., Brison, D. R. and Kimber, S. J. (2012). Comparison of gene expression in fresh and frozen-thawed human preimplantation embryos. *Reproduction* 144, 569–582.

Shi, X., Ni, Y., Zheng, H., Chen, S., Zhong, M., Wu, F., Xia, R. and Luo, Y. (2011). Abnormal methylation patterns at the IGF2/H19 imprinting control region in phenotypically normal babies conceived by assisted reproductive technologies. *European Journal of Obstetrics & Gynecology and Reproductive Biology* 158, 52–55.

Smallwood, S. A., Tomizawa, S., Krueger, F., Ruf, N., Carli, N., Segonds-Pichon, A., Sato, S., Hata, K., Andrews, S. R. and Kelsey, G. (2011). Dynamic CpG island methylation landscape in oocytes and preimplantation embryos. *Nature Genetics* 43, 811–814.

Smith, Z. D. and Meissner, A. (2013). DNA methylation: Roles in mammalian development. *Nature Reviews Genetics* 14, 204–220.

Son, W. Y., Park, S. E., Lee, K. A., Lee, W. S., Ko, J. J., Yoon, T. K. and Cha, K. Y. (1996). Effects of 1,2-propanediol and freezing-thawing on the in vitro developmental capacity of human immature oocytes. *Fertility and Sterility* 66, 995–999.

Song, S., Ghosh, J., Mainigi, M., Turan, N., Weinerman, R., Truongcao, M., Coutifaris, C. and Sapienza, C. (2015). DNA methylation differences between in vitro- and in vivo-conceived children are associated with ART procedures rather than infertility. *Clinical Epigenetics* 7, 41.

Sutcliffe, A. G., Melhuish, E., Barnes, J. and Gardiner, J. (2014). Health and development of children born after assisted reproductive technology and sub-fertility compared to naturally conceived children: data from a national study. *Pediatric Reports* 6, 5118.

Tan, Q., Frost, M., Heijmans, B. T., von Bornemann Hjelmborg, J., Tobi, E. W., Christensen, K. and Christiansen, L. (2014). Epigenetic signature of birth weight discordance in adult twins. *BMC Genomics* 15, 1062.

Tan, S. L., Doyle, P., Campbell, S., Beral, V., Rizk, B., Brinsden, P., Mason, B. and Edwards, R. G. (1992). Obstetric outcome of in vitro fertilization pregnancies compared with normally conceived pregnancies. *American Journal of Obstetrics & Gynecology* 167, 778–784.

Tierling, S., Souren, N. Y., Gries, J., Loporto, C., Groth, M., Lutsik, P., Neitzel, H., Utz-Billing, I., Gillessen-Kaesbach, G., Kentenich, H. *et al.* (2010). Assisted reproductive technologies do not enhance the variability of DNA methylation imprints in human. *Journal of Medical Genetics* 47, 371–376.

Tobi, E. W., Heijmans, B. T., Kremer, D., Putter, H., Delemarre-van de Waal, H. A., Finken, M. J., Wit, J. M. and Slagboom, P. E. (2011). DNA methylation of IGF2, GNASAS, INSIGF and LEP and being born small for gestational age. *Epigenetics* 6, 171–176.

Tomizawa, S., Kobayashi, H., Watanabe, T., Andrews, S., Hata, K., Kelsey, G. and Sasaki, H. (2011). Dynamic stage-specific changes in imprinted differentially methylated regions during early mammalian development and prevalence of non-CpG methylation in oocytes. *Development* 138, 811–820.

Toth, T. L., Baka, S. G., Veeck, L. L., Jones, H. W., Jr., Muasher, S. and Lanzendorf, S. E. (1994a). Fertilization and in vitro development of cryopreserved human prophase I oocytes. *Fertility and Sterility* 61, 891–894.

Toth, T. L., Lanzendorf, S. E., Sandow, B. A., Veeck, L. L., Hassen, W. A., Hansen, K. and Hodgen, G. D. (1994b). Cryopreservation of human prophase I oocytes collected from unstimulated follicles. *Fertility and Sterility* 61, 1077–1082.

Turan, N., Ghalwash, M. F., Katari, S., Coutifaris, C., Obradovic, Z. and Sapienza, C. (2012). DNA methylation differences at growth related genes correlate with birth weight: A molecular signature linked to developmental origins of adult disease? *BMC Medical Genomics* 5, 10.

Turan, N., Katari, S., Gerson, L. F., Chalian, R., Foster, M. W., Gaughan, J. P., Coutifaris, C. and Sapienza, C. (2010). Inter- and intra-individual variation in allele-specific DNA methylation and gene expression in children conceived using assisted reproductive technology. *PLoS Genetics* 6, e1001033.

Uyar, A. and Seli, E. (2014). The impact of assisted reproductive technologies on genomic imprinting and imprinting disorders. *Current Opinion Obstetrics Gynecology* 26, 210–221.

Vergouw, C. G., Kostelijk, E. H., Doejaaren, E., Hompes, P. G., Lambalk, C. B. and Schats, R. (2012). The influence of the type of embryo culture medium on neonatal birthweight after single embryo transfer in IVF. *Human Reproduction* 27, 2619–2626.

Vermeiden, J. P. and Bernardus, R. E. (2013). Are imprinting disorders more prevalent after human in vitro fertilization or intracytoplasmic sperm injection? *Fertility and Sterility* 99, 642–651.

Walls, M. L., Hunter, T., Ryan, J. P., Keelan, J. A., Nathan, E. and Hart, R. J. (2015). *In vitro* maturation as an alternative to standard in vitro fertilization for patients diagnosed with polycystic ovaries: a comparative analysis of fresh, frozen and cumulative cycle outcomes. *Human Reproduction* 30, 88–96.

Wang, N., Le, F., Zhan, Q. T., Li, L., Dong, M. Y., Ding, G. L., Xu, C. M., Jiang, S. W., Huang, H. F. and Jin, F. (2010). Effects of in vitro maturation on histone acetylation in metaphase II oocytes and early cleavage embryos. *Obstetrics and Gynecology International* 2010, 989278.

Wellons, M. F., Fujimoto, V. Y., Baker, V. L., Barrington, D. S., Broomfield, D., Catherino, W. H., Richard-Davis, G., Ryan, M., Thornton, K. and Armstrong, A. Y. (2012). Race matters: A systematic review of racial/ethnic disparity in Society for Assisted Reproductive Technology reported outcomes. *Fertility and Sterility* 98, 406–409.

Wells, D. and Patrizio, P. (2008). Gene expression profiling of human oocytes at different maturational stages and after in vitro maturation. *American Journal of Obstetrics & Gynecology* 198, 455 e451–459; discussion 455 e459–411.

Whitelaw, N., Bhattacharya, S., Hoad, G., Horgan, G. W., Hamilton, M. and Haggarty, P. (2014). Epigenetic status in the offspring of spontaneous and assisted conception. *Human Reproduction* 29, 1452–1458.

Wisborg, K., Ingerslev, H. J. and Henriksen, T. B. (2010). In vitro fertilization and preterm delivery, low birth weight, and admission to the neonatal intensive care unit: A prospective follow-up study. *Fertility and Sterility* 94, 2102–2106.

Zhu, J., Lin, S., Li, M., Chen, L., Lian, Y., Liu, P. and Qiao, J. (2014). Effect of in vitro culture period on birthweight of singleton newborns. *Human Reproduction* 29, 448–454.